Physics of High-Density
Z-Pinch Plasmas

Springer

New York
Berlin
Heidelberg
Barcelona
Hong Kong
London
Milan
Paris
Singapore
Tokyo

Michael A. Liberman   John S. De Groot
Arthur Toor   Rick B. Spielman

Physics of High-Density Z-Pinch Plasmas

With 156 Illustrations

 Springer

Michael A. Liberman
P.L. Kapitsa Institute for Physical Problems
Russian Academy of Sciences
ul. Kosygina 2, 117334 Moscow
Russia
and
Physics Department, University of Uppsala
Box 530, S-75121 Uppsala
Sweden

John S. De Groot
Department of Applied Science
University of California
Davis, CA 95616
USA

Arthur Toor
V Division
Lawrence Livermore National Laboratory
PO Box 808
Livermore, CA 94550
USA

Rick B. Spielman
Sandia National Laboratories
Albuquerque, NM 87175
USA

Library of Congress Cataloging-in-Publication Data
Physics of high-density Z-pinch plasmas / Michael A. Liberman . . . [et
al.].
     p.   cm.
   Includes bibliographical references and index.
   ISBN 0-387-98568-9 (alk. paper)
   1. Pinch effect (Physics)   2. Magnetohydrodynamics.   I. Liberman,
M.A. (Mikhail Andreevich), 1942–
QC718.5.P45P48   1998
530.4′4—dc21   98-9925

Printed on acid-free paper.

Production managed by A. Orrantia; manufacturing supervised by Thomas King.
Camera-ready copy prepared by Bartlett Press, Inc., Marietta, GA.
Printed and bound by Edwards Brothers, Inc., Ann Arbor, MI.
Printed in the United States of America.

9 8 7 6 5 4 3 2 1

ISBN 0-387-98568-9 Springer-Verlag New York Berlin Heidelberg   SPIN 10692260

Preface

This is presumably the first textbook devoted to high-density Z pinches. Our objective was to provide an experimental motivation and a theoretical background for on-going research in this field. Many of the problems are similar to those encountered in inertial-confinement fusion. Most of the analysis is sufficiently general to apply to dynamical plasmas, in general, including those encountered in laser and particle-beam fusion. At the same time, we did not aim to produce another textbook on general plasma theory, since good introductory monographs on the subject are already available both in English (Krall and Trivelpiece, 1973; Chen, 1984) and in Russian (Golant et al., 1977).

A great deal of progress has been made in recent years in research on dynamic, fast Z pinches and imploding plasma liners. These impressive advances, conducted on multi-megaampere 100-ns drivers, has revitalized the field of Z-pinch physics. In 1997 researchers at Sandia National Laboratories conducted dramatic experiments at 20 MA on the Z machine achieving total x-ray powers greater than 200 TW and x-ray energies of 1.8 MJ with tungsten-wire-array Z pinches. For many years the focus of research efforts in fast Z-pinch implosions has centered on optimizing the power output in order to provide an x-ray source in the 1–5 keV region for lithography, x-ray microscopy, simulation of nuclear weapons effects, x-ray lasers, etc. The new experimental Z-pinch capabilities are certain to expand the applications of Z pinches. Most of these applications depend on the extent to which the instabilities inherent to the plasma acceleration can be controlled. Hydromagnetic Rayleigh–Taylor instabilities and cylindrical load symmetry are critical limiting factors in determining the assembled plasma that can be produced. For this reason most of this book is devoted to dynamics of Z-pinch plasmas and stability during the implosion. A number of techniques including external frozen-in magnetic fields can be quite effective in stabilizing a dynamic pinch. In addition, ultra-high magnetic fields produced in Z-pinch implosions can have many interesting applications.

Although *Physics of High Density Z-Pinch Plasmas* is mainly theoretical, we discuss the operating principles of the main types of Z-pinch facilities, recent experiments on the Saturn and other accelerators at Sandia National Laboratories as well as some of the relevant experimental results. All mathematical expressions are given in a closed analytic form convenient for practical use. We tried to present our material so as to make it useful for the whole pulsed power research community, including theoreticians, experimentalists, engineers, and students.

During the preparation of this book we have benefited from discussions with many of our colleagues, and we take the opportunity to express our gratitude to them all. Our deep gratitude is due to Drs. R. Baksht, A. Bud'ko, M. K. Matzen, J. Hammer, S. Golberg and A. I. Kleev for their helpful comments during our work on the book. We acknowledge the valuable collaboration of Prof. H. Herold and Dr. N. V. Filippov who made their original schemes of plasma focus devices available for reproduction.

Michael A. Liberman
John S. De Groot
Arthur Toor
Rick B. Spielman

This book has a long history. At the very beginning, in 1989–1990, the book was planned by M. Liberman and A. Velikovich as a book on dynamics and stability of Z pinches and plasma liners. Later N. Rostoker joined the project but unfortunately the project was never completed. Nevertheless, during the preparation of the present book, the authors have benefited from the first version of the manuscript dating from 1991–1992 by M. Liberman, A. Velikovich and N. Rostoker, and we would like to thank A. Velikovich and N. Rostoker for their valuable contributions to the first version of the manuscript. We also appreciate comments made by N. Pereira in 1992 regarding the first version of this manuscript.

We take the opportunity to express our thanks to many of our colleagues for their help, comments, criticism, and useful discussions. Our deep gratitude is due to Drs. Rina Baksht, Anatolii Bud'ko, Boris Kadomtsev, Keith Matzen, Jim Hammer, Laurent Jacquet, Philippe Arnault, Sergey Goldberg, Andrey Kleev and Charles Hartman for their helpful comments during our work on the book. We acknowledge the valuable collaboration of Prof. H. Herold and Dr. N.V. Filippov who made their original schemes of plasma focus devices available for reproduction.

Contents

1

Introduction

Physics of High-Density Z-Pinch Plasmas is the first book to provide a comprehensive and complete view of the field of Z-pinch research. This field has a long history and has recently received renewed vigor by dramatic advances in pulsed-power technology, computational and modeling capabilities, and plasma physics understanding.

A Z pinch in its simplest form is a column of plasma in which current is driven in the axial (**z**) direction by an electrical power source producing an azimuthally directed magnetic field that tends to confine the plasma. The response of the Z-pinch plasma to that applied current is at first glance quite simple but, upon more careful study, complex beyond belief. The entire Z-pinch phenomenon evolves rapidly into nonlinear hydrodynamic behavior. The atomic, ionization, and radiative physics that are inherent in Z-pinch implosions add yet another layer of complexity. There are few fields in physics like plasma physics in which a simple assembly of relatively common components gives results that are difficult to understand qualitatively let alone quantitatively.

Z pinches have wide application because they are a highly efficient and cost effective technique to heat a small mass to a very high temperature. Z pinches have been used widely to produce x-ray sources (Exploding Wires, 1959–1968; Burtsev et al., 1990). More recently, interest has been renewed in the possible use of Z pinches as a controlled thermonuclear fusion reactor. Surprisingly, theoretical understanding of Z pinches has lagged their practical applications.

Our objective is to give an overview of Z-pinch research and to develop a physical understanding of the important processes that determine the behavior of current-carrying dense plasmas confined in a Z-pinch geometry. Our approach is to review the extensive Z-pinch literature, including papers that are not widely available, to compare theoretical, computational, and experimental results, and to develop simple models to aid the reader's understanding. *Physics of High-Density Z-Pinch Plasmas* is intended as a resource for advanced undergraduate and graduate students, researchers, and engineers in the field.

Z pinches are at once deceptively simple and extremely complex. Equilibrium configurations, using the static ideal-MHD model, are developed in Chapter 2. The dynamic evolution of Z-pinch-confined plasmas is developed in Chapters 3,

4, 5, and 6. The dynamical behavior is critical not only because simple Z-pinch confined plasmas are always dynamically unstable, but also because the kinetic energy of Z-pinch plasma motion can be comparable with or even larger than its internal thermal energy. Thus, these plasmas can rapidly become turbulent, breaking simple cylindrical symmetry and introducing three-dimensional dynamics. Understanding the transition to this unstable behavior is critical for practical applications of Z pinches. Physics issues connected with higher-Z plasmas, rarely discussed in plasma physics texts, are addressed. Radiation production and transport, which strongly affect these simple equilibria, are also discussed.

The full set of MHD approximations is discussed in Chapter 3. These full MHD approximations are required, since the plasma density and temperature vary by orders of magnitude spatially and temporally during the Z-pinch event. The initially cold material is ionized, heated, and compressed to form a dense hot plasma during the Z-pinch event. Several self-similar models are developed to extend understanding of these complex devices. Finally, it is shown that some features of Z-pinch plasmas require models beyond the MHD approximation.

The stability of quasi-steady-state Z-pinch-confined plasmas is developed in Chapter 4. This field is important in the application of Z pinches to controlled thermonuclear fusion.

Many Z-pinch applications require the implosion of an initially cold plasma to form a hot compressed plasma. The accelerated plasma is unstable to the hydromagnetic or magneto Rayleigh–Taylor instability as discussed in Chapters 5 and 6. Mitigating this instability is the subject of much of present Z-pinch research. A simple planar model is developed in Chapter 5 to further the understanding of this difficult topic. Several models of the Rayleigh–Taylor instability of the implosion of cylindrical plasmas are presented in Chapter 6. Computer simulations are required to understand and describe the multidimensional features of the nonlinear evolution of the imploding, unstable plasma. This rich source of present research is discussed in Chapter 6.9.

Finally, many of the applications of Z pinches are discussed in Chapter 7. These applications include the use of hot and dense Z-pinch plasmas as unique sources of x rays for lithography and x-ray microscopy, for material equation-of-state studies, for inertial-confinement fusion, for pumping x-ray lasers, for neutron production. Also, imploding Z pinches can produce ultrahigh magnetic fields, and can focus high-energy particles with the magnetic field profile formed in a Z pinch acting as a magnetic lens.

Throughout this book experimental data are included in order to connect many of these plasma physics topics to the problems found in the real world.

1.1 An historical perspective

In the late 1700s, electrostatic generators were used to create electrical breakdown in air, thereby producing the earliest versions of Z pinches. A large generator made by M. Van Marum in 1784 (Turner and Levere, 1973) was able to store $\sim 30\,\mathrm{kJ}$

of energy and produced a 60-cm-long spark in air (Finn, 1971). Van Marum's generator is on exhibition at Teyler's Musuem in the Netherlands, and a picture of it can be found at the website: http://www.teylersmuseum.nl/engels/ruimtes/instrument/start.html.

The earliest work on high current pinches in this century involved exploding wires which were used as soft-x-ray sources (Exploding Wires, 1959–1968; Burtsev et al., 1990). Typically, a voltage of tens of kilovolts from a charged capacitor was applied to a fine wire about 10–100 μm in diameter. At first the wire vaporizes and the current decreases (the "dunkel" pause). As the electric field increases and the vapor expands, breakdown takes place. A low-resistance plasma is created that is capable of carrying a large current, and the plasma pinches. With both ohmic and compressional heating, the plasma reaches a temperature of 10–100 eV and radiates a blackbody spectrum, since the absorption mean free path is much less than the diameter of the plasma. The plasma electron density after pinching is typically about 10^{21} cm^{-3}.

Early controlled thermonuclear fusion research programs were focused on microsecond-timescale, high-current, deuterium Z pinches with a plasma density after compression about 10^{18} cm^{-3}. These pinches were regarded as possible fusion systems based on a quasi-static magnetic confinement of the plasma. Analytic equilibrium models of Z-pinch plasmas suggested that fusion temperatures and densities with reasonable confinement times were possible. Both linear (Tuck, 1958) and toroidal (Butt et al., 1958) systems were studied during the period 1952–1960. Unfortunately, fast, magnetohydrodynamic instabilities were observed to disrupt plasma confinement. The theoretical basis for such magnetohydrodynamic instabilities (Rosenbluth, 1956) was developed. It was not possible to eliminate the instabilities either experimentally or theoretically, so the fusion community lost interest in Z pinches for many years. In recent years there has been a revival of interest in several forms of pinches—the reversed field pinch (RFP) (Alper, 1990; Prager et al., 1990) and the field reversed configuration (FRC), Tuszewski (1988). These forms have about the same plasma density, but improved MHD stability.

The development of high-voltage, fast-pulsed-power technologies in the last 30 years (see Martin et al., 1975, Baker et al., 1978; Hammel et al., 1984; Pereira et al., 1984; Turman et al., 1985; Bloomquist et al., 1987; Spielman et al., 1997) stimulated a renewed interest in high-density Z-pinch systems. Very intense x rays were observed from exploding single wires of various atomic compositions at the Naval Research Laboratory in 1969 (Shanny and Vitkovitsky, 1969, Mosher et al., 1975). Previous experiments had involved electrical powers of ~100's GW; this new pulsed-power technology made possible an electrical power of ~1 TW in these early experiments. As a result, it was possible to reach plasma temperatures where the plasma (about 1-mm diameter and a density of 10^{21} cm^{-3}) became optically thin and the temperature was therefore not limited by the radiation law of a black body.

The limitation of exploding single wires is primarily due to their high initial impedance, typically 1–2 Ω. Scaling these systems to higher currents to reach higher x-ray energies and powers would have required pulsed power systems of both extremely high voltage (> 10 MV) and high power (> 100 TW). This pulsed-power limitation lead to the reexamination and further development of dynamic Z pinches (Linhart et al. 1962, Turchi and Baker, 1973; Stallings et al., 1976; and others). It was noticed that a more effective way to produce a high-density and high-temperature plasma was to convert the kinetic energy of an imploding plasma after its stagnation on axis. In these configurations, the single wire is replaced with a cylindrical array of wires, a cylindrical liner, or a cylindrical gas puff. These loads have a low initial impedance and couple well with the low-impedance, high-current generators being developed at the time (1970s–1980s) by Maxwell Laboratories, Inc., Physics International Co., the Air Force Weapons Laboratory in Albuquerque, Sandia National Laboratories, Kurchatov Institute of Atomic Energy in Troitsk, and the Institute of High Current Electronics in Tomsk.

At present it is possible to produce current pulses with peak values as large as 20 MA and rise times on the order of 100 ns, the impedance of modern high-current pulsed-power systems being on the order of 0.1–0.25 Ω. The rate of current increase can be as high as 3×10^{14} A/s, and the energy delivered to the Z pinch can be optimized by choosing the initial Z-pinch geometry and mass, and improving plasma uniformity.

The results of recent experiments with fiber-initiated dense Z pinchesfiber-initiated dense Z pinches (Scudder, 1985; Hammel and Scudder, 1987; Sethian et al., 1987), with compressional (Haines, 1978a; Haines et al., 1988) and gas-embedded (Dangor et al., 1983; Hammel et al., 1984; Haines et al., 1986) Z pinches, with imploding gas-puff Z pinches (Shiloh et al., 1978, 1979; Matzen et al., 1986; Dukart et al., 1987; Felber et al., 1988a, b) and wire arrayswire arrays (Spielman et al., 1989; Sanford et al., 1996; Deeney et al., 1997) indicate that modern, fast Z-pinch systems are capable of greater stability and can provide higher compression, much greater energy density, and longer plasma confinement times than predicted by the conventional estimates from the 1950s and 1960s. A comprehensive review of the subject was given by Dangor (1986).

High-density Z pinches have since been developed as versatile x-ray sources that can radiate 1–2 MJ of radiation and have electron temperatures as high as a few keV. The conventional views on the applications of Z pinches, particularly those related to controlled thermonuclear fusion, are currently being seriously reconsidered.

1.2 Characteristics of modern Z-pinch systems

The design of a Z-pinch system is determined by both the available pulsed-power energy source and the initial load configuration; the latter can be represented by a gas column, a solid cryogenic fiber, a metal wire or a multiwire array, a laser-produced ionized channel in a gas, an annular liner made of a metal foil, or an annular gas jet. The actual configuration chosen depends on the limitations of the available pulsed power driver (the peak current and pulse width determine pinch mass) and the desired application. In most cases today the time scale of the current drive is in the 50–150 ns range to limit the growth of MHD instabilities (described later in this monograph). Larger drivers are capable of accelerating more massive Z pinches and produce more optically thick plasmas. Many of the engineering difficulties found in Z-pinch load fabrication with $\sim$ megaampere drivers are greatly reduced or disappear with higher currrent drivers.

Parameters of existing Z-pinch facilities vary over a broad range: the energy delivered to the load from 1 kJ to 1 MJ, the pinch current from 100 kA to 20 MA, the current rise time from 10 ns to 10 ms, the electrical power from 0.01 to 60 TW. In any case, the large energy density involved in a required high-temperature plasma pinch can be maintained for a relatively short time interval in the submicrosecond or nanosecond range. Hence most of the Z pinches are an integral part of pulsed power systems, and their behavior is to a large extent determined by their dynamics, often far from equilibria.

The conventional scenario of the development of a Z pinch consists of two stages. The first stage includes breakdown, rapid ionization, and heating of an initially cold gas, fiber, or wire material that is turned into gaseous plasma. The physical effects governing this stage are avalanche ionization in the strong electric field and related phenomena, essentially dependent on ionization cross sections and other individual characteristics of the atoms of the ionized medium. They are studied in the theory of breakdown based on the kinetic approach (e.g., see Vikhrev and Braginskii, 1980; Bruzzone and Vieytes, 1990) and will not be discussed here. The subject of this monograph is the next, or second, stage, which begins when a column of conducting plasma, at least singly ionized, is prepared and interacts with the magnetic field due to the current flowing in it. We are interested mainly in the macroscopic behavior of these Z-pinch plasmas, their gross dynamics and stability. The atomic processes—ionization, radiation emission, and transport— may be and often are equally important during this stage, but their influence now can be measured by their contribution to the macroscopic balance of plasma momentum and energy.

Typical Z-pinch plasma parameters of interest are: electron number density $n_e \sim 10^{21}$ cm^{-3}, temperature $T = 0.01$–10 keV, minimal plasma-column radius $r = 100\ \mu$m to 1 mm, and length L ~ 1–2 cm. Since a particle (ions and electrons) mean free path in a singly ionized plasma is on the order of $\sim \mu$m's, the plasmas in question are in most cases collisional, and the fluid approach is adequate. However, the mean free path may be comparable to some characteristic lengths such as the perturbation wavelength, so that kinetic effects can also be important, especially when the Z-pinch stability is concerned. The instabilities developing there can eventually produce, at the time of stagnation, small regions where the density and temperature significantly differ from those of the surrounding plasma — the "hot spots," which are the main sources of hard x rays and neutrons emitted by Z pinches (e.g., see Negus et al., 1979); some authors call them "bright spots" (Pereira and Davis, 1988). (The problem is that the main diagnostic for bright spots is x-ray emission so the bright spots could have high density and/or high temperature.) The theory describing the formation of these spots and related phenomena like acceleration of nonthermal particles in Z pinches is a subject of other review papers (e.g., see Vikhrev and Braginskii, 1980; Vikhrev, 1986; Trubnikov, 1986) and will not be treated here in any detail. Similarly, noncylindrical Z pinches (plasma focus) have certain specific features of their own, which are beyond the scope of the present monograph, having been reviewed elsewhere (Filippov et al., 1962; Mather, 1964; Schmidt, 1980; Bilbao et al., 1984, 1985; Herold et al., 1988; Kelly et al., 1989).

Here we discuss macroscopic dynamics and stability of cylindrical, linear Z pinches. Our goal is to describe under what conditions a current-carrying plasma column will implode, explode, oscillate around an equilibrium state, or tend to equilibrium asymptotically; how long can it remain stable, i.e., conserve its cylindrical symmetry, and what scaling laws are to be used to estimate the plasma parameters that can be obtained with the aid of a Z pinch. It turns out that the dynamics and stability of Z pinches are closely related. For instance, the most dangerous Rayleigh–Taylor instabilities develop when the pinch plasma is accelerated to the axis and may result in destruction of its cylindrical symmetry before an equilibrium steady state is achieved—thus conventional estimates based on the assumption of Bennett equilibrium may be quite erroneous, indeed, irrelevant. Although the problems in question have been studied for almost 50 years, a fully consistent theory of a Z pinch is not yet complete; below we stress the points that still need further elucidation. Because the development of this theory was nearly stopped in the 1960s, many of the important results of the early work on Z pinches have been forgotten (and sometimes independently reproduced afterwards by other authors). We have tried to find and cite the original papers.

1.3 The various types of Z pinches

This section describes the individual types of Z pinches that are used in research today. We have separated the various configurations into groupings based on one or more general physics criteria. Those groups are: the dynamic pinch, where the kinetic energy of the implosion dominates most other heating mechanisms at stagnation; the equilibrium pinch, where ohmic and compressional heating are most important; the dense plasma focus, a μs-implosion where the sheath is initiated by surface flashover and mass accretion and axial shear are important; and vacuum arcs, where kinetic effects and beams dominate.

1.3.1 Dynamic Z pinches

While there are a number of distinct embodiments of dynamic Z pinches, it is possible to describe the physical processes that occur during a generic Z-pinch implosion. A high-voltage pulse is applied to two electrodes from a pulse-forming line or capacitor bank. The current passes through a gas shell (liner or wire array) between the electrodes. As the gas breaks down (liner or wires heat and vaporize), a plasma is formed having an axial current channel. This resulting plasma is accelerated rapidly toward the axis by the pressure of the azimuthal magnetic field produced by the current flowing in the plasma. The dynamics of this self-constriction process depend mainly on the following parameters of the system: the length and initial radius of the plasma, the initial plasma mass, and the amplitude and rise time of the electric current. If the current grows slowly (relative to the plasma conductivity), or the above parameters give rise to anomalous resistivity so that the current density is distributed over the entire plasma, then a diffuse Z pinch is formed. If the current is concentrated in a skin layer, as in a snowplow pinch, the current sheet converges to the axis, sweeping and accreting the plasma ahead of it like a cylindrical piston. An equilibrium is only transitory in a dynamic Z pinch. At stagnation the pressure of the plasma is balanced by plasma inertia and by the magnetic pressure.

An implosion is a convenient way to provide effective energy coupling and power gain. The electrical energy stored in the energy source is converted first to the kinetic energy of the imploding plasma (mainly the ions), then to its thermal energy after its stagnation at the axis, and finally to radiation, which in most cases is the goal of the whole process. Some ohmic and compressional heating of the plasma takes place in the course of implosion and stagnation, but the main energy source is in the final thermalization of ion kinetic energy. Most of the energy is then radiated away from the plasma volume in the extreme ultraviolet (xuv) and x-ray spectral ranges for virtually all implosion regimes.

Isolated bright spots may appear in dynamic Z pinches for conditions where the implosion velocity is low—less than 5–10 cm/μs. For this condition the soft x rays are emitted mainly from these bright spots, the amount and spectra of this radiation being very sensitive to the dynamics of the implosion, amplitude of the current pulse, atomic number of the gas, etc. (Deeney et al., 1997, Sanford et al., 1996) The dependence of the plasma liner uniformity, determined from spectral measurements, on the rate of current rise has been studied by Mehlman et al. (1986) and Stephanakis et al. (1986).

X-Ray Production

In present pulsed-power accelerators as much as 20 MA flow through centimeter-scale Z-pinch plasmas, accelerating them to implosion velocities as high as 10^8 cm/s. The hot, dense plasma that is generated when this imploding plasma stagnates and thermalizes at the cylindrical axis of symmetry can be an efficient source of x radiation. The efficiency of converting the electrical energy stored in the capacitor banks of these large pulse-power generators into radiation has been demonstrated to be greater than 10% for high-Z plasmas (Spielman et al., 1985a; Matzen, 1997; Spielman et al., 1997).

For many years the focus of research efforts in fast Z-pinch implosions has centered on optimizing the x-ray energy and power output in the 1- to 5-keV spectral region in order to provide an x-ray source for radiation–material interaction studies. In addition to this traditional application, recent work at Sandia National Laboratories (SNL) has emphasized the generation of softer x rays (< 1 keV) that can be thermalized into a near-Planckian x-ray source by containing them within a large, high-Z-lined cylindrical hohlraum (Matzen, 1997). (See Fig. 1.1.) These large volume ($\sim$ 6 cm^3), long lived ($\sim$ 20 ns) near-Planckian radiation sources provide a well-characterized x-ray drive for ablator physics and radiation symmetrization experiments relevant to the Inertial Confinement Fusion Program, as well as for astrophysical opacity and shock physics experiments. Since higher x-ray intensities and fluences enhance the utility of these Z-pinch-driven x-ray sources for all of these applications, the source-development research at Sandia National Laboratories has emphasized increasing the x-ray energy and power output (Matzen, 1997; Sanford et al., 1996; Spielman et al., 1997; Deeney et al., 1997).

The radiation that is generated upon stagnation depends on the plasma atomic ionization state, Z, density, and temperature, which are determined, in turn, by the convergence ratio of the pinch and the ability to convert kinetic energy per particle into internal energy. Compact, uniform pinches on axis produce the high plasma densities and temperatures that are needed in order to achieve a large x-ray power. These conditions can be difficult to reach if the imploding plasma shell is disrupted in any manner. Analytical estimates and computations show that the hydromagnetic Rayleigh–Taylor instability will ultimately limit the stagnation densities and temperatures that can be achieved (see Chapters 5 and 6). This instability that results from the inward radial acceleration of the plasma by the magnetic field, forms

FIGURE 1.1. A photograph of a Z-pinch hohlraum. The hohlraum is 2-cm tall and 2.5-cm diameter. The stainless steel walls are covered with $\sim$ 5μm of gold.

at the plasma–vacuum interface. For large diameters and high-velocity implosions, the Rayleigh–Taylor instability can grow rapidly, causing severe deformation of the imploding shell through bubble and spike formation. The density and temperature inhomogeneties that are associated with such deformations both increase the sheath thickness and inhibit the rapid, efficient assembly of a hot, dense plasma on axis. Therefore, mitigating this instability is important for optimizing the x-ray energy and power output. The first step in improving implosion quality is to minimize the initial axial and azimuthal nonuniformities in the load itself. This symmetry requirement, coupled with the additional requirement of low mass (and low density) in order to match the current levels and optimal implosion times of existing pulsed-power accelerators, has led to the choice of gas puffs, thin foils, metal vapor jets, low-density foams, and wire arrays as the loads in experiments on existing pulsed-power accelerators.

Typically, ~ 0.1- to 10-mg masses can be efficiently imploded on short-rise-time electrical generators such as the 25-TW Saturn accelerator (see Fig. 1.2) and the 50-TW Z accelerator. Z (see Figs. 1.3 and 1.4) with currents as large as 20 MA in the load, can efficiently implode masses of several mg/cm (Spielman et al., 1989; Spielman et al., 1997). The optimal, experimentally produced total x-ray energy output from several generations of pulsed-power accelerators at Sandia is plotted as a function of load current in Fig. 1.5. From these experimental data, which were obtained using gas-puff, annular-foil, and wire-array loads, the optimal x-ray energy output is observed to scale approximately as I^2 (Matzen, 1997). Basically, this is a statement of energy conservation where the energy initially stored in the magnetic fields scales as I^2. This measured x-ray energy output is somewhat higher than the radial kinetic energy calculated from the simple zero-dimensional scaling laws when we use the observed $\sim 10:1$ convergence ratio. Two-dimensional calculations of these fast Z pinches (Peterson et al., 1996) show that the additional energy comes from coupling of the $\mathbf{J} \times \mathbf{B}$ force into the finite-thickness sheath during the stagnation process. Since these calculations and experiments imply that total x-ray energy output is determined by the current, the x-ray pulse width must be decreased in order to optimize the x-ray power on any given pulsed-power accelerator.

The goals of the recent Z-pinch-driven x-ray-source development research were to perform reproducible experiments that could be compared quantitatively to detailed numerical simulations, to study the x-ray power optimization, and to provide an x-ray source that can be reliably fielded in a hohlraum configuration (Matzen, 1997). Wire arrays have been studied extensively. An annular array of wires used on Z, as shown in Fig. 1.6, is an attractive load configuration because the initial conditions (mass and radius) can be accurately determined and easily varied. A schematic of the experimental arrangement and the equivalent circuit are shown in Fig. 1.7. Low-density metal-doped foams offer similar advantages and are planned to be used in future, higher-current experiments. Thin annular foils (or liners) offer the potential of excellent axial and azimuthal uniformity. However, within the mass range that can be imploded on Saturn and Z, they must be on the order of a few hundred nanometers thick and are difficult to fabricate and handle without introducing

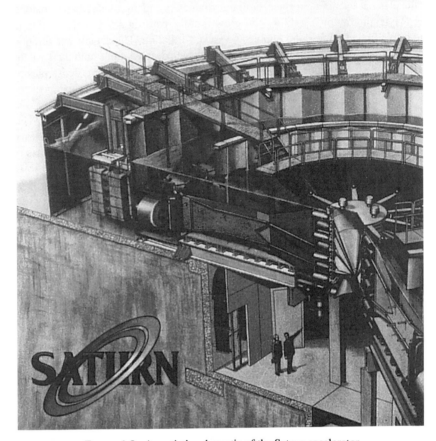

FIGURE 1.2. An artist's schematic of the Saturn accelerator.

significant azimuthal and axial perturbations. Z-pinch experiments with gas-puff loads are difficult to compare quantitatively to theoretical predictions since the mass distribution varies both radially and axially, is time dependent, often suffers from fluid turbulence, and is difficult to quantify. Metal vapor jets provide a uniform annular load for smaller pulsed-power accelerators, but do not provide enough mass for high-current drivers such as Saturn or Z. In addition, neither gas-puff nor metal-vapor-jet loads are easy to field in a hohlraum configuration.

Wire Array Loads

It was found that wire-array loads can provide excellent initial axial uniformity. Recent research with aluminum and tungsten wire arrays at Sandia National Laboratories on Saturn and Z, where the number of wires in cylindrical wire-array loads have been greatly increased from about 12 (See Fig. 1.8) up to 300 wires led to record x-ray power generation in the laboratory. The best obtained energy

FIGURE 1.3. A photograph of Z showing the Marx generators, the water pulse-forming section, the diagnostic work platforms, and the vacuum section.

and power of x-ray output are 2 MJ and 250 TW (Spielman et al., 1998). The initial circumferential gap between adjacent wires was varied from a maximum of ~ 6 mm to a minimum of ~ 0.4 mm. As shown in Fig. 1.9 (Sanford et al., 1996) for initial wire-array radii of both 8.6 and 12 mm, there is a dramatic increase in the peak x-ray emission power when the initial interwire gap was decreased below ~ 1.5 mm. This x-ray power increase was a result of both an increase in total energy of the emitted x rays and a decrease in the x-ray pulse width. Time-resolved x-ray pinhole pictures show that the diameter of the emission region decreases with increasing wire number, consistent with a more stable implosion and a higher density stagnation. One possible explanation for these results is suggested from preliminary one- and two-dimensional radiation-hydrodynamics calculations (Sanford et al., 1996; 1997). At an interwire gap of $\leq$ 1.5 mm, one-dimensional radiation-hydrodynamics calculations using the measured prepulse on Saturn as the input current drive predict that the expansion of the individuals wires is large enough to merge with the expanding plasma from the adjacent wires prior to the time at which the annular sheath begins to implode. At larger initial interwire gaps, the increasing current of the main drive pulse recompresses the individual wires (self-pinch), and they implode as a collection of unstable individual wires [see Fig. 1.10(A)]. With

FIGURE 1.4. Photograph showing the primary diagnostic line-of-sight for Z located on LOS 5/6. The diagnostics fielded on LOS 5/6 include a five-channel XRD array, a six-channel PCD array, a three channel nickel-bolometer array, a time-integrated convex-KAP spectrometer, and a time-resolved convex-KAP spectrometer. All of the diagnostics are individually alignable and have their own vacuum pumping system. A double-wall screen box is located immediately adjacent to the LOS in order to record high-bandwidth data. The cables shown are 20-ns long and have a 3-dB bandwidth of > 5 GHz. The digitizers are Tekronix TDS684A's and have a bandwidth of 1 GHz.

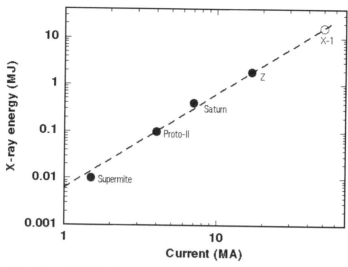

FIGURE 1.5. The optimized total x-ray output from several generations of pulsed-power accelerators at Sandia National Laboratories. The filled circles are measurements and the open circle is extrapolated. The dashed line indicates that the x-ray energy scales approximately as I^2 (Matzen, 1997).

FIGURE 1.6. Photograph of the wire-array load used on Z.

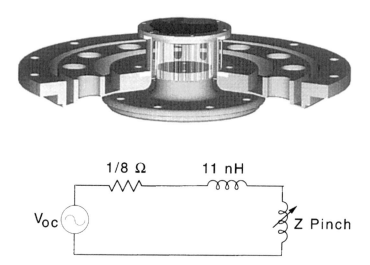

FIGURE 1.7. Schematic of experimental arrangement and the equivalent circuit for the wire-array load.

FIGURE 1.8. An early wire-array load used on Saturn with $\sim$ 24 wires.

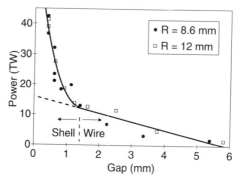

FIGURE 1.9. Total x-ray power versus wire gap.

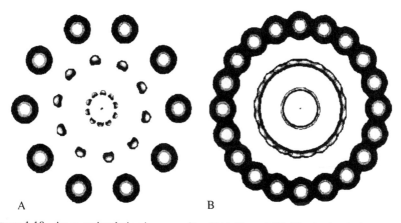

FIGURE 1.10. An x, y simulation in mm units of (A) 10- and (B) 40-wire implosions at 86, 11, 3, and 0 ns before stagnation for small radius load.

a large number of wires (smaller interwire gap), the current per wire is too small to recompress the individual wire plasmas, the plasmas from adjacent wires merge, and the two-dimensional (x-y) calculations predict the implosion of an annular plasma sheath [Fig. 1.10(B)]. Although these calculations provide a potential explanation for the dramatic increase in x-ray power output at smaller gaps, they are not definitive. For example, the low-temperature plasma resistivities are not well known, which greatly affects the one-dimensional calculations of the wire expansion. The two-dimensional (x-y) calculations do not account for sausage or kink instabilities in the axial direction and use a very simple radiation loss model. In any case, these experiments with wire arrays conclusively demonstrated the necessity of using a very large number of wires in a wire array in order to optimize the x-ray energy and power output from Z-pinch implosions, and represent breakthrough Z-pinch implosion load configurations for large pulsed-power generators.

Cylindrical Liners

One of the first dynamic load configurations proposed was the annular liner (Linhart et al., 1962). Experiments conducted at the Air Force Weapons Laboratory on the Shiva-Star facility demonstrated the utility of aluminum liners for microsecond-time-scale Z-pinch experiments (Turchi and Baker, 1973; Burns et al., 1977; Baker et al., 1978). Later experiments on Proto-II (5 MA, 50 ns) and Saturn (10 MA, 50 ns) at Sandia National Laboratories confirmed the utility of aluminum liners for Z-pinch x-ray sources. These experiments also showed that the development of the magneto-Rayleigh–Taylor instability was enhanced by poor liner quality. Aluminum liners with thickness of $\sim$ 200–300 nm had large azimuthal and axial wrinkles. Fabrication data suggest that until the current levels on pulsed-power drivers exceed 40–60 MA the quality of liners will be worse than the quality of equivalent-mass wire arrays.

Gas Puff Configurations

Gas-puff Z pinches represent an important class of dynamic Z-pinch device where plasma implosion is used (Shiloh et al., 1978; Stallings et al., 1979; Burkhalter et al., 1979, Baksht et al., 1981, Marrs et al., 1983). A typical scheme of a gas-puff Z-pinch device is shown in Fig. 1.11. The imploding annular gas sheet is a supersonic jet emitted by a gas-injection nozzle in the cathode, directed toward the anode; the gas is ionized by the main current pulse or preionized before it. To avoid initial nonuniformities in the gas flow, the anode is a honeycombed transparent structure, so that the gas jet propagates through the anode plane without reflection (Wessel et al., 1986). This is a very convenient way to obtain a relatively

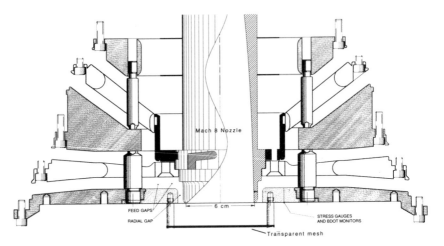

FIGURE 1.11. Schematic of a gas-puff Z-pinch device.

dense plasma sheet of macroscopic thickness and adjustable line mass. Typical number densities of the neutral gas in the gas puff are $\sim 10^{17}$–10^{18} cm^{-3}. Gas-puff Z pinches are used mainly as xuv and x-ray radiation sources. A gas-puff Z pinch has been used as a flashlamp for photopumping an x-ray laser (Maxon et al., 1985; Spielman et al., 1985a). Note, the application of gas-puff Z pinches to very-high-current devices can be limited because the required initial mass leads to very high required gas densities that are not readily available from gas-puff sources.

1.3.2 Equilibrium Z pinches

There are a number of Z-pinch configurations that are not characterized by a sizable implosion and high velocities. These configurations are included in the general class of equilibrium Z pinches. They include exploding wires, fiber pinches, gas-embedded Z pinches, and capillary discharges.

Exploding wires (Gol'din and Kalitkin, 1970; Mosher et al., 1975; Baksht et al., 1983; Aspden, 1985; Aranchuk et al., 1986; Graneau, 1987; Burtsev et al., 1990) are mainly used as light sources or compact x-ray sources. By passing a short current pulse through a fine metal wire, one can produce a high-density Z pinch the line density of the plasma thus being well-defined. The rapid ohmic heating in the wire results in its thermal explosion. To describe the dynamics of exploding wires, one should take into account the equation of state of the fiber material (Gol'din and Kalitkin, 1970); a macroscopic description of phase transformation and other hydrodynamic processes following the wire expansion is usually needed.

Attractive possibilities for fusion applications resulted from the recent experiments involving frozen deuterium fibers (Hammel et al., 1984; Scudder, 1985; Sethian et al., 1987; Scudder et al., 1987; Kies, 1988; Decker and Kies, 1989; Hammel, 1989; Ishii et al., 1989; Sethian et al., 1989). Single dielectric fibers of polyethylene or deuterated polyethylene were used as exploding wires in many previous experiments (e.g., see Young et al., 1977; Stephanakis, 1989); nevertheless, the results of the experiments with fiber-initiated high-density Z pinches (HDZP) turned out to be surprising in many respects. Typical experimental conditions of the above experiments are a peak voltage of ~ 2 MV, applied to a fiber made of solid deuterium whose initial diameter is ~ 20–100μm, and a length of ~ 5 cm. A very dense plasma is produced ($n_e \sim 10^{21}$–10^{22} cm^{-3}), but as the current grows very rapidly (the rise time being of order 100 ns) the plasma does not explode but seems to be confined by the azimuthal magnetic field of the current. These experiments demonstrated stability of the pinch plasma to $m = 0$ and $m = 1$ perturbations and high neutron yields observed with moderate currents (from ~ 250 kA to ~ 640 kA). The plasma column was found to be stable for about 100 ns, which is about 100 Alfvén transit times, in apparent contrast with the predictions of the conventional MHD stability theory. An intriguing feature of the fiber-initiated HDZP (stressed by Sethian et al., 1987) is that they appear stable as long as the current is rising, whereas the stability is lost after the current peak.

One possible explanation that seems to be reasonably correct is that the stability of HDZP before the current peak is due to evaporation of material from the dense core.

Dynamics and stability of HDZP were studied using both numerical simulation and analytical self-similar solutions (Coppins et al., 1988; Ikuta, 1988; Culverwell and Coppins, 1989; Lindemuth et al., 1989; Lindemuth, 1989a, b; Rosenau et al., 1989; Glasser, 1989; Glasser and Nebel, 1989; McCall, 1989; Meierovich and Sukhorukov, 1989; Bud'ko and Liberman, 1989a, b; Bud'ko et al., 1990b). There have been many attempts to explain the unexpected properties of the HDZP, first of all their anomalous stability (see Sec. 3.5), but at present no explanation is fully satisfactory, and further research is required. Typical parameters of powerful pulsed Z-pinch facilities are given in Table 1.

For a gas-embedded Z pinch, the pressure of the external gas is added to the magnetic pressure. To produce a gas-embedded Z pinch, a narrow plasma channel is created in a dense gas by a laser pulse, which is sometimes followed by additional ohmic preheating prior to the main current pulse (see Haines, 1981; Jones et al., 1981; Dangor et al., 1983; Hammel et al., 1984; Haines et al., 1986).

Another intriguing configuration is the capillary discharge (Lee and Ziegler, 1988; Marconi and Rocca, 1989). A $\sim 500\text{-}\mu m$ diameter hole, $\sim 1\text{--}3\text{-cm}$ long is fabricated in an insulator such as plastic or ceramic. A voltage of $\sim 10\text{--}100$ kV is applied across the material, and a surface discharge forms in the hole. The plasma that is evaporated from the sides of the hole sustains the discharge. The discharge is confined by the capillary and a high-density, nearly local thermodynamic equilibrium (LTE) plasma is formed. This type of plasma has been used to produce collisionally pumped vacuum ultraviolet (vuv) lasers (Wernsman et. al., 1990).

1.3.3 The dense plasma focus

Plasma-focus devices have been studied by many authors. Owning to the special design of the electrodes in a plasma focus, the initial axially nonuniform shape of a pinch in a plasma focus device provides favorable conditions for generation of a strong electron beam, as a result of plasma stagnation near the axis. Thus, the action of a plasma focus is determined by the dynamics of the current sheet. Figures 1.12 and 1.13 illustrate the design principle of the two main types of plasma focus devices: the Filippov type (Filippov et al., 1962) (Fig. 1.12) and the Mather type (Mather, 1964; Schmidt, 1980) (Fig. 1.13). The main difference between them is in the aspect ratio of the anode diameter, D, to its height, L: $D/L > 1$ and $D/L < 1$ for Filippov and Mather types, respectively. An interesting modification of the Mather-type plasma focus is one with an annular anode (e.g., see Schmidt et al., 1980). The typical parameters of a plasma focus are energy stored in the capacitor bank from 1 kJ to 1 MJ, voltage from 10 to 300 kV, current ~ 1 MA, and initial gas pressure from 2 to 5 Torr. The discharge is initiated at the lower end

TABLE 1.   Typical Z-Pinch Installation Parameters

Installation	Location[a]	Current pulse amplitude $I_m (MA)$	Current pulse duration (ns)	Power (TW)	Typical radius R of plasma column (cm)	Typical ion density n (cm^{-3})
Module Angara 5-1	KI	2	100	0.5	0.5	
Angara 5-1	KI	4.5	70	6	1	
SNOP-1	IHCE	0.5	70		0.7	10^{17}
SNOP-3	IHCE.	1.5	100		0.7	8×10^{17}
I-4	IHCE	2.5	100		0.7	2×10^{18}
Blackjack 3	ML	1	100	1	2	0.2×10^{19}
Blackjack 5	ML	4.6	100	10	2	0.4×10^{19}
Double Eagle	PI	3	100	8	2	0.4×10^{19}
PROTO-II	SNL	7.5	40	10	1	10^{19}
Saturn	SNL	11.5	50	25	1	10^{19}
Z	SNL	20	100	50	2	2×10^{19}
Fiber Experiment	LANL LANL	0.25 0.8	125 100	0.15	1.5×10^{-3} $(3-4) \times 10^{-3}$	5×10^{22} 5×10^{22}
Fiber Experiment	NRL	0.64	130		4×10^{-3}	5×10^{22}

[a] KI: Kurchatov Institute, Moscow and Troitsk, Russia; IHCE: Institute of High Current Electronics, Tomsk, Russia; ML Maxwell Laboratories, San Diego, CA; PI: Physics International, San Leandro, CA; SNL: Sandia National Laboratory, Albuquerque, NM; LANL: Los Alamos National Laboratory, Los Alamos; NM; NRL: Naval Research Laboratory, Washington, DC.

of the electrodes, where a current sheet is formed with electron number density $n_e \sim 10^{16}-10^{17}$ cm^{-3}.

The propagation of the current sheet to the axis before the final pinch compression takes a considerable time interval (3–5 μs). The motion of the current sheet is accompanied by the accretion of mass in the sheet and, depending on the geometry, significant axial mass motion. During this interval, part of the energy stored in the capacitor bank is transformed into the magnetic field energy in the anode–cathode gap behind the current sheath, and partially converted into the kinetic and thermal energy of the imploding plasma. The current sheet converges to the axis, driving

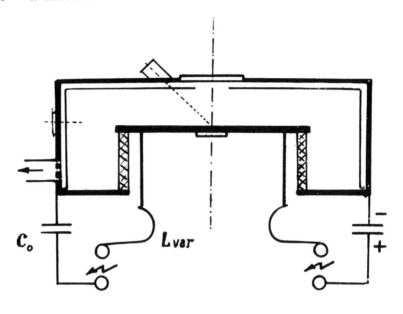

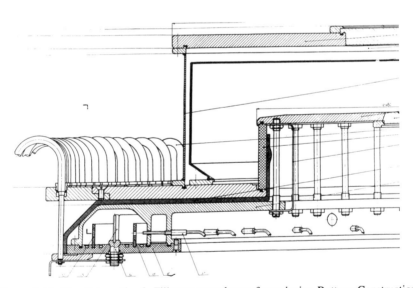

FIGURE 1.12. Top: Schematic of a Filippov-type plasma-focus device. Bottom: Construction of the MT device (courtesy of N.V. Filippov and T.I. Filippova). Here $C = 576$ mF; L is an adjustable inductance chosen to match the current sheath dynamics and the LC time of the circuit (here 6 ms); $U = 20$ kV. Dimensions of the chamber are determined by the anode diameter ($D = 66$ cm). The walls of the chamber and its bottom serve as a cathode.

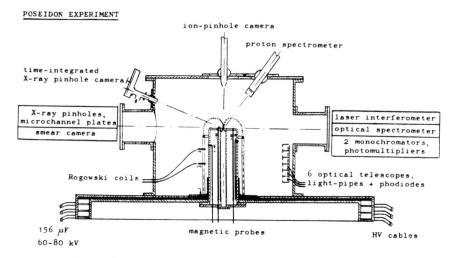

FIGURE 1.13. Schematic of a Mather-type plasma-focus device.

an imploding shock wave, which heats neutral gas and involves it in the inward motion. Most of the kinetic energy is thermalized at stagnation; some of the kinetic and magnetic energy is transformed into the energy of an accelerated particle beam, the amount being determined by the regime of the plasma focus action. The amount of energy available for the plasma compression is sensitive to many factors, such as, the geometry of the electrodes, the pressure and atomic number of the gas (or composition of the gas mixture), the material of the anode and of the insulating walls. Typical plasma parameters at the end of compression are: $n_e \sim 10^{20}$ cm^{-3}, $T_i \sim 0.3$–9 keV, $T_e \sim 0.3$–2 keV. Thus, a plasma-focus device can be used not only as a powerful x-ray source, but also as a neutron source, the observed neutron yield per pulse being as high as 10^{12}.

The dense-plasma pinch formed at the final stage of compression has much in common with a conventional cylindrical Z pinch. However, it exhibits violent instabilities leading to plasma turbulence, which is established both by electrical and laser scattering measurements (Bernard et al., 1979). In particular, development of filamentation instabilities is observed both during the implosion of the current sheet and at the final stage (Sadowski et al., 1984; Bruzzone and Fischfeld, 1989). The dynamics of the plasma-focus devices were studied numerically by many authors (e.g., see Bazdenkov et al., 1978; Gureev et al., 1975; Gureev, 1978; Gerusov and Imshennik, 1985; Gerusov, 1988).

1.3.4 Plasma arcs

The vacuum spark devices (Cilliers et al., 1975; Negus and Peacock, 1979; Korop et al., 1979) are low-voltage, low-energy (below ~ 20 kV and 2 kJ, respectively) diodes conventionally used as sources of xuv and x-ray radiation. These devices are similar to Z pinches but the plasma kinetics are different. In a vacuum spark, the current flows through a rarefied gas ($n_0 \sim 10^{16}$ cm^{-3}), and the plasma is nonequilibrium (non-Maxwellian), with temperatures in the range $\sim$ 1–30 keV. Most of the radiation comes from the bright spots with tens of microns dimensions, where the plasma density can exceed the average value by orders of magnitude. Line radiation is characterized by the presence of K_α lines suggesting that the radiation production mechanism is through inner-shell ionization from electron beams. The plasma density is low enough that the plasma is essentially collisionless, so kinetic instabilities and anomalous transport determine the behavior of the vacuum-spark plasma, whereas the characteristic dimensions of dense Z pinches are typically much greater than a mean free path, so the plasma is collisional.

1.4 Pulsed-power drivers

The purpose of any pulsed-power system is to transform the electrical energy from the rather low powers ($>$ MW) and long timescales (seconds) that can be obtained from an electrical outlet to the high powers ($>$ TWs) and short times ($\sim$ 10 ns–10μs) and small spatial scales (cm) that are required to drive a Z-pinch load. A simple, imploding Z pinch is unstable to magneto Rayleigh–Taylor modes (see Chapter 5) that degrade the plasma density at stagnation (see Chapter 6). The total number of e-foldings that characterize the instability growth during the implosion can be reduced by reducing the implosion time (again see Chapter 5). Thus, pulsed-power techniques that reduce the pulse width of the electrical power delivered to the load are important for reducing the growth of magneto-Rayleigh–Taylor instabilities and therefore the performance of the Z pinch.

A number of pulsed-power systems are typically used to drive Z-pinch loads. The simplest is a primary energy-storage device, usually a capacitor bank, a high-power low-inductance switch, and a low-impedance transmission line connected to the load. This system has the advantage of low cost and small size but is limited to pulse widths of a few microseconds for interesting stored energies ($\sim$ MJ).

A Marx generator coupled with a pulse-forming line is the most widely used, high-power system today. The erected Marx generator slowly ($\sim \mu$s) charges an intermediate storage capacitor, which subsequently is rapidly (10–100 ns) discharged through the load. Systems that utilize pulse-forming lines have the advantage of simplicity but at the expense of efficiency.

Inductive energy storage rather than capacitive energy storage can be used to drive Z pinches. The vacuum inductor is charged on a microsecond time scale using capacitors (usually in a Marx configuration) as the primary energy store. A conducting plasma is placed between the storage inductor and the load. Plasma erosion or hydrodynamic motion partially interrupts the current flowing in the inductor switching the current to the load. Inductive energy storage concepts can be very inexpensive and efficient, with the limitation being the effectiveness of the plasma opening switch (POS). These different systems are discussed below.

1.4.1 Primary energy storage device

Chemical explosives or high-energy-density capacitors are used as primary energy devices in pulsed-power systems (Knoepfel, 1970). The cost per joule of chemical explosives is very low ($\sim$ \$0.01/J), but the generator must be replaced for each Z-pinch event. Chemical explosives are typically used for infrequent experiments requiring very large stored energies ($\geq$ 50 MJ) where the high capital costs of capacitors can be prohibitive. The advantage of capacitors is that they are low cost ($\sim$ \$0.1/J), relatively long-lived, and highly efficient ($\sim$ 90%). The development of higher energy-density capacitors continues to reduce the size and cost of MJ-class Z-pinch drivers. Modern capacitors can store megajoules of energy in $\sim$ 10 m^3.

1.4.2 Slow (microsecond) pulsed-power drivers

The simplest pulsed-power system has a capacitor bank as the primary energy-storage device, a fast low-inductance closing switch, and a low-inductance transmission line connected to the load. The transmission line is fabricated with two or more closely spaced parallel conducting plates that are separated by a very thin dielectric sheet, such as Mylar, to prevent breakdown. The advantage of this simple pulsed-power system is that the total generator cost and the laboratory space requirement are low. The disadvantage is that the pulse width of the power delivered to the Z-pinch load is limited to the microsecond range. This is because the charge voltage of available energy storage capacitors is limited ($V_{max} \sim$ 100 kV) and it is difficult to reduce the generator inductance to below a minimum of $L_{min} \sim$ 10 nH. A very good approximation of the equivalent circuit for the simple pulsed-power system is a capacitor of capacitance C, and an equivalent series inductor of inductance L. The equivalent inductance includes the inductance of the capacitors, the transmission line, and the switch. Typically the switch inductance dominates the inductance of the load (although this is not always so for modern switches). The generator pulse width is $\tau_{1/4} \sim (LC)^{1/2} = (2LE)^{1/2}/V$, where V is the capacitor charge voltage and E is the electrical energy stored, $E = CV^2/2$. A simple way to estimate the rise time is to use and $V = L(dI/dt)$, $I \approx (V/L)\Delta t$, and we see that $\tau_{min} \sim (MA \cdot nH)/MV \sim$ 1 μs for stored energy, $E \sim$ MJ, and V $\sim$ 100 kV. As can be seen the rise time of capacitive-storage pulsed-power systems increases as the stored energy increases.

The rise time of such direct drive schemes can be somewhat improved by placing the capacitors in a simple Marxed configuration (see the next section). This has been done on the Atlas driver (See Fig. 1.14) being constructed at Los Alamos National Laboratory (Parsons et. al., 1997).

1.4.3 Fast ($\sim$ 10–100 ns) pulsed-power systems

Short-pulse, high-voltage power generators were developed in the United States and in Russia after the pioneering efforts in the early 1960s by J.C. Martin (Martin et al., 1996) at the Atomic Weapons Establishment in England. The basic

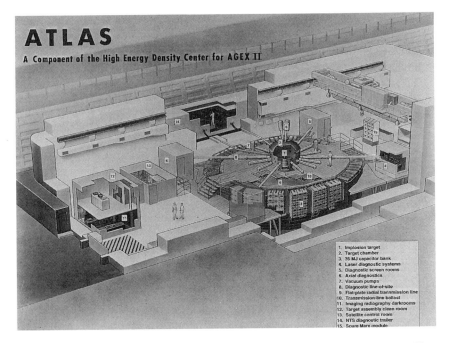

FIGURE 1.14. An artist's schematic of the Los Alamos National Laboratory Atlas Facility is shown.

idea is to use capacitors configured as a high-voltage Marx generator to slowly ($\sim$ microsecond) charge a very-low-inductance high-voltage intermediate-storage capacitor that is subsequently rapidly ($\sim$ 10–100 ns) discharged through a transmission line to the load. A high-power low-inductance switch isolates the load from the intermediate storage device until it is fully charged (Nation, 1979; Miller, 1982). The intermediate storage device is typically a cylindrical pulse-forming line (PFL). By staging a series of these PFLs the final pulse generated can be as short as 10 ns. Modern generators use ultra-pure water as the dielectric medium.

A more recent and exciting development in the field of fast, pulse-forming systems is the use of inductive energy storage (IES) rather than capacitive energy storage in the PFLs (Sincerny et al., 1993; Kim et al., 1995). IES systems make use of the fact that a higher energy density can be stored in magnetic fields than in electric fields. Thus, IES schemes are potentially more compact and less expensive than their capacitive storage predecessors. In IES devices the primary energy storage is still capacitive (usually in a Marx configuration, see the next section). The vacuum inductor is charged on a submicrosecond time scale. Figures 1.15 and 1.16 show two typical IES systems in which a Marx generator charges a coaxial inductor with current in less than 1 μs. A plasma, injected into the vacuum gap (sometimes aided with applied magnetic fields), carries the current. After a conduction time determined by the plasma mass and, possibly, the applied magnetic

FIGURE 1.15. A photograph of the Decade Module-2 IES/POS testbed located at Physics International Co., San Leandro, CA. Energy is first stored in the coaxial inductor on the left-hand-side of the photograph, switched with the POS shown in the middle, and flows to an electron beam at the right-hand-side of the device. (Photograph courtesy of PRIMEX/Physics International Co.)

field the current flowing in the plasma is interrupted and energy is switched into the load. A conducting plasma is injected between the anode and the cathode of the inductor at a location between the storage inductor and the load. Sufficient plasma must be injected to withstand the conduction time or the time it takes to charge the storage inductor. It is clear that shorter charging or conduction times reduce the plasma quantity needed in the POS. At low plasma densities plasma erosion, aided by the Hall effect, "opens" the circuit. At high densities, typically found in high-current, long-conduction time systems, hydrodynamic motion interrupts the current flowing in the inductor by moving the plasma out of the gap. In this concept the ability of the "plasma opening switch" (POS) to open, while withstanding the high voltage developed across the Z-pinch load, is the key technological challenge.

The Marx Generator

A Marx generator is a number (N) of capacitors that are charged in parallel and discharged in series by a number of switches (see Fig. 1.17). The basic idea is to

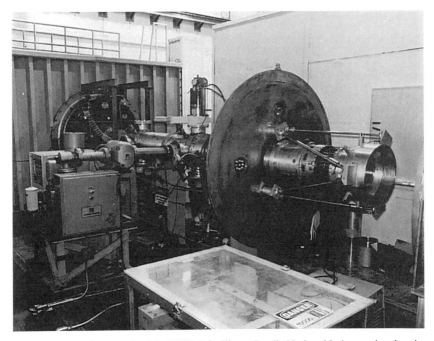

FIGURE 1.16. A photograph of the TESLA facility at Sandia National Laboratories showing the coaxial inductive energy store and one version of a magnetically controlled plasma opening switch (right-hand side).

increase the voltage that drives the load without increasing the individual capacitor voltage requirement. The electrical switches in Marx generators are usually gas-filled spark gaps of fixed dimensions, with the operating (breakdown) voltage range set by varying the gas pressure (typically SF_6 or a mixture of SF_6 and air). The charge and trigger resistors are fabricated of flexible plastic tubing filled with a copper sulfate ($CuSO_4$) solution of the appropriate resistivity. For output voltages greater than about 1 MV, the Marx generator is usually immersed in transformer oil to prevent flashover. The advantage of the Marx generator is the erected capacitance (C_{stage}/N) together with the series inductance of the Marx ($N L_{stage}$) gives a rise time nearly the same as that of a single stage. There is no stored energy rise-time penalty!

Pulse-Forming Lines

The Marx generator is used to charge a PFL. Typically, a coaxial transmission-line PFL is used with either simple or double (Blumlein, 1948) transmission lines. (A simple transmission line is a cylindrical capacitor.) The line is typically short, compared to the transit time of an electrical pulse down the line. Thus, the line

FIGURE 1.17. A photograph of a modern low-inductance Marx generator on the Sandia Z accelerator.

acts as a lumped capacitor and the Marx generator and PFL act as a capacitive divider. The peak charge voltage of the PFL, when the current in the circuit is zero, is $V_0 = 2V_m C_m/(C_m + C_1)$, where V_m and C_m are the voltage and capacitance of the Marx generator, respectively.

During the discharge through the load, the PFL can be considered a transmission line with a characteristic impedence that is well-approximated by $Z_0 = (L_1/C_1)^{1/2}$. If the load impedance is matched to the PFL impedance, i.e., $Z_0 = Z_1$, then there are no reflections at the load and the energy is extracted in a time, 2τ, where $\tau = (L_1 C_1)^{1/2}$. The voltage at the load is $V_0/2$.

A. D. Blumlein (1948) developed an improved PFL in which the voltage into a matched load is equal to the charge voltage, V_0. A simple Blumlein circuit consists of three coaxial cylinders with the intermediate cylinder charged by the Marx generator (Fig. 1.18). The center cylinder is connected through a charging inductor to the outer cylinder. The charging inductor is shorted during the slow charging of the PFL and open during the fast discharge of the PFL through the load. After the outer and intermediate cylinders are charged, a Blumlein switch shorts the inner and intermediate cylinder together, so that a voltage of $2V_0$ and a pulse width of 2τ is developed across the charging inductor, which is effectively an open circuit. The Z-pinch load is connected through a high-power switch in parallel with the charging inductor. The load switch is closed at time τ after the Blumlein switch closes. A voltage of V_0 and pulse width of 2τ are developed across the load.

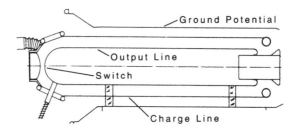

FIGURE 1.18. Schematic of a simple Blumlein circuit.

Hermes II a 10 MV and 100 kA accelerator at Sandia National Laboratories was one example of a large Blumlein machine (see Fig. 1.19).

One of the problems with these PFL designs is that a prepulse is generated in the load as the PFL is charged. Typically, the pulse width of the prepulse is quite long (microseconds). Because the prepulse can be deleterious for some designs, methods have been developed to control the amplitude and pulse width of the prepulse. A prepulse switch can be used prior to the load to reduce the prepulse.

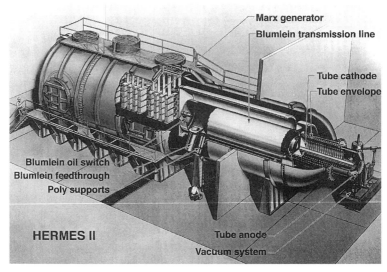

FIGURE 1.19. A schematic of the Hermes-II facility at Sandia National Laboratories is shown.

1.4.4 Electrical switching

Normally-Open Switches

The key to many pulsed-power designs is the requirement for normally open, high-power switches. The switches are basically an open circuit that is switched to low impedance by ionization (breakdown) of the switch medium. Water, oil, and high-pressure gas are all possible switch media. The switches must have low inductances and low resistances to minimize electrical reflections and losses, and the capability of synchronization to nanosecond accuracy. The switches operate in a self-break or triggered mode. The self-break switches switch when the voltage across the switch reaches a predetermined value.

The most common triggered-switch configurations are the trigatron, field-distortion gaps, and laser-triggered configuration. The trigatron switch has a trigger pin in one of the electrodes. A fast, high-voltage pulse is applied to the pin, resulting in an arc between the pin and the electrode, which causes breakdown of the switch medium. A field-distortion switch is triggered by applying a high voltage to a mid-plane conductor. The applied voltage distorts the electric field created by the primary applied voltage and initiates a breakdown. A laser can be used to trigger switches by creating an ionized channel between the conducting electrodes of a switch. In all cases, low-jitter (1- to 3-ns) performance has been demonstrated.

Normally-Closed Switches

POS's are normally closed, high-power switches that are used to decrease the pulse width of pulsed-power generators operating in the inductive-energy-storage mode. The basic idea is that cold ($T_e \sim 1$ eV) plasma is injected into one end of a vacuum-transmission line/inductor and shorts the transmission line. The magnetic pressure of the electromagnetic field in the tranmission line and/or complex plasma erosion processes due to current in the plasma acts to move or erode the plasma away from the transmission line, and after some time, the transmission line becomes an open circuit, i.e., the POS opens. A POS must open after conducting a MA-level current flowing through the inductor for times approaching a microsecond. The combination of a long conduction time at high current represents the significant technical risk of this concept.

1.4.5 Magnetically-insulated vacuum transmission lines

The simultaneous requirements of high voltage and low inductance on large Z-pinch drivers eventually causes the electric field on transmission line cathodes to exceed the self-emission threshold, typically ~ 250 kV/cm. A current of electrons will then flow to the anode at the unipolar Child–Langmuir limit. Given time this current will reach the anode of the transmission line, heat the anode material, and through a number of processes generate an anode plasma. When this plasma expands and reaches the cathode, the conditions for a catastrophic arc are met.

FIGURE 1.20. The lower two MITLs from Z are shown. The MITLs are fabricated from aluminum at large diameter and stainless steel inside a diameter of 1 m.

The use of low-impedance PFLs coupled with good vacuum transmission line design allows the self magnetic field of, first, the electron loss current and, later, the current through the load to inhibit this electrical loss. Vacuum power feeds that take advantage of this process are called self-magnetically insulated transmission lines (MITLs). At the simplest level, when the magnetic field is such that the Larmor radii of the emitted electrons is significantly less than the anode–cathode gap, losses of electrons to the anode cease. The Z accelerator at Sandia National Laboratories has 3-m diameter MITLs that operate at a peak voltage of 3 MV with anode–cathode gaps as small as 1 cm (see Fig. 1.20).

2

Equilibria of Z-Pinch Plasmas

2.1 Steady-state equilibria of Z-pinch plasmas

2.1.1 MHD equilibria of a Z pinch – The Bennett equilibrium

The Z pinch represents the simplest one-dimensional (1-D) model of the toroidal configuration. The parameters of Z-pinch plasmas can be estimated in a very simple way, if a steady-state equilibrium is assumed (Bennett, 1934; Blackman, 1951). This simplifying assumption rarely can be justified fully, taking into account the intrinsically dynamic nature of most Z-pinch experiments. In reality, the estimates based on this assumption are often correct only within an order of magnitude. Nevertheless, they are convenient and, therefore, conventionally used.

Setting the velocity and all time derivatives to zero in the ideal magnetohydrodynamic (MHD) equations (3.44) and in Maxwell's equations reduces them to what are known as magnetostatic equations describing the steady-state equilibria of perfectly conducting plasmas:

$$\nabla P - \frac{1}{c} \mathbf{J} \times \mathbf{B} = 0, \tag{2.1}$$

$$\nabla \times \mathbf{B} = \frac{4\pi}{c} \mathbf{J}, \quad \nabla \cdot \mathbf{B} = 0. \tag{2.2}$$

Some immediate consequences of (2.1) and (2.2) can be obtained by taking the divergence of (2.2) and the scalar product of (2.1) with $\mathbf{B}$:

$$\nabla \cdot \mathbf{J} = 0, \quad \mathbf{B} \cdot \nabla P = 0. \tag{2.3}$$

(Note, the result that the divergence of $\mathbf{J}$ is zero also follows naturally from ideal MHD and the assumption of charge neutrality.)

Z pinches represent the simplest class of unbounded (in one dimension) plasma equilibria. The second equation (2.2) is trivially satisfied because of cylindrical symmetry: all quantities are functions only of the radial coordinate r. A detailed analysis of steady-state plasma equilibria is given by Shafranov (1957, 1963) (see also Freidberg, 1987). In particular, Shafranov proves that no plasma equilibrium can exist that is bounded in space and not supported by a magnetic field from external conductors. Shafranov also proved the remarkable property that the simple Z pinch ($B_r = B_z$, $B_\varphi \neq 0$,) is the only possible plasma equilibrium.

Eliminating $\mathbf{J}$ between (2.2) and (2.1) and substituting $\mathbf{J}$ from Ampere's law into the momentum equation in cylindrical coordinates,

$$ J_z = \frac{c}{4\pi r} \frac{d}{dr}(r B_\varphi), $$

yields

$$ \frac{d}{dr}\left(P + \frac{B_\varphi^2}{8\pi} + \frac{B_z^2}{8\pi}\right) + \frac{B_\varphi^2}{4\pi r} = 0. \tag{2.4} $$

Equation (2.4) is the basic radial pressure-balance relation for a Z pinch. The terms in the derivative represent the particle (kinetic) pressure and the magnetic pressure. The last term in (2.4) represents the tension force generated by the curvature of the B_φ magnetic field lines that provides radial confinement of the plasma. For the pinch plasma, equilibrium is made up of the tension force that acts inward, thereby providing radial pressure balance, and an outward force due to the plasma pressure and any compression of the field B_z. There is only one equation, (2.4), for three functions, P_r, $B_\varphi(r)$, and $B_z(r)$, so two free functions are available to specify equilibrium; (2.4) can be satisfied with any two of them (or any two relations between them) taken arbitrarily. Since the plasma density $n(r)$ and temperature $T(r)$ do not enter (2.4) separately, one of them can be chosen arbitrarily for any given pressure profile. This results in a great variety of the possible equilibrium profiles. In a simple Z pinch [without an external magnetic field $B_z(r)$], we have one degree of freedom less. For instance, assuming an uniform current density J_z in the plasma column, we find

$$ B_\varphi(r) = \frac{2I}{cR}\frac{r}{R}, \quad P(r) = (I^2/\pi R^2 c^2)(1 - r^2/R^2), \tag{2.5} $$

where I is the total current flowing through the pinch and R is the radius of the plasma column. If the heat conductivity of the plasma is sufficiently high so that the temperature is uniform, then the ion-density profile has the same parabolic shape:

$$ T_w(r) = (I^2/2Nc^2) = \text{constant}, \quad n(r) = (2N/\pi R^2)(1 - r^2/R^2), \tag{2.6} $$

where $T_w = T_i + \overline{Z}T_e$ is the weighted plasma temperature, $\overline{Z}$ is the average ion charge, and N is the ion line number density

$$ N = 2\pi \int_0^R n(r)r\,dr. \tag{2.7} $$

Integrating (2.4) for a simple Z pinch, one easily finds the Bennett condition, i.e., the radially integrated form of the pressure-balance condition:

$$ NT_B = I^2/2c^2(1 + \overline{Z}), \tag{2.8} $$

where a weighted (by density) mean temperature, the Bennett temperature, T_B, is

$$T_B = \frac{2\pi}{N(1 + \overline{Z})} \int_0^R P(r)r\,dr. \qquad (2.9)$$

In general, the balance between the inward force and outward forces is represented by the Bennett relation (2.8), which is an integral equilibrium condition valid for any profiles of $J_z(r), n(r)$, and $T(r)$. For an isothermal Z pinch, where $\overline{T}_e = \overline{T}_i = T_B$, Eq. (2.9) can be presented in a form convenient for practical use:

$$T_B(\text{keV}) = 3.12(I/\text{MA})^2(1 + \overline{Z})^{-1}(N/10^{18}\text{ cm}^{-1})^{-1}. \qquad (2.10)$$

The lower the line density is, the higher the plasma temperature will be, hence, the better for most applications. However, there is an important limitation on N from below, stemming from the fact that the average electron-drift velocity, $\overline{v}_{De} = |J_z|/en_e$, should not exceed the ion thermal speed, $\overline{v}_{thi} = (T_i/M_i)^{1/2}$. Otherwise, current-driven plasma microinstabilities (e.g., Mikhailovskii, 1977) would develop in the pinch plasma. Using the Bennett relation (2.8), we express this condition on the average values of $\overline{v}_{De}$ and $\overline{v}_{thi}$ in the pinch via N:

$$\overline{v}_{De}/\overline{v}_{thi} = 2^{3/2}(1 + \overline{Z})^{1/2}\overline{Z}\Pi^{-1/2} < 1, \qquad (2.11)$$

where the dimensionless parameter, Π, is defined as (Braginskii, 1963)

$$\Pi = 4\overline{Z}^2 e^2 N/M_i c^2 = 6.1 \times 10^2(\overline{Z}^2/\overline{A})(N/10^{18}\text{ cm}^{-1}), \qquad (2.12)$$

and $\overline{A}$ is the average ion atomic weight. To satisfy (2.11) in a DT (deuterium–tritium) plasma, N should not be smaller than $N_0 = 6.1 \times 10^{15}\text{ cm}^{-1}$. Note that the actual values of v_{De} and v_{thi} can differ significantly from the average ones [e.g., $v_{De} \to \infty$ near the pinch boundary for the profiles (2.5) and (2.6)], so that near the periphery of the pinch, where n_e and $\overline{Z}$ may be low and J_z finite, this instability can develop even with $N > N_0$.

It is often useful to define the efficiency of plasma confinement by the magnetic field as the ratio of the averaged plasma pressure to the averaged magnetic pressure (Freidberg, 1987). Using (2.4), it is straightforward to calculate the basic pinch-plasma parameter, the so-called Bennett Pinch Relation, $\beta = (4\pi/I^2) \int_0^R P\,r\,dr$. Using (2.4), one finds $\beta = 1/c^2$, which is valid for any simple Z-pinch profile. Similarly, the MHD safety factor $q(r)$ is given by $q(r) = 2\pi r B_z/L B_\varphi$, and for the kink safety factor we obtain $q_* = q(R)$.

The characteristic velocity of small perturbations propagating radially in a steady-state Z pinch is that of a fast magnetosonic wave. The latter can be estimated by the Alfvén velocity, $V_A = B_\varphi(4\pi\overline{\rho})^{-1/2}$, where $B_\varphi = 2I/Rc$ is the azimuthal magnetic field near the outer boundary of the pinch and $\overline{\rho} = Nm_i/\pi R^2$ is the average mass density:

$$V_A = I/c(Nm_i)^{1/2} = 7.6 \times 10^7(I/\text{MA})(\overline{A}N/10^{18}\text{ cm}^{-1})^{-1/2}\text{ cm/s}. \qquad (2.13)$$

The average ion thermal velocity, $\bar{v}_{thi} = (2\overline{T}_i/m_i)^{1/2}$, with the average ion temperature, $\overline{T}_i$, being given by (2.9), is expressed via V_A:

$$\bar{v}_{thi} = (1 + \overline{Z})^{-1/2} V_A. \tag{2.14}$$

The time-scale characteristic of the Bennett equilibrium is determined by the Alfvén transit time

$$\tau_A = R/V_A. \tag{2.15}$$

When the parameters of a Z pinch depend on time (e.g., time-varying current or an imbalance between ohmic heating and radiation losses), time-dependent Bennett equilibria can be considered if the time scale of this variation is much greater than τ_A, so the Z pinch can adjust to its slowly changing parameters.

In general, the magnetostatic equations are very difficult to solve with three-dimensional (3-D) geometry. Fortunately in many cases, experimental devices have a certain degree of symmetry, so the dimensionality of the equations is reduced. The important case in practice is that of axisymmetric toroidal geometry describing plasma equilibrium in tokamaks. Analytical solutions for a noncircular plasma were considered by Solov'ev (1975).

2.1.2 Anisotropic Z-pinch equilibria

In the ideal case, a plasma would be in equilibrium for times that are appreciably greater than the mean electron–ion collision time. Under these conditions, the plasma pressure is a scalar quantity and the equilibrium conditions are given by (2.1) and (2.2). For modern pinches operating at high currents ($I \geq 1$ MA), the pinch plasmas may be collisionless: resistivity is unimportant, pressure is not isotropic, and, therefore, ideal (and non-ideal) MHD is invalid. It is also possible that pressure anisotropy of the collisionless pinch plasma affects the instability growth rate (Faghini and Scheffel, 1987; Coppins and Scheffel, 1989, 1992).

The simplest description of a collisionless plasma, which includes anisotropy, is the Chew–Goldberg–Low (CGL) single-fluid model (Chew et al., 1956). Equations (2.1) and (2.2) are still valid for a steady-state equilibria, but with a pressure tensor instead of a scalar pressure. The pressure is assumed to be locally isotropic in the plane perpendicular to magnetic field, $P_\perp$, while the pressure parallel to **B** has an independent value, $P_{||}$. In cylindrical coordinates, the pressure-balance relation for an anisotropic Z pinch can be obtained in the same way as (2.4) and takes the following form:

$$\frac{d}{dr}\left(P_\perp + \frac{B_\varphi^2}{8\pi}\right) + \frac{B_\varphi^2 + (P_\perp - P_{||})}{4\pi r} = 0. \tag{2.16}$$

Different classes of anisotropic equilibria were discussed by Coppins and Scheffel (1992). They found that the effect of equilibrium pressure anisotropy in the collisionless small-Larmor-radius regime cannot provide significant stabilization, but the growth rate decreases with increasing anisotropy.

2.2 Equilibria of radiating Z pinches

2.2.1 The Pease-Braginskii equilibrium

Plasma radiation is the most important energy-loss mechanism in Z pinches, from both physical and practical points of view. A steady-state equilibrium of a Z pinch is possible, if the total radiative energy losses do not exceed the energy released by resistive (ohmic) heating. An excess of energy released by ohmic heating, if any, can be absorbed by the electrodes due to axial heat conductivity. The radius of a quasistatic pinch is determined by the balance between ohmic heating and radiative cooling: if the heating exceeds the radiation, the pinch will expand, and vice versa. For optically thin plasmas bremsstrahlung energy losses, which dominate in hydrogen and deuterium plasmas, the power density of radiated energy is

$$Q_r = (32/3)(2/\pi)^{1/2} e^6 h^{-1} (m_e c^2)^{-3/2} \bar{g} \bar{Z}^2 n_e n_i T_e^{1/2}$$
$$= 1.33 \bar{g} \bar{Z}^2 n_e n_i (T_e/\text{keV})^{1/2} \text{ kJ cm}^{-3}\text{ns}^{-1}. \tag{2.17}$$

where $\bar{g}$ is the Gaunt factor averaged over the Maxwellian distribution (in most cases $\bar{g}$ is close to unity, and below we assume $\bar{g} = 1$), n_e and n_i are expressed in cm^{-3}. Assuming the plasma to be transparent to the radiation in the whole range of parameters of Z pinches, this term represents bremsstrahlung radiative losses from the plasma volume. Integrating over the plasma volume (per unit length) the radiative losses term (2.17) and the ohmic heating term, J^2/σ, where $\sigma = \sigma_0 T_e^{3/2}$(eV) for the classical Coulomb plasma conductivity ($\sigma_0 = 8.86 \times 10^{12}/\ln \Lambda$), and with the aid of the Bennett condition (2.8), we obtain the condition of power equilibrium formulated above

$$P_j \equiv 2\pi \int_0^R (J^2/\sigma) r \, dr = P_r \equiv 2\pi \int_0^R Q_r r \, dr, \tag{2.18}$$

where P_j and P_r are ohmic heating power and radiated power per unit length of the Z pinch, respectively. About 40 years ago, Pease (1957) and Braginskii (1957b) (see also Braginskii and Shafranov, 1958) independently predicted a critical current I_{PB}, at which a simple Z pinch would have an energy balance between the ohmic heating and bremsstrahlung radiation losses. Taking into account the pressure balance given by the Bennett condition (2.8) for force equilibria, this gives a limitation on the current $I < I_{PB}$, where

$$I_{PB} = \left(\frac{3\pi}{8}\right)^{1/2} \frac{h^{1/2} m_e c^{7/2}}{e^2} \frac{\bar{Z}+1}{\bar{Z}} (\lambda \ln \Lambda)^{1/2}$$
$$\approx 0.685 \frac{\bar{Z}+1}{\bar{Z}} [\lambda \ln(\Lambda/10)]^{1/2} \text{MA}, \tag{2.19}$$

In Λ is the Coulomb logarithm, and the factor λ in (2.19) is a numerical factor depending on the temperature, density, and current density profiles in the Z pinch:

$$\lambda = (4/3) \left[\int_0^1 \widetilde{J}^2(x) \widetilde{T}^{-3/2}(x) \, dx \right] \cdot \left[\int_0^1 \widetilde{n}^2(x) \widetilde{T}^{1/2}(x) \, dx \right]^{-1} ; \qquad (2.20)$$

here $x = r^2/R^2$, $\widetilde{J} = J(\pi R^2/I)$, $\widetilde{T} = T_e(1 + \widetilde{Z})/\widetilde{T}$, and $\widetilde{n} = n(\pi R^2/N)$. For the profiles (2.5) and (2.6), $\lambda = 1$, and in most cases λ is close to unity.

Existence of the critical current I_{PB} becomes clearer if one integrates the energy equation over the pinch cross section, taking into account (2.8), assuming thermal isolation outside the pinch radius $R(t)$, a uniform current density and temperature, and a parabolic density profile. The result is

$$\frac{d}{dt}(I^2 R^{2(\gamma-1)}) = \frac{16(\gamma - 1)N^{3/2}}{\pi \sigma_0 I R^{4-2\gamma}} \left(1 - \frac{I^2}{I_{PB}^2} \right). \qquad (2.21)$$

Equation (2.21) can alternatively be derived from the first law of thermodynamics (Haines, 1989). All these derivations consider an equilibrium state of a pinch plasma, while the main limitations stem from the dynamic nature of the pinches.

In equilibrium, the current I determines the plasma temperature. Using (2.8) to estimate the resistive electric field E applied to the pinch

$$E = I/\sigma \pi R^2 = \text{const} \cdot N^{3/2}/R^2 I^2, \qquad (2.22)$$

E decreases with increasing I (Vikhrev and Braginskii, 1980). However, for most experimental conditions, the electrical impedance of the plasma column is determined mainly by its inductance and not by resistivity.

The maximum average temperature possible in an equilibrium corresponds to $I = I_{PB}$ and equals,

$$\overline{T} = I_{PB}^2/2c^2N = 3.05 \frac{\overline{Z} + 1}{\overline{Z}} (N/10^{18} \text{ cm}^{-1})^{-1} \text{ keV}. \qquad (2.23)$$

At currents above I_{PB}, radiation losses exceed ohmic heating, and the thermal pressure begins to fall. The pinch contracts more rapidly under the excess magnetic pressure, and as the density increases, the radiation rate increases correspondingly. There is no Bennett equilibria with $I > I_{PB}$, and the pinch will undergo a spontaneous contraction (Shearer 1976, Vikhrev 1978) that will be stopped only when additional effects come into play, such as opacity or the onset of degeneracy in the electron gas (Meierovich and Sukhorukov, 1975; Robson, 1989). Equation (2.22) shows that further increase in the resistive electric field results only in additional compression of the pinch, leading to increased radiative losses for constant current and temperature. Note that there are no such limitations if the plasma flow in the Z pinch is not in force balance.

The issue of radiative collapse has remained relatively ignored until recently, when interest was revived by advances in pulsed-power technology, which make it feasible to drive currents much greater than I_{PB}. The question remains to what extent is it realistic to attain a high density through compression by the self-magnetic field. Some authors believe that it is possible to produce degenerate laboratory plasmas with densities similar to white dwarf stars by the process of radiative collapse (Meierovich and Sukhorukov, 1975; Chittenden et al., 1989). Obviously, long before the pinch becomes a black hole, the plasma will have become strongly coupled and the MHD approximations will have broken down. For a realistic study of the problem of the modification of the equation of state, transport coefficients and radiation losses are necessary. The simple expression (2.19) for the limiting current of an equilibrium Z pinch, which is virtually independent of the plasma temperature and density, is valid only for a classical (Spitzer) plasma conductivity, $\sigma \propto T^{3/2}$, and bremsstrahlung losses, $P_r \propto n^2 T^{1/2}$. In this case, the Pease–Braginskii current is lower for high-Z ions, so that collapse should be easier to obtain in high-Z ions. In contrast to the case of pure Spitzer resistivity, the Pease–Braginskii current becomes a function of the line density and the pinch radius, if the resistivity is enhanced due to an anomalous mechanism. Robson (1991) discussed the consequences for radiative collapse when the resistivity is enhanced by an anomalous component arising from current-driven microinstabilities. Many authors considered generalizations of the simple model, taking into account the difference between electron and ion temperatures, radiation of atoms (in the case of a partially ionized plasma), radiation of high-Z impurity ions, etc. (see Braginskii, 1957b; Haines, 1960, 1989; Shearer, 1976; Miyamoto, 1987; Chittenden et al., 1989; Robson, 1988, 1989, 1991). Since these mechanisms provide greater radiative power losses than bremsstrahlung if the plasma temperature is not too high ($T < 1$ keV), the corresponding values of the Pease–Braginskii current are smaller. In particular, $I_{PB} \approx 0.65$ MA for a deuterium Z pinch with carbon impurities, $I_{PB} \approx 0.11$ MA for a Z pinch in argon (Shearer, 1976).

Steady-state equilibria of radiating Z pinches with classical conductivity, bremsstrahlung radiative losses, and additional end-energy losses can exist with arbitrary density, temperature, and current-density profiles provided that (2.1) and (2.4) are satisfied. Taking into account other mechanisms of energy loss and dissipation, like resonant line and recombination radiation, heat conduction, etc., the possible equilibrium profiles are further specified (see Braginskii, 1957a,b, 1958; Rukhadze and Triger, 1968; Rosanov et al., 1968; Shearer, 1976; Dagazian and Paris, 1986; Bobrova and Razinkova, 1987; Bobrova et al., 1988). For instance, assuming a flat plasma-conductivity profile, we find (2.4) for the magnetic field and plasma pressure profiles. But the temperature and density profiles corresponding to this pressure profile will now be determined by the radiative losses term. In particular, when both the end losses and the radial heat conductivity are neglected, equilibrium implies a local balance between ohmic heating and radiative losses. The temperature and density profiles are easily found from this condition. The plasma equilibria, in the presence of radiative losses, are known to exhibit considerable inhomogeneities (Field, 1965; Chuideri and Van Hoven, 1979), and the same is true for the equilibria of radiating Z pinches, which turn out to be quite heterogeneous, as demonstrated by Bobrova and Razinkova (1987) and confirmed by the numerical calculations of Bobrova et al. (1988) made for a deuterium plasma with 1% xenon impurity and $I = I_{PB}$.

Another limitation results from the fact that in a realistic system the pinch current is driven by an external voltage source. During contraction, the pinch resistance and inductance rises dramatically, causing the current to fall rapidly. However, the current remains finite, since the large inductance of the driver balances the voltage drop across the pinch with a large negative $L_g(dI/dt)$ term, where L_g is the generator inductance. Possible limitations and processes terminating radiative collapse in a equilibrium Z pinch, including the limitation due to the external circuit, the effect of opacity, and degeneracy of the electron pressure, were considered (Haines, 1989; Chittenden et al., 1989; Robson, 1988, 1989, 1991).

Equilibria of radiating Z pinches have been observed (Spielman et al., 1994) in high-power Z pinches. To make this possible, the Z pinch should remain close to the steady state for times greater than the characteristic times of radiative loss and magnetic field diffusion. In most experiments these conditions are not fulfilled. Thus, the assumption of plasma equilibrium (or local heat balance, which is more demanding) may imply density and temperature profiles that are not realistic. For instance, the model of heterogeneous equilibrium, used by Bobrova et al. (1988) to explain the results of the experiment of Aranchuk et al. (1986) with exploding copper wire, predicts a temperature profile with a minimum near the axis. This is hardly compatible with the absence of skin effects, which is characteristic of the experiments with wires exploded by microsecond current pulses.

A fully consistent approach requires a study of Z-pinch dynamics, which are certainly substantially affected by radiative energy losses. When physically meaningful steady-state equilibria exist, the dynamic Z pinches should approach them, thus realizing some of the many *a priori* possible equilibrium structures. Otherwise the temperature, density, and magnetic field profiles of a Z pinch would be essentially dynamic. However, the existence of steady-state equilibria does not mean that such a state can be approached in the course of contraction. The instabilities that are developed during the collapse of the pinch will play an important role. Two-dimensional numerical simulations indicate (Neudachin and Sasorov, 1991) that the prolonged existence of heterogeneous Z pinches is due to their stability against small perturbations and a possible nonlinear stabilization of MHD instabilities in heterogeneous Z pinches. However, if the development of kink instabilities causes the formation of even a small axial magnetic field, then this magnetic field will terminate radiative collapse, since the axial magnetic field is frozen in the pinch plasma. Therefore, the pressure of the axial magnetic field grows faster during the collapse than does the pressure of the self-magnetic field.

Another example of a process that can terminate the radiative collapse is leakage currents outside the collapsing Z pinch. The model of radiative collapse assumes that the current density in the pinch is constant and that all the current into the diode flows through the pinch. Leakage current outside the Z pinch was discussed by Terry and Pereira (1991) and by Sasorov (1991). It can be caused by non-neutral electron flow in the initial stage of the pinch or neutralized current flow in a tenuous plasma that may be present in the pinch periphery (Terry and Pererira, 1991). Another reason can be turbulent spreading of a superheated skin layer of a Z pinch, which arises due to the MHD interchange instability skin layer. This increase in the size of the Z-pinch corona results in a recapture of the current by the low-density plasma corona (Sasorov, 1991).

Leakage currents outside the pinch are difficult to confirm experimentally, but presumably were discovered as an "intense evaporation" of plasma from the surface of the pinch in experiments on the Angara-5 facility (Batjunin et al., 1991). Shorter pulses reduce the influence of pinch instabilities; unfortunately, the faster current rise results in more current flowing as a skin current. This results in the magnetic pressure being at the highest near the nozzle in the gas-puff Z pinches. For a constant mass per unit length, the pinch will collapse first near the nozzle and then along the remaining pinch length. This is the characteristic zipper effect (Stallings et al., 1979; Hussey et al., 1986). The limitation of radiative collapse in gas-puff implosions was studied experimentally (Stallings et al., 1979) and using 2-D modeling (Deeney et al., 1993).

Radiation emission dramatically affects the development of the pinch compression. Radiation is emitted by the plasma at all frequencies, but the bulk of the emission at the time of highest compression is in the soft-x-ray region of the spectrum (10–5000 eV). Three mechanisms are largely responsible for soft-x-ray emission: bremsstrahlung (free–free) radiation, recombination (free–bound) radiation, and line emission (bound–bound). Radiation emission results in plasma cooling. The effect of this cooling on the dynamics of the plasma compression can be negligible, as in the case of pure bremsstrahlung from unseeded discharges, or dominant, as in the case of heavily seeded or high-Z discharges. The influence of radiative cooling on the dynamics of a Z-pinch plasma was studied in Bailey et al. (1986), where the time history of implosion $R(t)$, plasma density, and electronic number density were recorded for a gas-puff Z pinch. The experiments were made with pure helium (He) and a mixture of He with 0.5% and 1.5% krypton (Kr), which results in a significant increase in radiative energy losses. The dynamics of radiative cooling observed in this experiment are shown in Fig. 2.1. We see that the minimum pinch radius and maximum temperature correspond to higher compression for the mixture than for pure He: the plasma column is compressed from the initial radius of 1 cm to 0.07 cm and 0.14 cm in the former and in the latter case, respectively. The effects of radiation emission on the dynamic development of a neon (Ne)-seeded dense-plasma focus was studied experimentally in Venneri et al. (1990). They show the occurrence of a much denser, cooler compression in the seeded plasma focus when high-Z seed gas (more than 20% Ne in weight) was added to the discharge. They found that the radiative cooling of the pinch plasma can be responsible for more than a factor of 10 increase in the pinch density.

The influence of radiative cooling on the dynamics of imploding Z pinches was studied by Clark et al. (1986), Boiko et al. (1988), and Gasilov et al. (1989) with the aid of 1-D numerical codes. The results of the calculations show that dynamic behavior of radiating Z pinches depends critically on nonequilibrium physical processes like ionization kinetics, formation of shock waves, radiative energy transport in resonant lines, etc.

The detailed description of radiative losses from mid-Z and high-Z plasmas and the impact of those losses on pinch equilibria is complicated by the effect of plasma

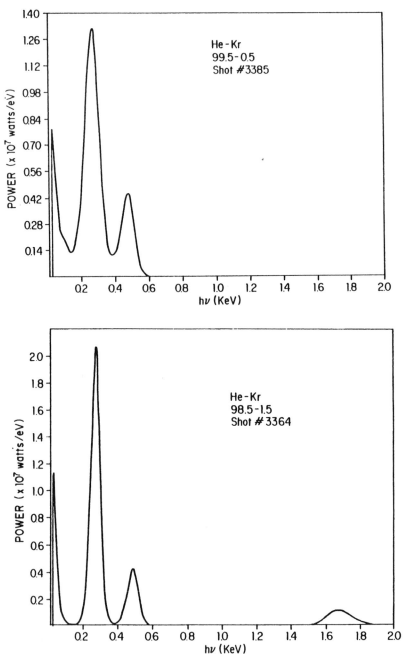

FIGURE 2.1. Two graphs showing the spectral power emitted by a helium/krypton mixture. The first graph has 0.5% Kr by number and the second 1.5% Kr by number with the balance He. Tripling the Kr fraction increases the total radiated x-ray energy > 50%.

opacity and the requirement for radiation transport models. As the Z pinch heats and compresses the plasma, the opacity for many of the bound–bound transitions can reach the point where the plasma is effectively optically thick. Radiation emitted from the interior regions cannot rapidly escape. The transition from an optically thin to an optically thick plasma state can terminate the collapse of the pinch.

2.3 Bright spots

Many authors believe that bright spots—these extreme plasma states produced in vacuum sparks—are already observable as bright spots seen in the experiments with plasma focus and Z pinches (Kononov et al., 1980; Gol'ts et al., 1986; Vikhrev et al., 1982, 1989; Koshelev et al., 1989), sometimes also called "micropinches." Some experimental evidence supports this point of view. In particular, in the experiment of Kononov et al. (1980) with a low-inductance vacuum spark, the electron density in the bright spot was estimated from the observed redistribution of intensities of resonant He-like calcium (Ca) and titanium (Ti) ions. The ionization states were assumed to be in a Saha equilibrium with temperature $T \sim 1.5$ keV. In the experiment of Gol'ts et al. (1986) with a 1-MA Z pinch in Kr, an estimate of the electron temperature in the bright spot, T_e, from 3 up to 5 keV, was based on the presence of the Kr XXXV ion lines in the x-ray spectrum. The dimensions of the spot were determined from an x-ray pinhole photograph, and an electron density, $n_e \sim 10^{23}$ cm^{-3}, was estimated from the Bennett condition, the measured values of temperature and radius of the micropinch. However, one cannot be quite sure that in these experiments either the Saha equilibrium or the Bennett equilibrium (with all the pinch current flowing via the neck) are achieved. For instance, estimating the inductance of the neck, one can see that only $\sim 2\%$ of the total current flows in it, which is consistent with the measurements of Hares et al. (1985). Further experimental research is certainly required.

Vikhrev et al. (1982, 1989) developed a model of the formation of a micropinch from a single neck. The model allows one to estimate the steady-state radius of the micropinch for a given current from the balance of ohmic heating due to anomalous resistivity and radiative energy losses. For a 150-kA Z pinch in an iron plasma, the radius of the micropinch is estimated by Vikhrev et al. (1982) as $R \leq 1$ mm, with $N_i \sim 5 \times 10^{15}$ cm^{-1}, $n_e \geq 10^{23}$ cm^{-3}, and $T \sim 1$ keV.

The physics of bright spots in Z pinches is not limited to ideas related to radiative collapse and micropinches. This was a very traditional field, opened in the 1950s, when it was first established that these spots are sources of nonthermal electrons and ions with energies from 20 keV to 1 MeV, hard x rays, and neutrons. As a rule, the bright spots appear near the moment of maximum compression, correlating with a spike in dI/dt. Mechanisms for the production of nonthermal electrons and ions were studied in a large number of works (see Trubnikov, 1958, 1986; Bernstein, 1970; Imshennik et al., 1973; Gary and Hohl, 1973; Gary, 1974; Haines et al., 1984; Deutsch et al., 1986; Marnachev, 1987; Deutsch and Kies, 1988a, b; Herold et al., 1988).

The first explanation, due to Trubnikov (1958), suggests that the nonthermal particles are accelerated by the high-induction electric field in the neck during the nonlinear stage of the sausage ($m = 0$ mode) instability development. A 2-D numerical simulation, taking into account the plasma axial outflow from the neck, shows that the calculated value of the induction electric field agrees with the observed spectra of nonthermal ions and the total value and anisotropy of the observed neutron yield. A detailed description of this mechanism is given by Trubnikov (1986).

Other physical mechanisms that can be responsible for the production of fast electrons and ions were discussed by the authors cited above. In particular, the generation of toroidal vortices accompanying the formation of a bright spot was considered by Marnachev (1987), the vortices being regarded as regions where the electrons are accelerated. An interesting mechanism for the acceleration of ions in plasma-focus Z pinches was suggested by Bernstein (1970), Gary (1974), and Deutsch and Kies (1988a, b). A beam of fast ions with energies ~ 100 keV is supposed to be produced when the current sheath converges to the axis and the ion mean free path is greater than the pinch radius, as a result of multiple reflections of ions from the radially accelerated current sheath that acts as a magnetic piston (a gyro-reflection acceleration). The decrease in the neutron yield with increased initial gas density in plasma-focus experiments is important, though indirect, evidence in favor of this mechanism. Experimental results on the dynamics of bright spots and the production of nonthermal particles in plasma-focus studies are discussed in detail by Herold et al. (1988).

Deeney et al. (1989) have presented experimental results (see Fig. 2.2) that show electron-beam-induced Ni-K_α emission prior to and during a rapid localized heating of the plasma and formation of intense bright spots. These bright spots radiate x rays with energies far higher than the thermal temperature plasma. Deeney et al. point out that this behavior occurs only when the kinetic energy of the implosion is small compared with the relevant ionization energies. Their conclusion is that bright spots are not formed by $m = 0$ instabilities and that they are always preceded by electron-beam-excited characteristic lines.

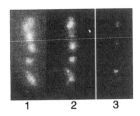

1   2   3

FIGURE 2.2. A time-resolved x-ray pinhole photograph of an nickel wire-array implosion on Double Eagle showing the presence of very localized emitting spots. The dimensions of the radiating regions are much smaller than the 100 μm resolution of the instrument.

It should be noted that the nature of the bright spots and the physical mechanisms responsible for their formation and for particle acceleration in Z pinches are not yet firmly established and are often discussed by different authors from contradictory points of view (see Trubnikov, 1986; Vikhrev, 1986). No doubt the development of the neck at the nonlinear stage of the sausage instability (the development of the RT instability, which considerably degrades the symmetrical implosion of the cylindrical collapse), as well as radiative energy losses play a very important role in all this. A simple model of a bright spot implosion to study the dynamics and stability of a stagnating bright spot was considered in Murakami et al. (1995).

3

Dynamics of Z-Pinch Plasmas

3.1 Formation of Z-pinch plasmas: Theoretical modeling

A fully consistent approach requires a study of Z-pinch dynamics. We emphasize the dynamics of Z-pinch plasmas or imploding plasmas, since dynamics certainly substantially affects Z-pinch parameters. While the pulsed-power generator characteristics strongly affect pinch parameters, depending on the goal of applications (fusion, neutron production, x-ray sources, etc.), the canonical diode consists of a cylindrical rod as the cathode, placed a certain distance (approximately the radius of the cathode) from the anode. Typically in dense pinches, the electric current is collisionally dominated, while current conduction in a vacuum spark, for example, may be dominated by plasma microinstabilities. Modern experimental facilities routinely produce Z-pinch plasmas for which the ideal MHD model may be inadequate at certain conditions. For a dense Z pinch, the current-driven plasma kinetic instabilities do not appear, if the average electron drift velocity $\bar{v}_{De}$, is less than the ion thermal velocity, v_{thi}. Then, we obtain the condition given by (2.11). It is convenient to express this condition in terms of the pinch current, I, taking into account, $I = Zev_{De}N$. The limiting line number density of a dense pinch is

$$N > 2 \times 10^{17} \text{ cm}^{-1}(I/\text{MA})\sqrt{M_i/(T/\text{keV})}/Z. \tag{3.1}$$

The Z-pinch load breaks down during the initial phase of the Z-pinch event. The load becomes plasma early into discharge in all cases of high current pinches. However, the initial mass distribution differs for the various phases and for the various pinch compressions, and this may be an important factor in pinch behavior. If the load is conductive (e.g., metal wires, thin metallic cylindrical foil, or plasma), the current starts to flow as soon as the electric pulse arrives, and the load voltage is largely inductive. However, if the material is nonconductive (e.g., a neutral gas, or plastic wires), the load voltage builds up rapidly until the material breaks down. Self-breakdown causes sparks that evolve into random current channels, which usually affect the implosion symmetry. Preionization of the nonconductive material by external means (usually UV flashboards) improves the symmetry of the implosion (Felber et al., 1988a, 1988b). Also, preionization may be superfluous, if the pulse-power generator has a sizable prepulse. Data suggest that higher-Z gases at higher densities in general require less preionization (Hsing and Porter, 1989). In this case, the preionization from the prepulse appears to have time to spread through the gas, as evidenced by sufficiently symmetric behavior of the implosion. Additional control over the initial condition of the pinch is possible. For example, control can be provided by injecting low-energy electrons along the diode axis, through a ring in the anode. The effect is seen experimentally in a reduced instability amplitude and improved implosion compression (Ruden et al., 1987). Accordingly, a suitable theoretical model can be used to describe Z-pinch dynamics and stability. The purpose of this chapter is to describe the alternative models used to describe Z-pinch plasma dynamics. We do not consider here the initial stage of Z-pinch plasma formation with detailed ionization dynamics, electrical breakdown processes, etc. (see, for example, Vikhrev and Braginskii, 1980). The ultimate goal, to predict the pinch characteristics in detail from first principles, including ionization dynamics, growth of instabilities, and radiation energetics, remains a very difficult unsolved problem today. However, when the pinch remains cylindrical, the simple hydrodynamic models do a credible job in correctly predicting the pinch parameters.

3.2 Zero-dimensional models of dynamic Z pinches

3.2.1 Thin-shell model

The gross dynamical properties of the pinch implosion, such as the implosion time or the implosion kinetic energy, can be computed reasonably well in a zero-dimensional approximation. The simplest model of an imploding gas-puff Z pinch or a foil liner is an infinitely thin annular shell imploded by the pressure of an azimuthal magnetic field produced by the axial current flowing in it. In this model, the imploding plasma is assumed to remain an infinitely thin cylindrically symmetric shell, and the radial position of the plasma shell as a function of time should be consistent with the acceleration by the $\mathbf{J} \times \mathbf{B}$ force and a constant mass per unit length. Indeed, such pinch dynamics have been corroborated in many experiments. The comparison is favorable: the measured radius $R(t)$ is found to be in a good agreement with the acceleration determined from the force $I^2(t)/R(t)c^2$ and constant mass per unit length. It is typical in Z-pinch modeling to consider the mass as an adjustable parameter, since the mass evolved in a pinch implosion is very hard to measure (the situation is different for wire arrays). The equation of motion of the shell of radius $R(t)$ is

$$\mu \frac{d^2 R}{dt^2} = \frac{I^2(t)}{Rc^2},$$

(3.2)

where μ is the mass per unit length of the shell.

Equation (3.2) is easily integrated, and for constant current, $I = I_{max}$, and with initial conditions $R(0) = R_0$, and $\dot{R}(0) = 0$, we obtain

$$\left(\frac{dR}{dt}\right)^2 = \frac{R_0^2}{\tau_A^2} 2 \ln\left(\frac{R_0}{R(t)}\right) \qquad (3.3)$$

and

$$t/\tau_A = \left(\frac{\pi}{2}\right)^{1/2} \text{erf}\left\{\left[\ln\left(\frac{R_0}{R(t)}\right)\right]^{1/2}\right\}, \qquad (3.4)$$

where $\text{erf}\{\ldots\}$ is the error function (Abramowitz and Stegun, 1964), $\tau_A = c\mu^{1/2}R_0/I_{max}$ is the Alfvén transit time [see (2.15)].

Equations (3.3) and (3.4) predict collapse of the shell: its radius goes to zero; and its velocity, pinch energy, and magnetic field go to infinity in a finite time interval. This is evidently not realistic, owing both to the instabilities growing in the course of the implosion and to the increasing shell inductance with decreasing shell radius. However, Eqs. (3.2) and (3.4) demonstrate the very weak dependence of the velocity and implosion time on the actual value of the radial compression ratio $\alpha(t) = R(t)/R_0$, provided that α is sufficiently large. To avoid the divergence in the pinch energy, velocity, and magnetic field, one needs to cut off the calculation when the plasma sheath reaches an experimentally determined value, typically about 0.1 of the initial radius. Taking $\alpha = 0.1$ in (3.3), we obtain reasonable estimates for the implosion velocity and implosion time (Hussey et al., 1980)

$$v_{imp} = \frac{dR}{dt} \cong 2\frac{R_0}{\tau_A} = \frac{2I_{max}}{c\mu^{1/2}} = \frac{2 \times 10^7 I_{max}(MA)}{\sqrt{\mu(10^{-4}\,\text{g/cm})}} \text{ cm/s}, \qquad (3.5)$$

$$\tau_{imp} \cong (\pi/2)^{1/2}\tau_A = 125 R_0 \text{ (cm)} \left[\mu(10^{-4}\text{g/cm})\right]^{1/2} (I_{max}/MA)^{1/2}, \qquad (3.6)$$

and for the kinetic energy per unit length of the liner, available for further transformation into thermal and radiative energy:

$$\varepsilon_k = E_k/L = 2.3 I_{max}^2/c^2 = 2.3 I_{max}^2(MA) \text{ kJ/cm}, \qquad (3.7)$$

where L is the pinch length. For instance, substituting into (3.5)–(3.7) the parameters of the fast Z-pinch experiments with the 30 TW Saturn accelerator, one of the most powerful drivers to date operating at Sandia National Laboratories (Spielman et al., 1989), $I_{max} = 10$ MA, $\mu = 5 \times 10^{-4}$ g/cm, $R_0 = 1.75$ cm, $L = 2$ cm, we obtain $v_{imp} = 9 \times 10^7$ cm/s, $\tau_{imp} = 50$ ns, $E_k = 460$ kJ, which are in reasonable agreement with the experimental data.

A more realistic current pulse shape in a pulsed high-power driver is

$$I(t) = I_{max} \sin(\pi t/2\tau_{max}), \quad 0 \le t \le 2\tau_{max}. \qquad (3.8)$$

The implosion regime in this case is dependent on the dimensionless parameter

$$a \equiv \tau_{max}^2/\tau_A^2 = I_{max}^2\tau_{max}^2/\mu c^2 R_0^2, \qquad (3.9)$$

which determines the optimization of the implosion, the connection between the mass line density of the shell, and the current pulse amplitude and duration. For a given value of $\alpha = \alpha_{\min}$ one can find the optimal value of $a = a_{opt}$ corresponding to the maximum kinetic energy delivered to the liner imploded from $R = R_0$ to $R = R_{\min}$ by the sinusoidal current pulse. In particular, for $\alpha_{\min}$ between 0.05 and 0.15, the value of a_{opt} varies from 5.5 to 5.3. With $a = a_{opt}$ the sinusoidal current pulse of amplitude $I_{\max}$ is almost as effective in imploding the thin shell as a constant current $I = I_{\max}$. For instance, taking $\alpha_{\min} = 0.1$ and $a = a_{opt}(\alpha_{\min}) = 5.3$, we find that the kinetic energy delivered to the liner by the current pulse (3.8) is only 7.5% less than that given by (3.7). The normalized implosion radius $\alpha = R(t)/R_0$ of a thin shell imploded by a constant current (dashed lines) and by an optimized sinusoidal current pulse (solid lines) are compared in Fig. 3.1.

During the rise time of the pulse, the current is roughly proportional to time, $I(t) = (dI/dt)_0 t$. Then the characteristic time of the implosion is

$$\tau_A = \mu^{1/4} \left(\frac{R_0 c}{(dI/dt)_0} \right)^{1/2}, \tag{3.10}$$

and stagnation corresponds to $t/\tau_A \cong 3$.

All the estimates given above are valid also for implosion of a wire array consisting of N_w parallel wires mounted symmetrically and coaxially between the anode and cathode plates. Then, in the equation of motion (3.2), μ is the total mass/length of the wire array and I is the total current flowing in it. The compact compression of a wire array by a current pulse of a given amplitude $I_{\max}$ was observed. Each wire moves to the axis as a separate plasma column (Baksht et al., 1989), if the characteristic rise time of the pulse, $\tau_f = I_{\max}/(dI/dt)_0$, exceeds the collapse

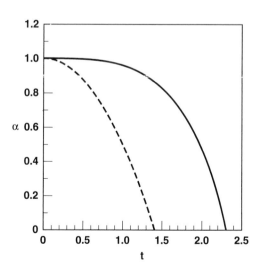

FIGURE 3.1. Dynamics of a thin-shell implosion by a current pulse: $I = I_{\max} \sin(\pi t/\tau_{\max})$, solid line; $I = I_{\max} = \text{const.}$, dashed line.

time $3\tau_A$, so that the collapse occurs during the phase of linear current rise. The optimal energy input in the wire array, measured by maximal output of radiative energy, is obtained when the moment of collapse corresponds to the peak current, $\tau_f \cong \tau_A$, in agreement with the preceding discussion.

Consider now a thin shell pinching an externally applied axial magnetic field. This geometry corresponds to the experiments of magnetic flux compression by imploding gas-puff Z pinches and electromagnetically imploded foil liners. The axial magnetic field B_{z0} exists initially both inside and outside the shell. As the shell implodes, it compresses the axial field inside while the axial field outside remains unchanged. Assuming magnetic flux conservation, we obtain instead of (3.2) the equation of motion in the form

$$\mu \frac{d^2 R}{dt^2} = \frac{I^2(t)}{Rc^2} + \frac{B_{z0}^2 R(t)}{4} \left(1 - \frac{R_0^4}{R^4(t)}\right). \tag{3.11}$$

Now there is no infinite compression: the pinch implodes to a turnaround radius $R_{\min} = \alpha_{\min} R_0$, and then expands. For constant current $I = I_{\max}$, the value of $\alpha_{\min}$ is given by

$$\ln(1/\alpha_{\min}) = b(\alpha_{\min} - 1/\alpha_{\min}), \tag{3.12}$$

where

$$b = (cR_0 B_{z0}/2I_{\max})^2 \tag{3.13}$$

is a dimensionless characteristic of the implosion. High magnetic flux compression corresponds to a small value, $b \ll 1$. (The experiments on high and ultrahigh magnetic field generation on gas-puff Z-pinch devices were made with $b < 10^{-2}$.) Here again, when the duration of the current pulse is finite, the implosion regime can be optimized by choosing an appropriate value of a, by adjusting the mass of the gas puff for a given value of b, which has the same affect. The turnaround radius has a shallow minimum as a function of a. For the range of values of b between ~ 0.01 and 0.1, and the current pulse shape (3.8), the maximum compression occurs for $a = a_{\mathrm{opt}} \approx 4$ (Felber et al., 1988b). For smaller values of a, the shell is too massive and slow, reaching turnaround well after the peak of the current pulse and the pinching force has passed. For larger values of a, the shell mass is too low, and it bounces off the compressed magnetic field early in the current pulse (Liberman et al., 1987; see Fig. 3.2). With $a = a_{\mathrm{opt}}$, the optimal implosion time to reach minimum radius is within a few percent of $1.1\tau_{\max}$, just after the current peak. The peak axial magnetic field $B_{z\,\max} = B_{z0}\alpha_{\min}^{-2}$ is only a few percent less than that estimated from (3.12) for constant current $I = I_{\max}$.

In most cases, the implosion is designed to produce a high density and temperature stagnation on axis. High temperature of the imploding plasma is generally not desirable, since this results in rapid expansion, lower density, and a less-effective stagnation. The optimal implosion is supersonic, even hypersonic, with an implosion velocity that is much greater than the characteristic sound velocity in the

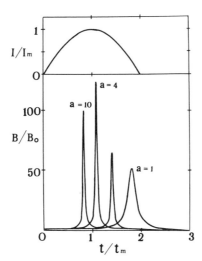

FIGURE 3.2. Dynamics of a thin shell imploded by a sine current pulse onto an axial magnetic field.

imploding shell. The ratio of thermal to kinetic energy of the liner is thus an important characteristic of implosion efficiency. A simple estimate can be made for an imploding wire array. Assuming each wire to be in Bennett equilibrium with current $I_w = I_{max}/N_w$ (a very rough approximation), we obtain from (3.6) and (3.7) for the whole array:

$$E_t/E_k = 3/(8N_w) \qquad (3.14)$$

e.g., for a six-wire array only about 6% of the energy of the current pulse is spent on heating the wires.

Equation (3.14) indicates that the thermal energy approaches zero as the wires approach a shell ($N_w \to \infty$). However, we cannot use the Bennett relation for a shell, so another estimate is needed. Consider a small element of the shell whose thickness, ΔL, is much less than its radius, R, the area of the element being ΔA. The shell is accelerated by the magnetic pressure, $B_\varphi^2/8\pi$, the current density in it thus being $J = (c/4\pi)(B_\varphi/\Delta L)$. The ohmic heating released in the element is

$$dQ = (J^2/\sigma)\Delta A\Delta L\, dt = (c^2/2\pi\sigma\,\Delta L)(B_\varphi^2/8\pi)\Delta A\, dt = (2/R_M)\, dE_k, \qquad (3.15)$$

where dE_k equals the work done by the magnetic pressure force $(B_\varphi^2/8\pi)\Delta A\, dR$, $dR = u\,dt$, u is the shell velocity, and the magnetic Reynolds number here is R_M (see Sec. 3.3). Equation (3.15) shows that the main heating of the shell corresponds to the acceleration phase, when the velocity is small and the value of the magnetic Reynolds number is low. Neglecting the increase in plasma density during this phase, one can integrate (3.15) and estimate the temperature corresponding to the given shell velocity. Assuming that the shell heating ends when the velocity reaches the characteristic value $2R_0/\tau_A$, we obtain

$$T = \left[\left(\frac{20}{3} \right) m_i m_e^{1/2} c^2 e^2 (\ln \Lambda)(Z/(1+Z)) \frac{R_0}{\tau_A} \right]^{2/5}$$

$$= 130 \left[\bar{A}(\ln \Lambda/10)(Z/(1+Z)) R_0(\text{cm})/\tau_A(\text{ns}) \right]^{2/5} \text{ eV}. \quad (3.16)$$

Two factors are not taken into account in (3.16)—adiabatic heating due to compression of the accelerated shell and ionization energy losses—partly compensate each other. For instance, for the experiment (Felber et al., 1988a) with imploding Ne gas-puff Z pinch on the PROTO-II accelerator ($R_0 = 1$ cm, $\tau_A = 50$ ns, $Z = 3$–4, $\bar{A} = 20$) the estimate (3.16) yields $T = 80$ eV, which is close to the temperature 100 eV measured before a pinch has moved appreciably. If the shell contains heavy ions emitting strong radiation in bound–bound resonant lines, which is not taken into account in (3.16), then the temperature of the shell is considerably lower, in most cases not greater than 10–20 eV, and (3.16) represents an upper estimate for the temperature. Note that the total ohmic heating of the shell would be greater in this case, due to the higher resistivity (higher ionization state, Z) of the plasma.

3.2.2 Snowplow model

The snowplow model, one of the simplest, commonly used models, predicts the implosion time from the generator parameters, the load mass, and initial radius. In this model (see Rosenbluth et al., 1954; Colgate, 1957; Shafranov, 1958b), an ideally conducting cylindrical shell is assumed to remain infinitely thin and cylindrically symmetric in the course of the implosion. The model suggests that all the mass is swept up by the shell as if the magnetic pressure were a snowplow. The snowplow model first reported in the 1950s, is a widely used picture of the formation of a pinch in an initially cold, nonconducting gas or plasma column. The idea was that the nonconducting gas was ionized as it was snowplowed into the sheath.

Consider a cylindrical plasma liner, which is collapsing toward the axis under the pressure of an azimuthal magnetic field, B_φ, produced by a pinch current, $I(t)$, carried by the imploding Z-pinch shell along the z axis. The plasma liner is imploding toward the axis into the inner space filled by unperturbed resting plasma. The sharp increase in the pinch current drives the implosion of the current sheath with a shock wave propagating ahead of the current sheath at the beginning of the implosion. We consider first the snowplow model that takes into account only the increase in the mass of the collapsing shell. In the spirit of the snowplow model, the equation of motion is

$$\frac{d}{dt} \left(dM \frac{d\mathbf{r}}{dt} \right) = -\mathbf{n} \, P \, da, \quad (3.17)$$

where $P = B_\varphi^2/8\pi$ is the pressure of the azimuthal magnetic field, $B_\varphi = 2I(t)/cR(t)$, generated by the pinch current, $I(t)$; $R(t)$ is the radius of the shell, and dM is the total mass of an element of area da of the shell, together with the mass of the plasma swept up by the area. We have

$$dM = \left(\sigma_M - \int_0^t \rho \mathbf{u} \cdot \mathbf{n} \, dt \right) da, \qquad (3.18)$$

where $\sigma_M(t)$ is the mass per unit area of the shell, $M_s = 2\pi R(t)\sigma_M(t) = \text{const}$ is the mass line density of the shell, ρ is the density of the resting plasma (gas) inside the shell, and $M = \pi \rho R_0^2$ is the mass line density of the gas column, $\mathbf{u}$ is the velocity, and $\mathbf{n}$ is a unit vector normal to the shell surface. Introducing cylindrical coordinates with the z axis directed along the Z-pinch axis and dimensionless variables

$$\alpha(\tau) = R(t)/R_0, \quad \tau = t/t_A, \quad \mu = M/(M_s + M), \qquad (3.19)$$

where $R_0 = R(0)$ is the initial radius of the shell, $t_A = R_0 c \sqrt{M_s + M}/I_{\max}$ is the characteristic time of the implosion, which coincides with the Alfvén transit time (2.15) for a constant mass line density, and $I_{\max}$ is the amplitude of the pinch current, $I(t) = I_{\max} \Phi(\tau)$, the equation of motion for the cylindrically symmetric implosion takes the form

$$\frac{d}{d\tau} \left[(1 - \mu\alpha^2) \frac{d\alpha}{d\tau} \right] = -\frac{1}{\alpha} \Phi^2(\tau). \qquad (3.20)$$

In this form, the equation of motion is universal, being dependent on the unique dimensionless parameter μ. Equation (3.20) can be solved exactly for constant current, $\Phi(\tau) = 1$, $I = I_{\max}$, with initial conditions $R(0) = R_0$, and $\dot{R}(0) = 0$. The radial velocity of the imploding shell is given by

$$u \equiv -\left(\frac{R_0}{\tau_A} \right) \frac{d\alpha}{d\tau} = \left(\frac{R_0}{\tau_A} \right) \frac{\sqrt{2\ln(\alpha^{-1}) + \mu(\alpha^2 - 1)}}{1 - \mu\alpha^2}, \qquad (3.21)$$

and the radial position of the shell, $\alpha(\tau)$, can be obtained by inverting the integral

$$\tau = \int_{\alpha(\tau)}^1 \frac{(1 - \mu\alpha^2) d\alpha}{\sqrt{2\ln(\alpha^{-1}) + \mu(\alpha^2 - 1)}}. \qquad (3.22)$$

The shell implosion computed from (3.20) for different pinch-current waveforms and a different value of μ is shown in Fig. 3.3.

Comparing (3.3) and (3.21) we see that higher implosion velocity corresponds to the same compression ratio $\alpha(t) = R(t)/R_0$ in the snowplow regime, which is not surprising since in the early stages of compression, the average accelerated mass in this case is smaller. Hence collapse, when $R(t)$ goes to zero, is reached sooner for $t/\tau_A = 0.92 < (\pi/2)^{1/2}$. In the case of a linearly rising current, collapse corresponds to $t = 1.5\tau_A$, where τ_A is given by (2.15) (see Krall and Trivelpiece, 1973).

The most important practical point is an estimate of the peak temperature and density that could be produced with an imploding Z pinch. One cannot obtain a reasonable estimate directly from the simple snowplow model given above because

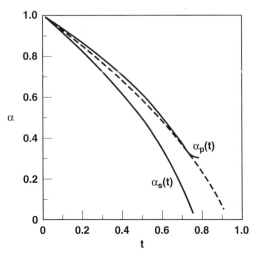

FIGURE 3.3. Dynamics of a Z-pinch implosion by a constant current according to a simple (dashed line) and snowplow model modified by Potter (1978); $\alpha_s(t)$ and $\alpha_p(t)$ represent the converging shock front and the magnetic piston, respectively.

it is based solely on momentum conservation (the energy balance is not taken into account). To do that, the snowplow model must be modified to include some additional physics determining the snowplow effect. The conventional model (Allen, 1957; Leontovich and Osovets, 1956; Shafranov, 1958b; Potter, 1978) assumes that a MHD shock wave is driven ahead of the converging current sheath that turns the unperturbed cold gas into plasma and involves it in the implosion. Other physical mechanisms are also possible, in particular for the case of a collisionless plasma: the current sheath represents a sharp field discontinuity, whereas some of the ions are reflected by the moving "potential wall," many of them are trapped by it, exhibiting a sort of snowplow behavior (Oliphant, 1974). The subsequent thermalization of the ions was considered in detail by Freidberg et al. (1972). The 0-D snowplow models developed by Potter (1978) and Miyamoto (1984) aimed to describe the shock compression and heating of the plasma by the imploding current sheath, which is more relevant for dense Z-pinch experiments.

The energy delivered to the plasma by a strong shock wave is divided equally between the kinetic and thermal energy of the shock-compressed plasma (Zel'dovich and Raizer, 1967; Liberman and Velikovich, 1986a). As the shock front velocity is greater than the velocity of a piston driving it, the flow evidently evolves to a 1-D pattern, though initially it can be well described by a 0-D model, when the layer of the compressed plasma is sufficiently thin compared to its radius. This is the physical basis of the simple slug model, developed by Potter (1978), of the plasma between the shock and the piston. Suppose the shock-compressed plasma has uniform velocity, density, and pressure distributions satisfying the strong shock

Rankine–Hugoniot boundary conditions at the shock front:

$$u = \frac{2}{\gamma + 1} D, \quad \rho = \frac{\gamma + 1}{\gamma - 1} \rho_0, \quad P = \frac{2}{\gamma + 1} \rho_0 D^2 \tag{3.23}$$

($D = dR_s/dt$ is the shock velocity), which is correct in plane geometry. Using the adiabatic law, $P \propto \rho^\gamma$, to account for the change in the plasma volume due to convergence, the following equations relating the shock and piston radii, R_s and R_p, can be obtained (Potter, 1978):

$$R_P \frac{dR_P}{dt} - \frac{2}{\gamma + 1} R_s D + \frac{1}{\gamma D} (R_s^2 - R_P^2) \frac{dD}{dt} = 0, \tag{3.24}$$

$$D = -\left(\frac{\gamma + 1}{4\pi \rho_0} \right)^{1/2} \frac{I(t)}{R_P(t)c}. \tag{3.25}$$

For the simple case of constant current, (3.24) and (3.25) can be integrated, with initial conditions $R_P(0) = R_s(0) = R_0$. The result is

$$R_P = R_0 \left(\frac{\gamma}{\gamma + 1 - R_s^2/R_P^2} \right)^{\frac{\gamma}{\gamma - 1}}. \tag{3.26}$$

Thus, the minimum radius that the plasma column achieves when the shock wave converges at the axis is estimated as

$$R_{\min} = R_0 \left(\frac{\gamma}{\gamma + 1} \right)^{\frac{\gamma}{\gamma - 1}}. \tag{3.27}$$

Note that for an ideal gas (simple singly ionized plasma) with $\gamma = 5/3$, the final compression is weak, $R_{\min} = 0.31 R_0$. Thus, the average plasma density is $\bar{\rho} \approx 10.5 \rho_0$, and the calculated profiles of density and temperature in the same slug model (Potter, 1978) correspond to a minimum final density on axis [$\rho = 4\rho_0$, cf. (3.23)] and a maximum density $\rho = 16.4\rho_0$ at the piston. The maximum temperature on axis is given by

$$NT = \frac{\gamma - 1}{\gamma + 1} \left(\frac{\gamma + 1}{\gamma} \right)^{\frac{2\gamma}{\gamma - 1}} \frac{I^2}{2c^2}. \tag{3.28}$$

Note that the right-hand side of (3.28) for $\gamma = 5/3$ is 5.245, about five times greater than that given by the Bennett relation (2.8) for a singly ionized plasma. This is an important feature of dynamic Z pinches: the nonequilibrium plasma temperature can be considerably higher for the same pinch current. On the other hand, the density increase for shock compression is limited; here, by a value of order 16. This conclusion is valid not only for the simple slug model under consideration. Self-similar solutions for imploding plasmas, driven by converging shock waves for cylindrical geometry and $\gamma = 5/3$, yield a 6.9-fold compression in converging flow and a 3.3-fold additional compression in the flow containing the reflected shock wave (the latter cannot be accounted for in the slug model). The maximum compression, about 23 times, is achieved directly after the reflected shock wave and is followed by an immediate expansion. Since the reflection of the shock wave from the axis is followed by rapid plasma expansion, high compression, which would be necessary, in particular for producing fusion conditions required by the inertial confinement schemes, cannot be obtained.

The problem is, in effect, similar to the problem in laser-driven inertial confinement fusion (ICF) (Nuckolls et al., 1972; Kidder, 1976) and one possible solution is exactly the same. To produce high compression, it must be adiabatic or almost adiabatic, without any shock waves converging to the axis. To do this, a special programming of the pinch current is proposed by Potter (1978), just like programming the laser pulse, suggested by Kidder (1976). An alternative method of increasing plasma compression in a Z pinch is also based on a 0-D analysis. Averaging the ideal MHD equations over a thin layer of the imploding plasma yields a third-order ordinary differential equation [the snowplow energy (SPE) equation, Miyamoto, 1984]. A study of Z-pinch implosion regimes, based on the SPE equation, shows that the final compression can be significantly increased by a precurrent pulse that provides an initial inward velocity of the plasma layer before the main current pulse begins.

The estimates, based on the modified snowplow equations, appear to be quite reasonable when the thickness of the imploding plasma shell is smaller than its inner radius. When they become comparable, the flow ceases to be essentially 0-D, and an MHD analysis is required. Thus, the snowplow estimates of plasma pressure, density, and temperature for the final stage of compression are generally correct within an order of magnitude.

3.2.3 Time-dependent equilibria of radiating Z pinches

Dynamics of radiating Z pinches can be studied with the aid of a simple analytical 0-D model, if the characteristic rate of time variation is much smaller than the inverse Alfvén transit time (or, alternately, if the velocity of the pinch boundary, $|dR/dt|$, is much smaller than the Alfvén velocity, V_A), so that we deal with a time-dependent Bennett equilibria (2.8) relating the values of current and temperature. Then in effect, the Z pinch is described by the power-balance equation, which for a singly ionized plasma ($\bar{Z} = 1$) is (Robson, 1988)

$$3N \frac{dT}{dt} = P_j - P_r - \frac{I^2}{c^2 R} \frac{dR}{dt}, \qquad (3.29)$$

P_j and P_r being given by (2.18). Note that $P_j = I^2 \mathbb{R}$, where

$$\mathbb{R} = 1.03 \times 10^{-4} (\ln \Lambda)/(\pi R^2 T^{3/2}) \Omega \ \mathrm{m}^{-1}, \qquad (3.30)$$

is the pinch resistance per unit length. The third term on the right-hand side of (3.29) represents the work done by the magnetic pressure. For an isothermal, flat-current Z-pinch structure (2.5) and (2.6), and taking into account (2.22), we obtain the equation

$$\frac{2}{R} \frac{dR}{dt} + \frac{3}{I} \frac{dI}{dt} = \frac{2S}{R^2 I^3} \left(1 - \frac{I^2}{I_{PB}^2} \right), \qquad (3.31)$$

where $S = 5.31 \times 10^{-11} N^{3/2} (\ln \Lambda) A^{-3}$ cm^2 s^{-1}, describing the dynamics of time-dependent equilibria of radiating Z pinches.

Equation (3.31) is easily solved. It is convenient to introduce the time scale

$$\tau_{PB} = R_0^2 I_{PB}^3 / 2S = 150 (R_0/100 \ \mu m)^2 (N/10^{20} \ m^{-1})^{-3/2} \ ns, \qquad (3.32)$$

corresponding to the radial length scale, R_0, of the Z pinch. For instance, one can find from (3.31) the time dependence of the current, $I(t)$, required to increase the pinch current up to I_{PB}, maintaining the pinch radius constant (Haines, 1960; Hammel, 1976):

$$2t/3\tau_{PB} = -2I/I_{PB} + \ln[(I_{PB} + I)/(I_{PB} - I)], \qquad (3.33)$$

corresponding to asymptotic laws

$$I/I_{PB} = \begin{cases} (t/\tau_{PB})^{1/3}, & \tau_{PB} \gg t \\ 1 - 2\exp[-(2t/3\tau_{PB}) - 2], & \tau_{PB} \ll 1 \end{cases} \qquad (3.34)$$

(3.31) can be also used to find the time dependence, $R(t)$, for a given current waveform (Haines, 1989):

$$R(t) = R_0 \left(I(t) I_{PB} \right)^{-3/2} \left[\frac{t}{\tau_{PB}} - \int_0^t \left(\frac{I(t)}{I_{PB}} \right)^2 \frac{dt}{\tau_{PB}} + C \right]^{1/2}, \qquad (3.35)$$

where C is an integration constant. In both cases we have a problem at the initial time: the law, $I \propto t^{1/3}$, found for currents, $I \ll I_{PB}$, by Braginskii (1957b) requires infinite initial rate of rise of the current, i.e., $dI/dt = \infty$ at $t = 0$; Eq. (3.35) predicts infinite radius when $I = 0$. However, this is of no importance because the time-dependent equilibria do not describe the initial stage of the pinch development (breakdown, etc.), so that (3.33)–(3.35) are valid only for finite values of I and t. Adequate account of this initial stage in (3.35) is reduced to a proper choice of the constants, R_0 and C (see Haines, 1989).

3.2.4 Radiative collapse

The above model of time-dependent equilibria predicts a radiative collapse for $I > I_{PB}$, that is, compression of a Z pinch to a zero radius in a finite time. Note that the Pease–Braginskii current, I_{PB}, says nothing about the pinch radius because Joule heating and optically thin radiation vary with plasma density in the same way. Currents exceeding I_{PB} would not allow power balance, and a stationary pinch would not be possible. Instead, the pinch would suffer radiative collapse. This was first shown by Shearer (1976) for the case of constant current and temperature. Indeed, taking $I = \text{const} > I_{PB}$ in (3.35), we find that $R(t)$ goes to zero when $t = \tau_{PB} I^3 [I_{PB}(I - I_{PB})]^{-1}$. Of course, a step-function current waveform is not realistic, but (3.35) describes a radiative collapse for any law of the current rise. Of particular importance is the case of $dI/dt = \text{const}$, which was studied by Robson (1988) and Haines (1989). Linear current rise corresponds to a constant voltage applied to a highly inductive Z pinch ($R \ll R_w$, where R_w is the radius of the return conductor) and is typical for experiments with fiber-initiated HDZP.

Neglecting C in (3.35) for a sufficiently large t/τ_{PB}, we find that the maximum value of about $\sqrt{3}I_{PB}$ cannot be exceeded by a linearly rising current because of radiative collapse, and this result does not depend on the value of dI/dt (for $dI/dt = \text{const}$) (Robson, 1988; Haines, 1989; see also Robson, 1989; Chittenden et al., 1989).

Note that unlike the collapse predicted by a thin-shell model or a snowplow model, when a plasma shell comes close to the axis, both models simply become invalid. The radiative collapse appears to be not just a feature of a particular 0-D model, but a physical phenomenon, which in principle can be realized in a subsonic compression of a plasma column by a current greater than I_{PB}. Of course, there are factors limiting the compression, which are not taken into account in the 0-D model. First, if the compression described by (3.35) is infinite, it cannot remain subsonic: with dI/dt limited, the compression velocity dR/dt diverges as $R \rightarrow 0$, whereas $V_A \propto I/\sqrt{N}$ remains finite, so that the subsonic condition $|dR/dt| \ll V_A$ is violated. Second, a Z-pinch compression is always unstable, and diverging velocity and acceleration should produce very strong instabilities (See Ch. 4). Third, when the pinch collapses, its inductance per unit length, $L = 2 \times 10^{-7} \ln(R_w/R)H$ m^{-1}, becomes infinite. Thus, an unlimited compression is impossible because it requires infinite voltage and infinite power applied to the Z pinch. For sufficiently small R_w/R, the current waveform cannot be prescribed arbitrarily; it should be determined from the circuit equations, with the Z-pinch dynamics taken into account in a self-consistent way.

However, radiative collapse regarded as a high (though not infinite) subsonic compression of a Z pinch by a current greater than I_{PB} seems to be realistic. The 0-D model given above can be used to describe it, if the varying inductance is taken into account. This was done by Robson (1988, 1989) and Haines (1989). The pinch current I, voltage V, and pinch radius R as a function of time were calculated by Haines (1989) for an external circuit represented by a transmission line with the Z-pinch assembly as a load, the maximum line voltage being 4 MV. The current is shown to rise almost linearly, up to the value of about I_{PB}, with the transmission line maintaining constant maximum voltage of 4 MV. The rapidly increasing inductance, L, then results in decreasing current, its maximum value being $\sim 1.5I_{PB}$. Radiative collapse continues until the pinch current is decreased below I_{PB}. After that, the pinch expands sharply from the minimum radius of ~ 0.14 mm and oscillates, approaching its equilibrium radius, whereas the current also approaches an equilibrium value, $I = I_{PB}$. If the maximum voltage provided by the transmission line is not sufficiently high to support the equilibrium state, then the expansion continues after collapse, its asymptotic law being $R \propto \sqrt{t}$. Although the increase in the inductance of the collapsing plasma should prevent radiative collapse, it is interesting to know the final equilibrium state produced by an unrestricted radiative collapse. The final stage of the pinch plasma after radiation collapse was discussed by Meierovich and co-workers (1975, 1982, 1986) who showed that the steady state in question is a degenerate-electron Fermi gas, whose charge is partly neutralized by the ions, since only the pressure of a degenerate-electron gas can stop the collapse.

One-dimensional calculations performed for cylindrical plasmas that account for inertia and dissipation, both analytical and numerical (see Vikhrev, 1978; Meierovich, 1985), also indicate the possibility of radiative collapse. The processes terminating radiative collapse were studied by Chittenden and Haines (1990), on the basis of a 1-D Lagrangian resistive-MHD code for the conditions of relevance to the present generation of hydrogenic-fiber Z-pinch experiments. They conclude that the termination process is a hybrid of radiation transport and free-electron degeneracy effects, in agreement with the results of Meierovich and Sukhorukov (1975). A regime similar to radiative collapse seems to be possible during the non-linear stage of the sausage-instability development, when high density is produced near the neck of the sausage (Yan'kov, 1985).

How close are the theoretical models to real Z-pinch radiative collapse? Experimentally the pinches have no problem carrying the current delivered to them by the pulsed-power generator, which can exceed $100 I_{PB}$. The real situation can differ essentially compared to the ideal models: for dense pinches the current can be carried mainly by a small fraction of the mass outside a cool core, i.e., the mass and current-density profiles give a large correction factor. For low-density pinches, radiative cooling may be too slow for the collapse to occur during the pinch lifetime. Radiative collapse can be modified because of space charge (Meierovich, 1985), for example, or because the plasma is not in ionizing equilibrium (Faenov et al., 1985). It seems likely that the bright spots in Z pinches and micropinches can be treated as a local radiative collapse effect (Koshelev et al., 1989). The bright-spot model includes the effect of opacity on the radiation and the contribution of some anomalous resistivity model on the ohmic heating (Vikhrev et al., 1982). In this model, the Pease–Braginskii current depends on the pinch radius, and power equilibrium then defines an equilibrium pinch radius for any current. The micropinches form only in a limited range of line densities. When the initial line density is outside this range, the micropinches may form when the line density enters into the contraction regime. It was found (Vikhrev et al., 1982) that, for an iron pinch at the current 150 kA, the equilibrium radius has a minimum of about 1 μm at an ion line density, $N_i \cong 5 \times 10^{15}$ cm^{-1}, corresponding to a Bennett temperature $T \cong 1$ keV. The model assumed that Bennett equilibrium was valid. It is likely that the discrepancy occurs between Z-pinch parameters computed from the proposed models and the measurements because the dynamics of the Z-pinch plasma are not taken into account.

The general interest in radiative collapse is due to the fact that this regime appears to be capable of producing extreme states of matter, high temperatures ($\geq$ keV), extremely high densities ($\geq 10^{24}$ cm^{-3}), pressures (≥ 1000 Mbar), magnetic fields (≥ 100 MG), and x-ray emission (~ 1000 TW) that are very interesting from a physical point of view, let alone their possible applications. We see that two general methods produce these extreme states. One, which is similar to the traditional scheme of magnetic confinement, employs a cylindrically symmetric subsonic compression of a dense Z pinch, as described above. This corresponds to the experiments with frozen deuterium fibers. The other is based on an instability development, a complicated 3-D flow triggered by chaotic sausage perturbations, which is expected to produce much the same results (see Yan'kov, 1985).

3.3 Fluid models of Z-pinch plasmas

It is essential to understand the transport processes in plasma, such as particle diffusion and heat conduction across the magnetic field, electric resistivity, viscosity, and energy transfer to address the critical issues in Z-pinch fusion research. These processes occur as the result of plasma interactions, which may be described either in terms of binary collisions between particles or collective effects involving the scattering of particles by plasma waves. In general, the state of a fully ionized plasma of electrons and ions can be specified in terms of the velocity distribution functions, which satisfy kinetic equations for electrons and ions. In the limit of no collisions this equation is commonly referred to as the Vlasov or the collisionless Boltzmann equation. This description of ions and electrons in terms of distribution functions is often more detailed than is required, and when the plasma is close enough to equilibrium, it is sufficient to describe the plasma in terms of macroscopic variables such as density, mean velocity, pressure, temperature, etc.

The equations that describe the behavior of these macroscopic variables can be obtained by taking appropriate moments of the kinetic equations (with the binary collision term included). These moments result in fluid equations for the ions and electrons. This monograph will not study the kinetic equations in detail but simply points out the limitations of the fluid approximation and the amount of physics contained in the collision term in the kinetic equation.

The fluid equations obtained from the moments of the fluid equation are not *closed*, in that there are more variables than equations. This situation is addressed herein by the choice of the equation of state.

3.3.1 The one-fluid MHD model

The most widely used model of dense Z-pinch plasmas is a one-fluid, two-temperature MHD model developed by Braginskii (1957a, 1963) and Rosenbluth and co-workers (1954, 1956). The one-fluid MHD equations are obtained by combining the separate fluid equations belonging the ions and the electrons and by applying a number of approximations. (See Krall & Trivelpiece, 1973; T. G. Cowling, 1957; and others.) The model includes equations of mass and charge continuity

$$\frac{\partial \rho}{\partial t} + \nabla \cdot \rho \mathbf{u} = 0, \tag{3.36}$$

$$\frac{\partial \rho_q}{\partial t} + \nabla \cdot \mathbf{J} = 0;$$

the equation of motion

$$\rho \frac{d\mathbf{u}}{dt} = -\nabla(n_e T_e + n_i T_i) + \frac{\mathbf{J} \times \mathbf{B}}{c} - \nabla \cdot \widehat{\pi}'; \tag{3.37}$$

and the energy equations for electrons and ions

$$\frac{3}{2} n_e \frac{dT_e}{dt} + n_e T_e \nabla \cdot \mathbf{u} = -\nabla \cdot \mathbf{q}_e + \frac{J_\parallel^2}{\sigma_\parallel} + \frac{J_\perp^2}{\sigma_\perp} + \frac{\mathbf{J} \cdot \mathbf{R}_T}{e n_e} - Q_\Delta - Q_r,$$

$$\frac{3}{2} n_i \frac{dT_i}{dt} + n_i T_i \nabla \cdot \mathbf{u} = -\nabla \cdot \mathbf{q}_i + \pi'_{\alpha\beta} \frac{\partial u_\alpha}{\partial x_\beta} + Q_\Delta, \tag{3.38}$$

where $\rho = m_i n_i + m_e n_e$ is the plasma mass density, $\rho_q = q_i n_i + q_e n_e$ is the plasma charge density, and u is the material velocity. J is the current density; σ is the plasma conductivity (indices $\parallel$ and $\perp$ correspond to parallel and transverse directions with respect to the magnetic field B); R_T is the thermoforce; $q_{e,i}$ are electron and ion heat fluxes, respectively; $q_\Delta = (3m_e/m_i)(n_e/\tau_e)(T_e - T_i)$ is the electron-ion collisional energy exchange term (τ_e is the electron collisional time); $\widehat{\pi}^i \equiv \pi^i_{\alpha\beta}$ is the ion viscous-stress tensor; Q_r represents radiative energy losses; and $d/dt \equiv \partial/\partial t + (\mathbf{u} \cdot \nabla)$ is the co-moving derivative.

An equation for the electric field, E, can be obtained by multiplying the ion momentum equation by the average ion charge, subtracting the electron fluid equation, and solving for the electric field. This gives

$$\mathbf{E} = -\frac{1}{c} \mathbf{u} \times \mathbf{B} + \left(\frac{\mathbf{J}_\parallel}{\sigma_\parallel} + \frac{\mathbf{J}_\perp}{\sigma_\perp} \right) + \frac{\mathbf{J} \times \mathbf{B}}{e n_e c} + \frac{\mathbf{R}_T}{e n_e} - \frac{\nabla P_e}{e n_e} + \frac{m_e}{e^2 n_e} \frac{d\mathbf{J}}{dt}. \tag{3.39}$$

Equation (3.39) is called the generalized Ohm's law. The last four terms on the right-hand side are often neglected since these terms involve spatial or temporal derivatives and the assumption is made that for slow enough and long enough scale lengths these terms can be neglected (see Braginskii, 1963). The $\mathbf{J} \times \mathbf{B}$ term is the Hall term and can be neglected if $u/c \gg B/(L e n_e)$, i.e., for high density, long scale length phenomena. The Nernst and Ettinghausen terms are contained in the R_T term on the right-hand side. Neglecting the last four terms we obtain the simplified the Ohm's law

$$\mathbf{E} = -\frac{1}{c} \mathbf{u} \times \mathbf{B} + \left(\frac{\mathbf{J}_\parallel}{\sigma_\parallel} + \frac{\mathbf{J}_\perp}{\sigma_\perp} \right).$$

Combining this equation with Faraday's law of induction we obtain an equation for $\mathbf{B}$ and, with Ampere's Law, an equation for $\mathbf{J}$

$$\frac{\partial \mathbf{B}}{\partial t} = \nabla \times \left[\mathbf{u} \times \mathbf{B} - c \left(\frac{\mathbf{J}_\parallel}{\sigma_\parallel} + \frac{\mathbf{J}_\perp}{\sigma_\perp} \right) \right], \tag{3.40}$$

$$\mathbf{J} = \frac{c}{4\pi} \nabla \times \mathbf{B}. \tag{3.41}$$

A number of common simplifying approximations are made in the development of the MHD equations. Plasma quasineutrality is generally assumed over the scale lengths of interest ($n_e \approx \bar{Z} n_i$), electron viscosity and inertia ($m_i \gg m_e$) being neglected. The transport coefficients for a fully ionized plasma have been calculated by Braginskii (1963). These approximations must be remembered for what they are, approximations. They limit the scope of applicability for the resulting MHD equations. Often after tedious solutions or computer modeling of the MHD equations it is easy to forget the limitations of the resulting answers.

In some cases, equations describing the ionization kinetics or radiative energy transport should be added to the system. For very high currents or particle beams interacting with the Z-pinch plasmas when relativistic effects are important, the single-fluid approximation turns out to be inadequate, and two-fluid model of a non-neutral plasma should be used instead (Meierovich and Sukhorukov, 1975; Solov'ev, 1984). When low-density plasmas are considered, like those of RFPs or EXTRAP, other fluid models may be relevant: the perpendicular MHD model, the CGL model, etc. Some of the stability problems cannot be treated with a fluid model and require a kinetic approach. But for dense Z-pinch plasmas (3.36)–(3.41) are valid in most cases, and most of the present discussion is based on this model.

3.3.2 The ideal one-fluid MHD model

In many cases, the collisional terms can be neglected in equations (3.36)–(3.41) to obtain the system of ideal MHD equations, which consists of (3.36) and (3.41), and

$$\frac{\partial \mathbf{B}}{\partial t} = \nabla \times (\mathbf{u} \times \mathbf{B}), \tag{3.42}$$

$$\frac{d}{dt}(P\rho^{-\gamma}) = 0, \tag{3.43}$$

$$\rho \frac{d\mathbf{u}}{dt} = -\nabla P + \frac{\mathbf{J} \times \mathbf{B}}{c}, \tag{3.44}$$

instead of (3.37)–(3.40), respectively; here $P = n_e T_e + n_i T_i$. Equation (3.38) is written for a particular case of the value of the adiabatic exponent $\gamma = 5/3$, but it may be useful sometimes to do the estimates with arbitrary values of γ in (3.43).

We can estimate the conditions under which the ideal MHD model is valid. If the pinch plasma is collision-dominated (Freidberg, 1982), i.e., the Knudsen number

$$K_n = \ell_\varepsilon / L \ll 1, \tag{3.45}$$

where $\ell_\varepsilon = (m_i/m_e)^{1/2} v_{thi} \tau_i$ is the mean free path corresponding to electron-ion energy relaxation (τ_i is the ion collisional time), and L is the characteristic length scale. Assuming $L = R$, $T_e = T_i = T$, we express this condition as

$$K_n = \frac{(m_i/m_e)^{1/2} R T^2}{4e^4 \bar{Z}^4 N \ln \Lambda} = \frac{5\bar{A}^{-1/2} R(100\ \mu\mathrm{m})[I\,(\mathrm{MA})]^4 (T/T_B)^2}{\bar{Z}^4 (1+\bar{Z})^2 (\ln \Lambda/10)(N(10^{18}\ \mathrm{cm}^{-3}))^3} \ll 1, \tag{3.46}$$

where T_B is the Bennett temperature (2.9). The second condition is expressed via the Hall coefficient

$$C_H = R/\rho_i = \bar{Z}^{-1}(1 + \bar{Z})^{-1/2}(T/T_B)\Pi^{1/2} \gg 1, \qquad (3.47)$$

where Π is given by (2.12), ρ_i is the ion Larmor radius. The inequality

$$\Pi = 6 \times 10^2 \bar{Z}^2 \bar{A}^{-1} N(10^{18} \text{ cm}^{-1}) \gg 1 \qquad (3.48)$$

should be satisfied to avoid plasma microinstabilities [see (2.11)]. Equation (3.48) is known to be a condition of hydrodynamic fluid behavior of the plasma: when it holds, the macroscopic velocity of electrons is close to that of ions, due not only to the collisions, but also to the action of a self-consistent magnetic field (Braginskii, 1963). The third condition is small plasma resistivity, that is the magnetic Reynolds number,

$$R_M \equiv \frac{uL}{\nu_m} \gg 1, \qquad (3.49)$$

where u and $\nu_m = c^2/4\pi\sigma$ are the characteristic values of velocity and magnetic viscosity, respectively. A large value of the magnetic Reynolds number allows the simplification of Ohm's law giving (3.42). Taking $u = V_A$, $L = R$, and estimating for classical plasma conductivity $\nu_m = m_e c^2/4\pi e^2 n_e \tau_e$, we obtain

$$\begin{aligned} R_M = L_u &= \frac{3\sqrt{\pi} I^4 R(T/T_B)^{3/2}}{4\sqrt{m_e m_i} e^2 c^6 \ln\Lambda \bar{Z}(1 + \bar{Z})^{3/2} N^2} \\ &= 1.6 \times 10^4 \frac{R(100 \ \mu\text{m})(I(MA))^4 (T/T_B)^{3/2}}{\sqrt{\bar{A}}\bar{Z}(1 + \bar{Z})^{3/2}(\ln\Lambda/10)(N(10^{18} \text{ cm}^{-1}))^2} \gg 1, \end{aligned} \qquad (3.50)$$

where $L_u = V_A L/\nu_m$ is, by definition, the Lundquist number.

The Z-pinch plasma can be regarded as a collision-dominated fluid, if the inequalities (3.46)–(3.48) hold. If the condition (3.50) also holds, it is also ideally conducting, at least when length and time scales of order of R and τ_A are considered. In most cases, the thermoforce term is of the same order as the ohmic terms. To neglect plasma viscosity and heat conductivity, the Reynolds number, R_e, and the Péclet number, P_e, should be large:

$$R_e \equiv \frac{uL}{\nu} \gg 1 \quad \text{and} \quad P_e \equiv \frac{uL}{\chi} \gg 1, \qquad (3.51)$$

where ν and χ are kinematic viscosity and thermal diffusivity, respectively. For the transport coefficients corresponding to isotropic plasma viscosity, $\nu = T\tau_i/m_i$, and parallel electron heat conductivity, $\chi = T\tau_e/m_e$, we obtain

$$\begin{aligned} R_e &= ((\bar{Z} + 1)(T_B/T)(m_i/m_e))^{1/2} K_n^{-1} \\ &= 8.5 \frac{(1 + \bar{Z})^{5/2}\bar{Z}^4(\ln\Lambda/10)(N(10^{18} \text{ cm}^{-1}))^3}{R(100 \ \mu\text{m})(I(MA))^4(T/T_B)^{5/2}} \end{aligned} \qquad (3.52)$$

$$P_r \equiv P_e/R_e = \sqrt{2m_e/m_i}\bar{Z}^{-2} \ll 1, \qquad (3.53)$$

$$P_e = 2(1 + 1/\bar{Z})(T_B/T) \cdot (\Pi/L_u), \qquad (3.54)$$

where $P_r \equiv \nu/\chi$ is the Prandtl number. From (3.46) it follows that the condition (3.50) is always satisfied when the plasma is collision-dominated. In particular, (3.50) is valid when condition (3.51) holds [see also (3.53)]. The value of P_e given by (3.54) is in most cases large, but it could be of order unity. However, this does not mean that the effect of heat conductivity can be important, because electrons in Z-pinch plasmas are usually strongly magnetized, i.e., $\Omega_e \tau_e \gg 1$, Ω_e being the electron gyrofrequency, so that radial heat conductivity is damped by a factor, $(\Omega_e \tau_e)^2$ and the value of P_e, is increased by the same factor (Braginskii, 1963). The axial heat conductivity affects only perturbations of a cylindrical Z-pinch, and can be shown to be of minor importance also.

The Hall term [the $\mathbf{J} \times \mathbf{B}$ term in the right-hand side of (3.39)] can be neglected if the inequality (3.47) is valid. For a cylindrically symmetrical Z pinch, this term vanishes identically for any value of C_H. Indeed, the direction of the vector product, $\mathbf{J} \times \mathbf{B}$, is radial, depends only on r, and hence does not contribute to the curl in the right-hand side of Eq. (3.40). The same is true for the next term, which is proportional to $\nabla n_e \times \nabla T_e$, and represents a source for the magnetic field. Some effects related to the Hall and Nernst terms were considered for a noncylindrical Z pinch by Vikhrev and Gureev (1977) (see also Vikhrev and Braginskii, 1980; Shaper, 1983; Coppins et al., 1984).

The radiative losses can be neglected in the first approximation, if the corresponding dimensionless parameter is large:

$$C_r = n_i(1 + \bar{Z})T/Q_r\tau_A \gg 1. \qquad (3.55)$$

If the bremsstrahlung radiative energy losses dominate, then C_r is of order of τ_{PB}/τ_A where τ_{PB} is given by (3.32). Comparing (3.32) and (2.15), we see that for typical Z-pinch parameters, the time scale for bremsstrahlung energy losses is much greater than τ_{PB}, which makes possible a comparatively slow evolution of the time-dependent equilibria described in Sec. 3.2.3. When heavy ions emitting intense radiation in resonant lines are present, C_r can be of order unity; the radiative losses are thus important in determining the dynamics of the fast processes in a Z pinch (it also turns out to be an instability mechanism; see Sec. 4.2.1).

Calculation of the dimensionless parameters, K_n, L_u, R_e, P_e, and Π, for typical experimental conditions in dense Z pinches, shows that the ideal MHD model is applicable in most cases, at least for the sonic time scale τ_A (an example is given in Table 2). The dynamics of subsonic ohmically heated radiating Z pinches are determined mainly by dissipation and losses on the time scale τ_{PB}. In turn, this is possible due to their remarkable stability during time intervals much greater than τ_A. The ideal MHD model remains a reasonable model, even when the pinch current is about 1 MA and $T \sim 2$ keV, so that the plasma is collisionless, the ion mean free path being greater than R. Then, the classical expressions for the transport coefficients are no longer valid. Nevertheless, if the ions are magnetized, $\Omega_i t_i \gg 1$, and (3.48) holds, the collisionless plasma behaves as an ideal collisional fluid due to the action of transverse magnetic field (Braginskii, 1963; Chen, 1984). Dense Z pinches thus appear to be the best field for applications of the conventional

TABLE 2.   Typical Plasma Parameters for some Well-Diagnosed Z-Pinch Experiments

Plasma Parameters	Machine name	
	SNOP-3	I-4
Current pulse amplitude I_m (MA)	1.5	2.5
Plasma temperature T (keV)	0.04–0.05	0.07
Plasma line density N (cm^{-1})	10^{18}	3×10^{18}
Radius R of the plasma column (cm)	0.7	0.7
Plasma density n (cm^{-3})	5×10^{17}–5×10^{19}	10^{18}–10^{20}
Bennett temperature T_B (keV)	$9/(1 + \bar{Z})$	$10/(1 + \bar{Z})$
Parameter Π (2.12)	$5 \times 10^2 \bar{Z}^2/A$	$10^3 \bar{Z}/A$
Knudsen number Kn	$\dfrac{7 \times 10^{-2} A^{1/2}}{\bar{Z}^4(1 + \bar{Z})^2}$	$\dfrac{8 \times 10^{-2} A^{1/2}}{\bar{Z}^4(1 + \bar{Z})^2}$
Magnetic Reynolds number Rm	$\dfrac{2 \times 10^3}{A^{1/2}\bar{Z}(1 + \bar{Z})^3}$	$\dfrac{7 \times 10^3}{A^{1/2}\bar{Z}(1 + \bar{Z})^3}$
Reynolds number Re	$9 \times 10^3 \bar{Z}^4(1 + \bar{Z})^5$	$6 \times 10^3 \bar{Z}^4(1 + \bar{Z})^5$
Prandtl number Pr	$\dfrac{3 \times 10^2}{A^{1/2}\bar{Z}^2}$	$\dfrac{3 \times 10^2}{A^{1/2}\bar{Z}^2}$
Péclet number Pe	$3 \times 10^2 \dfrac{\bar{Z}^2(1 + \bar{Z})^5}{A^{1/2}}$	$2 \times 10^2 \dfrac{\bar{Z}^2(1 + \bar{Z})^5}{A^{1/2}}$

ideal MHD model, whose region of validity completely excludes other plasmas of fusion interest (cf. Freidberg, 1982; Coppins, 1988).

When perturbations of a cylindrically symmetric Z pinch are considered, it should be noted that the scale length and time, k^{-1} and, γ^{-1}, corresponding to a perturbation (k and γ being the wavenumber and the growth rate, respectively), can differ from those characteristic of the pinch, R and τ_A. In addition, the perturbations break the symmetry, allowing nonzero contributions of the Hall and $\nabla n_e \times \nabla T_e$

terms in (3.40), of parallel heat conductivity term in (3.38), etc. Thus, the fluid model describing the perturbations may not be generally the same as that needed to describe the dynamics of the unperturbed Z pinch.

For instance, an unperturbed Z pinch can be dissipative; the same is not true for the dominate instability mode. The latter can be sufficiently large-scale and develop fast enough to be described by an ideal MHD model. In particular, if a radiating dense Z pinch is believed to remain stable much longer than τ_A, then the absence of instability modes with growth rates of order τ_A^{-1} should be proved. Of course, the inverse is also possible. The unperturbed flow can represent a fast compression of an annular liner, the relevant time scale being of order of τ_A, the plasma resistivity not being as important in determining its dynamics, provided that condition (3.50) holds. However, the instability mode dominating during the linear stage of development can be determined just by the condition, $L_u \sim 1$, shorter wavelengths being damped due to resistivity and longer ones having lower growth rates (see Sec. 4.3).

3.4 Self-similar dynamics of an ideal MHD Z pinch

One-dimensional, self-similar solutions of the fluid equations are obtained assuming that all the flow variables are separable functions of time and that the radius is normalized to a properly determined length scale (Zel'dovich and Raizer, 1967; Barenblatt, 1979). The use of self-similar solutions to study the dynamics of Z pinches was first suggested by A. D. Sakharov (see, Braginskii, 1957b).

At present, while obtaining numerical solutions of fluid equations is not a major problem, the importance of self-similar solutions stems from their ability to represent, in some cases, the asymptotic states of quite complicated plasma flows, and to be attractors in the space of all possible solutions. Thus, a self-similar solution is not only a particular solution of the equations under consideration, it has a certain degree of generality; it exhibits some universal features of the flow structure, which are independent of the initial conditions. Of course, this statement is not a theorem that could be proved mathematically once and for all. Rather, it is a result of experience, and its validity in each case should be established through numerical study. In addition, this property of self-similar solutions depends largely on both the functional space where the solutions are sought and the type of flow.

Self-similar solutions can be (and often are) attractors for strictly 1-D solutions, whereas the introduction of 3-D perturbations can reveal instabilities driving the flow away from its cylindrically symmetrical structure. The self-similar solutions describing plasma expansion have generally a better chance to be attractors than those describing implosion; solutions describing subsonic flows with dissipation taken into account have a better chance than those constructed for ideal supersonic flows. The self-similar solutions are valuable for their qualitative insight and guidance into the relevant areas of Z-pinch plasma dynamics.

3.4.1 Self-similarity anzatz

Self-similar dynamics of Z pinches were discussed extensively by Liberman and Velikovich (1986b), Bud'ko and Liberman (1990a,b; 1991; 1992), and Bud'ko et al. (1993; 1994). Assuming cylindrical symmetry, the absence of rotation, i.e., $u = (u, 0, 0)$, and $T_e = T_i = T$, we can present (3.36), (3.42)–(3.44) in the form

$$\frac{\partial n}{\partial t} + \frac{1}{r}\frac{\partial}{\partial r}(rnu) = 0, \tag{3.56}$$

$$\frac{\partial B_\varphi}{\partial t} + \frac{\partial}{\partial r}(uB_\varphi) = 0, \tag{3.57}$$

$$\frac{\partial B_z}{\partial t} + \frac{1}{r}\frac{\partial}{\partial r}(ruB_z) = 0, \tag{3.58}$$

$$\frac{\partial T}{\partial t} + u\frac{\partial T}{\partial r} + (\gamma - 1)\frac{T}{r}\frac{\partial}{\partial r}(ru) = 0, \tag{3.59}$$

$$\rho\left(\frac{\partial u}{\partial t} + u\frac{\partial u}{\partial r}\right) + \frac{\partial P}{\partial r} + \frac{B_\varphi}{4\pi r}\frac{\partial}{\partial r}(rB_\varrho) + \frac{\partial B_z}{\partial r} = 0, \tag{3.60}$$

where $n \equiv n_i$, $\rho = m_i n$, $P = (1 + \bar{Z})nT$.

Consider solutions of Eqs. (3.56)–(3.60), which can be presented as separable functions of time and a self-similar coordinate

$$\xi = \frac{r}{R(t)}, \tag{3.61}$$

where $R(t)$ is a time-dependent length scale of the problem. Arbitrarily choosing any reference time, $t = t_0$, and a corresponding length scale, $R_0 = R(t_0)$, and introducing a dimensionless compression ratio

$$\alpha(t) = R(t)/R_0, \tag{3.62}$$

we can separate the variables ξ and t in (3.56)–(3.59) by choosing flow variables in the form

$$u(r, t) = \dot{R}(t)\xi U(\xi), \tag{3.63}$$

$$n(r, t) = n_0 \alpha(t)^{2\chi} N(\xi), \tag{3.64}$$

$$B_\varphi(r, t) = B_{\varphi 0}\alpha(t)^\mu H_\varrho(\xi), \tag{3.65}$$

$$B_z(r, t) = B_{z0}\alpha(t)^\nu H_z(\xi), \tag{3.66}$$

$$T(r, t) = T_0 \alpha(t)^{-2\lambda}\Theta(\xi), \tag{3.67}$$

where n_0, T_0, $B_{\varphi 0}$, and B_{z0} are reference number density, energy, azimuthal, and axial magnetic field; U, N, H_φ, H_z, and Θ are dimensionless functions of ξ, giving self-similar profiles of velocity, density, magnetic field, and temperature; χ, λ, μ, and ν are the self-similarity exponents (we assume N, H_φ, H_z, and $\Theta > 0$).

The equation of motion (3.60) becomes

$$\ddot{\alpha}U + (\dot{\alpha}^2/\alpha)(U - 1)\frac{d}{d\xi}(U\xi) + \alpha^{2(\mu-\chi)-2}\left[\frac{H_\varphi}{\xi^2 N}\frac{d}{d\xi}(\xi H_\varphi)\right]$$
$$+ \beta\alpha^{-2\lambda-1}\left[\frac{1}{\xi N}\frac{d}{d\xi}(N\Theta)\right] + b\alpha^{2(\nu-\chi)-2}\left[\frac{H_z}{\xi N}\frac{dH_z}{d\xi}\right] = 0,$$

(3.68)

where the unit of time is the Alfvén transit time $\tau_A = R_0(4\pi m_i n_0)^{1/2}/B_{\varphi 0}$;

$$\beta = 4\pi(1 + \bar{Z})n_0 T_0/B_{\varphi 0}^2, \quad b = B_{z0}^2/B_{\varphi 0}^2,$$

(3.69)

are dimensionless parameters of the problem.

There are two ways of separating the variables in (3.68). One of these methods requires a linear coordinate dependence of the velocity, i.e., homogeneous deformation (since the "rate of deformation" tensor, du_i/dx_k, does not depend on the coordinates) (see Sedov, 1959). Assuming

$$U(\xi) = 1,$$

(3.70)

we can satisfy (3.56)–(3.59) identically for arbitrary $N(\xi)$, $H_\varphi(\xi)$, $H_z(\xi)$, and $\Theta(\xi)$ by an appropriate choice of the self-similarity exponents in (3.64)–(3.67):

$$\chi = -1, \quad \lambda = \gamma - 1, \quad \mu = -1, \quad \nu = -2.$$

(3.71)

Then (3.68) would be also satisfied, if all the terms in square brackets are constant. This imposes three conditions on the functions $N(\xi)$, $H_\varphi(\xi)$, $H_z(\xi)$, and $\Theta(\xi)$. Hence, one function remains arbitrary: constructing the self-similar profiles, one can postulate either an arbitrary shape of one of these profiles or an arbitrary relation between them (see also Saiz et al., 1992). This arbitrariness can be also used to construct self-similar solutions, taking into account some physical mechanisms beyond the ideal MHD model like ohmic heating or radiative losses (see Sec. 3.4.3.). The time dependence of all flow variables is then found from (3.68).

If (3.70) and (3.71) are not valid, and hence $N(\xi)$, $H_\varphi(\xi)$, $H_z(\xi)$, and $\Theta(\xi)$ are not arbitrary, then the terms in the left-hand side of (3.68) cannot have a similar coordinate dependence. This would require eight conditions of the five functions U, N, H_φ, H_z, and Θ. Equation (3.68) can be separable a second way, if the variables in (3.68) and all the terms in (3.68) have a similar time dependence. This entails a power law or exponential dependence, $\alpha(t)$ (for simplicity, we omit the additive constant in this dependence), the exponent being determined by λ:

$$\alpha(t) = \begin{cases} \left|\dfrac{t}{t_0}\right|^{1/(1+\lambda)}, & \lambda \neq -1, \\ \exp\left(\pm\dfrac{t}{t_0}\right), & \lambda = -1. \end{cases}$$

(3.72)

The kinetic and magnetic pressures for any given ξ should be also proportional to each other, and from (3.64)–(3.67) we obtain

$$\mu = \nu = \chi - \lambda.$$

TABLE 3.   Comparison of Self-Similar Solutions of Different Classes

	Non-homogeneous deformation
Inertia	retained
Dissipation	neglected
Exponents c, l, m	form a 2-parameter family
Time dependence of length scale $R(t)$	power law; for an exceptional case, an exponential function
Time dependence of total current $I(t)$	current is finite only if a current sheath on the pinch boundary is introduced; then $I(t)$ depends on time, not by a power law
$U(x)$	is to be determined from the equation for the self-similar profiles
$N(x), Q(x), H(x)$	are to be determined from the equation of the self-similar profiles
A fluid particle with Lagrangian corrdinate q and Eulerian coordinate $R_q(t)$	passes via all phases of profiles (i.e., via all values of x); $R_q(t)$ is not expressed explicitly in terms of $R(t)$ and is not a power law
Weak discontinuity	can take place only along the limiting characteristic, for one definite value of x
Shock wave	can take place in a general case and is characterized by a fixed value of x

Consequently, self-similar solutions of this class in ideal MHD form a two-parameter family (χ and λ being the remaining free parameters), as do the self-similar solutions in gas dynamics (Guderley, 1942; Brushlinskii and Kazhdan, 1963; Zel'dovich and Raizer, 1967; Meyer-ter-Vehn and Schalk, 1982). This family includes the so-called self-similar solutions of the second kind, for which the self-similarity exponents cannot be found independently from dimensional considerations, conservation laws, etc. The types of self-similar solutions available for dynamic Z pinches are listed in Table 3, taken from Bud'ko et al. (1989c). Below, they will be considered separately.

TABLE 3.   *cont.*

	Homogeneous deformation	
	Ideal MHD	Dissipation-dominated subsonic
Inertia	retained	neglected
Dissipation	neglected	retained, but not all at once
Exponents c, l, m	unambiguously determined	form a 1-parameter family
Time dependence of length scale $R(t)$	determined from equation of motion; as a rule, no power law	power law; for an exceptional case
Time dependence of total current $I(t)$	constant current; it can change only if a current sheath on the pinch boundary is introduced; then $I(t)$ depends on time, not by a power law	an exponential function
$U(x)$	1	1
$N(x), Q(x), H(x)$	one of these functions or any relation between them can be chosen arbitrarily	are to be determined from the equations for the self-similar profiles
A fluid particle with Lagrangian corrdinate q and Eulerian coordinate $R_q(t)$	characterized by a definite value of x, $R_q(t)$ being proportional to $R(t)$	
Weak discontinuity	can take place for any (not necessarily one) value of x	cannot take place
Shock wave	can take place in a special case and is characterized by a time-dependent value of x	

3.4.2 Self-similar solutions with homogeneous deformation

The flow regime of homogeneous deformation studied first by Kulikovskii (1957, 1958), and then by Felber (1982) and Solov'ev (1984), may be realized when all the essentially nonuniform processes, such as propagation of shock waves, sonic perturbations, etc., are completed, and each fluid particle corresponds to a fixed point of the self-similar profiles (if the instabilities have not broken the initial cylindrical symmetry before that). Here, the self-similar coordinate ξ is simply a Lagrangian coordinate of a fixed-plasma particle. Though the radial flow velocity, u, may be less or greater than the local fast magnetosonic velocity, C_f, the motion is essentially subsonic. This means that the velocity of the self-similar profiles with respect to the plasma particles is zero. There is no exchange of waves between the plasma particles; the self-similar flow is organized in such a way that each particle moves consistently with all the others.

It follows from Eqs. (3.61) and (3.67) that this regime corresponds to a constant axial current, $I = I_{max} = $ const. If there is no axial magnetic field, the equation of motion (3.68) becomes

$$\ddot{\alpha} + \alpha^{-1} - \beta\alpha^{1-2\gamma} = 0, \tag{3.73}$$

the initial conditions being

$$\alpha(0) = 1, \quad \dot{\alpha}(0) = 0, \tag{3.74}$$

and the self-similar profiles satisfying

$$\frac{1}{\xi N}\frac{d(N\Theta)}{d\xi} = -1, \qquad \frac{H_\varphi}{\xi^2 N}\frac{d(\xi H_\varphi)}{d\xi} = 1. \tag{3.75}$$

In particular, choosing a uniform temperature profile, $\Theta(\xi) = $ const $= 1$, one can integrate (3.75) to obtain (Felber, 1982)

$$N = \exp\left(-\frac{\xi^2}{2}\right),$$

$$H_\varphi = \frac{2}{\xi}\sqrt{1 - (1 + \xi^2/2)\exp(-\xi^2/2)}. \tag{3.76}$$

The self-similar profiles for this case are plotted in Fig. 3.4.

Equation (3.73) has the same form as an equation of motion of a point mass in a potential field, $V(\alpha)$, α being an analog of a coordinate. The corresponding energy integral is

$$\frac{\dot{\alpha}^2}{2} + \ln\alpha + \frac{\beta}{2(\gamma - 1)}\left(\alpha^{-2(\gamma-1)} - 1\right) = 0. \tag{3.77}$$

Comparing (3.3), (3.73), and (3.77) we see that the equation of motion is in effect the same as provided by the 0-D model, the only difference being due to term proportional to β. [For $\beta = 0$, (3.3) is the solution of (3.73) describing collapse.] The value $\beta = 1$ corresponds (parameter β here is defined by Eq. (3.69), do not

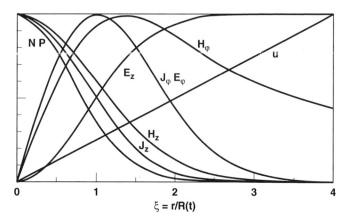

FIGURE 3.4. Self-similar profiles for a Z pinch with Gaussian density profile (Felber, 1982).

confuse with b in Sec. 2.1.1.) to the Bennett equilibrium: $\alpha(t) = 1 = $ const. For $0 < \beta < 1$, Eq. (3.73) describes an oscillatory motion of a particle in a potential well, $\alpha = 1$ being one turnaround point, and $\alpha = \alpha_{min} < 1$ another, the period of the motion being of order of τ_A. The implosion is stopped because the thermal pressure grows with increasing $\alpha(t)^{-1}$ faster than the magnetic pressure if $\gamma > 1$, as seen from (3.73). Then, the expansion of the plasma column to its initial radius $\alpha = 1$ follows, and the process continues. The peak compression corresponds to a maximum temperature, which is higher than the Bennett temperature corresponding to the same values of I and N by factor

$$T/T_B = C\beta\alpha_{min}^{-2(\gamma-1)}, \tag{3.78}$$

where C is a dimensionless factor of order unity, which depends on the self-similar profiles. For $\gamma = 5/3$ and β in the range 0.2–0.05 (which corresponds to radial compression α_{min}^{-1} in the range 6–30) $\alpha_{min} \approx 2\beta/3$, and the right-hand side of (3.78) is approximately equal to $1.7C\beta^{-1/3}$. We see that adiabatic heating, described by the present self-similar solutions, can be very effective and much larger than ohmic heating (in particular, because there is no limit on the current, like the Pease–Braginskii limit), provided that the imploded Z-pinch plasma is sufficiently low-beta and stable (large value of α_{min}^{-1} achieved without loss of cylindrical symmetry).

If an axial magnetic field compressed by the imploding Z pinch is present, then the corresponding term, $-\beta\alpha^{-3}$, and, $(\beta/2)(\alpha^{-2} - 1)$, should be added to the left-hand sides of (3.73) and (3.77), respectively, thus obtaining an analog of (3.3) and (3.11). The appropriate condition on the axial magnetic field profile is

$$\frac{H_z}{\xi N}\frac{dH_z}{d\xi} = -1, \tag{3.79}$$

which for a Gaussian isothermal pinch (3.76) yields a Gaussian profile of $H_z(\xi) = $ const $\cdot N(\xi)^{1/2}$. The case $\beta + b = 1$ corresponds to a Bennett equilibrium of a

Z pinch in the presence of axial magnetic field. For $0 < \beta + b < 1$, we obtain oscillations of the plasma column, as before. Now the pressure of an axial magnetic field contributes to stopping the implosion. Its influence is greater, the higher the compression ratio, because the compression of an axial field corresponds to an effective adiabatic exponent, $\gamma_{\mathrm{eff}} = 2$, which is greater than the value of the γ characteristic of the plasma. It turns out that an axial magnetic field is needed to make the oscillations of a Z pinch observable. In the experiments (Felber et al., 1988b) with a gas-puff Z pinch, up to four radial bounces of a plasma column were seen in an experiment with $b = 0.01$. When the axial magnetic field was absent, the other experimental conditions being the same, the oscillations could not be observed because instabilities destroyed the cylindrical symmetry within one compression–expansion cycle.

Equation (3.77) shows that self-similar solutions of this class cannot describe flows where the pressure gradient changes sign, like an annular plasma liner with zero pressure at $r = R_{\min}(t)$, $R_{\max}(t) = \mathrm{const} \cdot N(\xi)^{-1/2}$ and maximum pressure in between. However, the ideal MHD model permits a discontinuity of pressure and magnetic field on the surface of the liner, which means that the current is partly concentrated in the skin layer at the surface. This makes it possible to construct solutions describing annular liners (Liberman et al., 1987). The pinch current is not constant then; its time dependence is periodic, the period being the same as that of the oscillations of the plasma column. Another possibility corresponds to a cold plasma liner ($\beta = 0$) when the pressure term simply drops out from the equation of motion.

3.4.3 Self-similar solutions with energy losses

Constructing self-similar solutions of the ideal MHD equations with homogeneous deformation, one can use the functional arbitrariness mentioned in Sec. 3.3.2 to develop the model by including some additional physics in it. The simplest way to do this is to account for energy losses. Indeed, the variables ξ and t remain separable with the addition of an arbitrary term simulating energy losses into the right-hand side of (3.59), provided that this term is separable itself. The self-similarity anzatz (3.63)–(3.67) remains valid, and (3.56)–(3.58) are still identically satisfied by arbitrary N, H_φ, and H_z, if (3.70) holds and the values of χ, μ, ν are given by (3.71). (Hence, this family of solutions corresponds to a constant current, $I = I_{\max}$, if no discontinuities are present.) However, instead of (3.67) we have now

$$T(r, t) = T_0 \theta(t) \Theta(\xi), \qquad (3.80)$$

where $\Theta(\xi)$, as before, represents the self-similar temperature profile, whereas the time dependence of the temperature is given by the new function $\theta(t)$ to be determined. Below, we assume that no axial magnetic field is present. Separation of variables in the equation of motion is done exactly as in Sec. 3.4.2. In particular for the self-similar profiles, Eq. (3.75) remains valid, and its time-dependent part now takes the form

$$\ddot{\alpha} + \alpha^{-1} - \beta\alpha^{-1}\theta = 0. \tag{3.81}$$

The new equations, one for the self-similar profiles and one for $\theta(t)$, are obtained by separation of variables in the energy equation, with the term proportional to the power density of energy losses Q on the right-hand side:

$$\frac{\partial T}{\partial t} + u\frac{\partial T}{\partial r} + (\gamma - 1)\frac{T}{r}\frac{\partial}{\partial r}(ru) = \frac{\gamma - 1}{(1 + \bar{Z})n}Q. \tag{3.82}$$

The most important cases are bremsstrahlung radiative energy losses (Meierovich, 1985), recombination radiative energy losses, and energy losses in a model equation (Tendler, 1988) simulating the effect of finite energy-confinement time. They correspond to

$$Q = \begin{cases} \eta_o^{br}\bar{Z}^2n^2T^{1/3}, & \text{(a)} \\ \eta_o^{rec}\bar{Z}^4n^2T^{-1/2}, & \text{(b)} \\ 2nT/\tau_E, & \text{(c)} \end{cases} \tag{3.83}$$

respectively. Here the coefficient η_o^{br} is given by (2.17),

$$\eta_o^{br} = 5 \times 10^{-24} \text{ erg eV}^{-1/2} \text{ cm}^3 \text{ sec}^{-1}, \tag{3.84}$$

τ_E is the energy-confinement time. Separation of variables in (3.82) yields two equations, one of them coupling $\theta(t)$ and $\alpha(t)$, and the other determining the self-similar profiles together with (3.75).

In particular, for bremsstrahlung energy losses [case (a)], the additional dynamic equation is

$$\frac{d}{dt}\left[\ln(\theta\alpha^{2(\gamma-1)})\right] = -\delta\alpha^{-2}\theta^{-1/2}, \tag{3.85}$$

where $\delta = \tau_A[(\gamma - 1)\eta_o^{br}\bar{Z}^2n_0T_0^{-1/2}/(1 + \bar{Z})]$.

The self-similar profiles are

$$\Theta(\xi) = (1/3)(1 - \xi^2), \quad N(\xi) = \Theta(\xi)^{1/2},$$
$$H_\varphi(\xi) = \frac{2}{(675)^{1/4}\xi}[1 - (1 + 3\xi^2/2)(1 - \xi^2)^{3/2}]^{1/2}. \tag{3.86}$$

Equations (3.81) and (3.85) demonstrate oscillations of a radiating Z pinch around the equilibrium state (time scale of oscillations, $\sim \tau_A$) and a gradual decrease of the equilibrium radius with the characteristic time, τ_A/δ. This is illustrated by Fig. 3.5. Note that one can neglect the ohmic heating term in Eq. (3.78), taking at the same time the radiative losses term into account, if

$$I_{max} \gg (T_B/T)I_{PB}, \tag{3.87}$$

where T_B is the Bennett temperature corresponding to the current I_{max}.

Cases (b) and (c) in (3.83) demonstrate a similar time dependence of the pinch radius (decaying oscillations around a decreasing equilibrium radius) and correspond to the same parabolic temperature profile, whereas the density and magnetic field profiles differ.

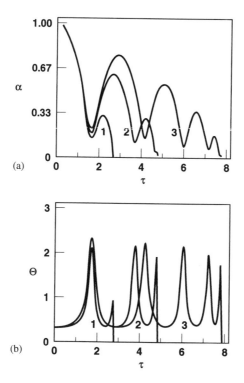

FIGURE 3.5. (a) Normalized pinch radius $\alpha(t)$ and (b) temperature $\Theta(t)$ for a self-similar solution with radiative losses (Meierovich, 1985).

3.4.4 Self-similar solutions of the second kind

Now we turn to the other family of self-similar solutions with a power-law time dependence of the pinch radius (3.72). For the case of a simple Z pinch with cylindrical symmetry and axial current, the solutions describe the collapse of shock waves and cavities (annular plasma liners) on axis, the outward propagation of shock waves reflected from the axis, and inverse Z-pinch flows and related phenomena.

The self-similarity anzatz (3.63)–(3.65), (3.67), and (3.72) yields four ordinary differential equations for the self-similar profiles: U, N, H_φ, and Θ. This system of equations can be reduced to a 3-D autonomous dynamic system by introducing new self-similar variables

$$S = \gamma\Theta/\xi(1 - U), \quad A = H_\varphi^2/\xi^2 N(1 - U), \tag{3.88}$$

related to the characteristic velocities. The local Mach numbers with respect to sonic, Alfvén, and fast magnetosonic velocities are expressed in terms of U, S, and A:

$$M_s^2 = u^2/v_s^2 = U^2/S(1 - U), \quad M_A^2 = u^2/V_A^2 = U^2/A(1 - U),$$

$$M_f^2 = u^2/C_f^2 = U^2/(S + A)(1 - U), \tag{3.89}$$

(Liberman and Velikovich, 1986b). The characteristic velocities are very important here because the self-similar profiles are formed by nonlinear waves, like shock or compression waves. The velocity of the self-similar profiles with respect to the plasma particles can be either less or greater than the local fast magnetosonic velocity (i.e., the waves in question may be either subsonic or supersonic), depending on whether the local value of $U + S + A$ is greater or less than unity. The plasma particles are involved in the self-similar motion by the waves that dominate the flow. A particle trajectory found by integrating the equation $dr_\xi/dt = u(r_\xi, t)$ (where ξ is a Lagrangian co-ordinate of a fixed plasma particle), is not, of course, given by a power law like (3.72).

The equations for U, S, A are

$$(U - 1)\frac{d(\ln S)}{d(\ln \xi)} + \gamma \left[\frac{dU}{d(\ln \xi)} + 2\left(U - \frac{\lambda + 1}{\gamma}\right) \right] = 0, \tag{3.90}$$

$$(U - 1)\frac{d(\ln A)}{d(\ln \xi)} + 2\left[\frac{dU}{d(\ln \xi)} + U - \lambda - 1 \right] = 0, \tag{3.91}$$

$$(U + S + A - 1)\frac{dU}{d(\ln \xi)}$$
$$+ \left[U(U - \lambda - 1) + 2S\left(U + \frac{\chi - \lambda}{\gamma}\right) + A(\chi - \lambda - 1) \right]$$
$$= 0, \tag{3.92}$$

and the continuity equation can be integrated together with (3.86), expressing N explicitly in terms of U, S, and ξ:

$$N = K_N S^{\frac{\chi+1}{\gamma-\lambda-1}} (\xi^2(1 - U))^{\frac{\lambda+\gamma\chi+1}{\gamma-\lambda-1}}, \tag{3.93}$$

where the positive integration constant K_N, determining the scale of N, is arbitrary to the same extent as n. When a solution of (3.86)–(3.88) is found, (3.89) can be used to find the profiles $N(x)$ and $H_\varphi(\xi)$ from the known profiles $U(\xi)$, $S(\xi)$, $A(\xi)$. The study of the solutions of (3.90)–(3.92) yields the following results.

1. No cylindrically symmetrical self-similar solutions of the ideal MHD equations are regular at $r > 0$. All the available solutions contain a singularity of $H_\varphi \propto 1/\xi$ at $\xi \to 0$, which represents a concentrated current flowing along the pinch axis. If the current along the axis is absent, the solutions are reduced to those of ideal hydrodynamics, with the magnetic pressure simulated by a boundary condition at the outer surface of the plasma column (strong skin effect). The solutions describing plasma implosion to the current-carrying axis represent collapse of a cavity (i.e., of a hollow, annular plasma liner) and of an ionizing shock waves converging to the axis.

2. At the moment of collapse, $t = 0$, the profiles of the flow variables are

$$u(r, 0) = \text{const} \cdot r^{-\lambda}, \quad n(r, 0) = \text{const} \cdot r^{2\chi},$$
$$B_\varphi(r, 0) = \text{const} \cdot r^{\chi-\lambda}, \quad T(r, 0) = \text{const} \cdot r^{-2\lambda}. \tag{3.94}$$

Such solutions make physical sense only for values of the self-similarity exponents λ and χ, which give, at $r = 0$, integrable densities of mass, current, and energy at the moment of collapse, i.e., for

$$\chi + 1 > 0, \quad \chi - \lambda + 1 > 0. \tag{3.95}$$

Note that if the power-law density profiles at $t = 0$ are integrable at the axis, they are inevitably not integrable at infinity. Thus, the solutions describe only the flow region near the axis. The cumulation of energy, producing infinite energy density at the axis where the plasma density itself may remain finite (for $\chi \geq 0$), is due to the fact that a decreasing amount of energy is distributed among the plasma particles, whose number is decreasing even faster.

3. As the implosion proceeds, the current is redistributed from the axis to the volume. At the moment of collapse, $t = 0$, no current flows along the axis.

Examples of the self-similar profiles of U, N, Θ, and H_φ are plotted in Fig. 3.6. This solution for $t < 0$ represents an implosion of a hollow plasma cavity (that is, of an annular plasma liner) onto a concentrated axial current bounded by a vacuum region (the parameters are $\lambda = 0.1$, $\chi = -0.4$); the same self-similar solution describes a reflected shock wave diverging from the axis for $t > 0$, though the profiles at this stage would be different. Figure 3.7 shows the time dependence of the concentrated current along the axis, $I_0(t)$, and the current flowing in the

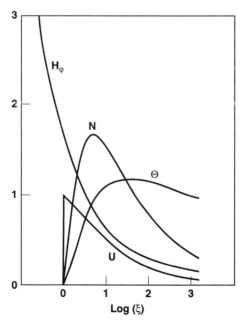

FIGURE 3.6. Self-similar profiles for an annular Z pinch imploded onto a concentrated current at the axis.

plasma liner, $I_p(t)$. It is important that that near the instant of collapse, $t = 0$, the main pinch current does not flow along the axis, but in the volume of the plasma liner.

4. The flow emerging after collapse remains self-similar and is described by the same equations. It contains a shock wave reflected from the axis. Since the reflected shock wave propagates into a heated plasma, it is characterized by finite Mach numbers and a temperature jump of order unity, unlike the strongly converging shock wave. If the plasma density is not singular at the moment of collapse, the total plasma compression after shock reflection is finite, just as in the case of conventional hydrodynamics. This confirms the conclusion made in Sec. 3.2.1 that, just as in the case of laser fusion, adiabatic implosion is preferable to obtain both high compression and considerable plasma heating.

5. Formation of a cumulative self-similar flow regarded in the context of a piston problem is characterized by "loss of control." Suppose that an external piston drives a converging shock wave or a collapsing cavity, so that that the plasma motion between the piston and the converging front corresponds to a self-similar solution. At some finite moment before the collapse, $t = t_1 < 0$, the self-similar coordinate, ξ_p, of the piston surface is equal to that of the so-called limiting characteristic. After that, neither characteristics nor waves emitted by the piston can affect the flow near the axis before the collapse: we can stop the piston or move it backwards, nevertheless, the flow near the collapse point remains self-similar only if the piston does not produce more converging shock waves. There are some indications that a 1-D flow near the converging front evolves to a self-similar regime, even when the piston motion does not correspond exactly to the self-similar solution. Therefore, after the loss of control, no 0-D approach remains valid. The transition between the 0-D and the essentially 1-D flow occurs when the thickness of the compressed plasma layer is of the order of its inner radius. Up to this moment, the stability of the cylindrically symmetrical flow is provided by the snowplow effect (see Sec. 6.3). After that, the flow near the axis is governed by two major factors: the inertia

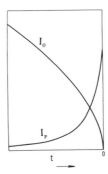

FIGURE 3.7. The axial current I_0 and the plasma current I_p versus time for a self-similar solution presented in Fig. 3.6.

of the converging plasma and the instability of a 1-D flow, which is characteristic of all the implosions.

6. The self-similar motion that exactly corresponds to the simple snowplow model is not an implosion but an explosion, namely, a plasma flow in an inverse Z-pinch configuration (the axial current flows through a plasma shell, and the return current is carried by a central conductor at the axis; the interaction of the two currents drives the plasma and the shock wave ahead of it from the axis). In this case, there is no loss of control. The diverging shock wave is driven by the magnetic piston that drives it all the time. The ratio of the shock wave radius to the inner radius of the free surface, which represents the magnetic piston, is constant in time, so that the current sheath can be regarded as "thin" at all times. The plasma velocity and pressure in this thin layer are virtually constant, thus justifying a 0-D approach, like the slug model (Potter, 1978). The self-similar solutions correspond to a power-law time dependence of the inverse Z-pinch current, $I(t) \propto t^s$, the shock and piston radii, $R_s(t)$ and, $R_p(t)$, being proportional to $t^{(s+1)/2}$ (see Greifinger and Cole, 1961; Liberman and Velikovich, 1986b). The slug model, applicable here, allows one to obtain a satisfactory solution for any $I(t)$.

7. The same problem can be studied for the case of magnetic flux compression by a plasma liner (Felber et al., 1988d). It is known from conventional gas dynamics that the cumulative compression of the density is limited, whereas the pressure increase is not [see (3.94)], where $\chi = 0$, and $\lambda > 0$. The axial magnetic field, being frozen-in, should be compressed to the same extent as the plasma density. On the other hand, $B_z^2/8\pi$ is an exact analog of pressure. Is the cumulative compression of an axial magnetic field thus limited? It was shown by Felber et al. (1988d) that an unlimited cumulative compression of B_z is possible, in principle, though it corresponds to a comparatively small negative degree in the power law, $B_z(t) \propto \alpha(t)^{-s}$, $s < 1$ as seen from (3.71) and (3.94) (cf. the case of adiabatic compression, $s = 2$). Here again, the adiabatic compression is more effective, though a cumulative compression regime can be used to increase the peak value of the axial magnetic field further after completion of the adiabatic implosion. A generalization of the above results for the case of a self-similar screw-pinch plasma implosion, with both axial and azimuthal currents present, has been given by Choe and Venkatesan (1990, 1992).

3.5 Self-similar solutions for time-dependent Z-pinch equilibria

3.5.1 Self-similar subsonic dissipation-dominated flows

The general interest in the fiber-initiated high-density Z pinches after the first successful experiments (Scudder, 1985; Sethian et al., 1987; Scudder et al., 1987) stimulated new active studies of the structure, dynamics, and stability of dense Z pinches that are close to Bennett equilibrium. The gross features of their comparatively slow evolution can be described by the 0-D models (see Sec. 3.2). However, the most intriguing feature of these experiments is the observed remarkable stability of the Z-pinch plasmas. A number of experiments performed during the last decade have demonstrated substantially improved global stability properties of pinch configurations. Experiments with gas-embedded Z pinches (Dangor et al., 1983), fiber-initiated dense Z pinches (Sethian et al., 1987) as well as with screw-pinch (pinch with an external axial magnetic field) configurations in ultra-low q (ULQ) and reverse field pinch (RFP) regimes (Kamada et al., 1989; Brunsell et al., 1990, 1991; Prager, 1992) indicate that pinch systems are capable of greater stability and can provide much higher energy density and plasma confinement times when compared to earlier investigations. The ULQ pinch operates at safety factor values, $0 < q(r) < 1$, where $q(r) = r B_z / R B_\varphi$ in the cylindrical approximation; B_z, B_φ are the toroidal (axial) and poloidal (azimuthal) components of the magnetic field, and R is the pinch major radius. Consequently, it represents an intermediate state between the tokamak $[q(a) > 1]$ and the RFP $[q(a) < 0]$, a being the minor radius of the plasma. This configuration is especially favorable from the reactor engineering point of view because of its high current density and beta value. The ULQ experiments in Toriut-6 and Repute-1 (Kamada et al., 1989), EXTRAP-L1 and EXTRAP-T1 facilities (Brunsell et al., 1990, 1991) reported pinch lifetimes, $t_0 \sim 10^2 \tau_A$ (τ_A is the Alfvén transit time), although operated at low $q(a)$ values.

To study the stability, a 0-D model is insufficient; at least a 1-D fluid model is needed. It has been found (Liberman and Velikovich, 1986b; Bud'ko and Liberman et al., 1990a, 1990b) that some simplified subsets of Eqs. (3.36)–(3.41) have self-similar solutions.

As the Z pinches in question appear to be stable for many Alfvén transit times, τ_A, they represent essentially subsonic plasma flows, their time scale τ_0 being much greater than τ_A. Thus, a slowly varying equilibrium state of the pinch plasma, which satisfies (2.4), can be considered and plasma inertia neglected. This does not mean that the velocity of the plasma flow is also neglected. As soon as the current, plasma temperature, and plasma radius depend on time, the velocity must be finite to satisfy the continuity and induction equations. However, the Mach number is indeed small (of order of τ_A/τ_0), so that the kinetic energy of the plasma is small compared to its thermal energy or magnetic energy, and therefore can be neglected.

But now, evidently, the criteria of validity for the ideal MHD model formulated in Sec. 3.3 [(3.50) in particular] are not satisfied because the characteristic time τ_0 is much greater than τ_A. The required fluid model of the Z-pinch plasma should take dissipation into account, because the time variation of interest is due just to the ohmic heating of the plasma column. Hence, we neglect inertia and viscosity in (3.37), but retain the dissipative terms in (3.38).

To obtain self-similar solutions, the anzatz (3.63) is used. The solutions in question are those with homogeneous deformation, i.e., (3.70) holds. The value of $\chi = -1$ is taken in (3.64) and (3.70) making (3.56) identically satisfied with an arbitrary density profile $N(\xi)$ and time dependence $\alpha(t)$. Separation of variables

(2.4) provides one equation for the self-similar profiles

$$\frac{d}{d\xi}(N\Theta + H_\varphi^2) + \frac{2}{\xi}H_\varphi^2 = 0, \tag{3.96}$$

and a relation between the self-similarity exponents μ and λ:

$$\mu = -\lambda - 1. \tag{3.97}$$

The variables in (3.39) can be separated when the transport coefficients are separable. The simplest way to do it is to assume a constant temperature profile, $\Theta(\xi) = \text{const}$, due to the high-electron heat conductivity, without explicitly taking into account the heat flow term in the energy equation. Then, the magnetic viscosity, $\nu_m = c^2/4\pi\sigma = \text{const} \cdot T^{-3/2}$, has a constant profile, too, and the resulting equations are easily solved (Braginskii and Shafranov, 1958). Of course, this is an *ad hoc* model, since the basic equations are not derived directly from (3.36)–(3.40). It was first shown by Velikovich et al. (1985) for a θ-pinch geometry (with only an axial magnetic field) that this can be done for two limiting cases: one corresponding to nonmagnetized plasma ($\Omega_e\tau_e \ll 1$), and the other to fully magnetized plasma ($\Omega_i\tau_i \gg 1$). For a Z-pinch geometry, the solution was obtained by Bud'ko and Liberman et al. (1990a, 1990b).

In both cases, the ohmic terms in (3.57) and (3.59) are among the dominating ones. The resulting time dependence of the pinch radius can be easily obtained as follows (Rosenau et al., 1989). Suppose a power-law time dependence of the pinch current

$$I(t) = I_{\max}(1 + t/\tau_0)^s, \tag{3.98}$$

where τ_0 is the time-scale characteristic of the current rise, $(I/\dot{I})_{t=0} = \tau_0/s$. The time dependence of temperature is determined by the Bennett equilibrium condition in (2.8): $T \propto I(t)^2$. On the other hand, the velocity depends on time as $\dot{\alpha}(t)$ [see (3.63)]. Thus, the ohmic term, $\nabla \times (\nu_m \nabla \times \mathbf{B})$, in (3.39) has the same time dependence as those retained in (3.42) if

$$\dot{B}_\varphi/B_\varphi = \mu\dot{\alpha}/\alpha \propto \nu_m/\alpha^2 \propto T^{3/2}/\alpha^2. \tag{3.99}$$

Integrating (3.99) with the aid of (3.98), we obtain a power-law time dependence for $\alpha(t)$

$$\alpha(t) = (1 + t/\tau_0)^{\frac{1-3s}{2}}, \tag{3.100}$$

and the values of the self-similarity exponents, μ, λ:

$$\mu = -\frac{5s - 1}{3s - 1}, \qquad \lambda = \frac{2s}{3s - 1}. \tag{3.101}$$

The inverse is also true: if the variables ξ and t are separated, then the time dependence of the current is given by (3.98), etc.

Note that the relation between $\alpha(t)$ and $I(t)$

$$\alpha(t) = \frac{\sqrt{(1 + t/\tau_0)}}{I(t)^{3/2}} \tag{3.102}$$

is independent of s and reproduces (3.35) for $I \ll I_{PB}$, which is quite natural because radiative energy losses are not taken into account here. We see that the self-similar subsonic Z pinch expands for $s < 1/3$, contracts for $s > 1/3$, and constant radius is maintained if $s = 1/3$. Plasma pressure decreases with time for $s < 1/5$, remains constant for $s = 1/5$, and increases for $s > 1/5$.

It should be noted also that for two values of s ($1/3$ and $-1/3$), when the pinch radius is either constant or a linear function of time, $\ddot{\alpha} \equiv 0$ [Eq. (3.60) where $B_z = 0$ is exactly reduced to (2.4)]. The self-similar solutions considered here are thus exact solutions of (3.56) and (3.60) and the corresponding induction and energy equations (Coppins et al., 1988). In all the other cases, they represent a first approximation with respect to the small parameter, τ_A/τ_0.

We see that the present results are somewhat consistent with those of Sec. 3.2. However, to make 0-D estimates, one must arbitrarily postulate some profiles of density, temperature, and magnetic field like those given by (2.5) and (2.6). The study of the self-similar solutions reveals an important fact: the profiles are not independent of the dynamics of the Z pinch. On the contrary, the dynamics determined by the time dependence of the current is the major factor in shaping the profiles.

3.5.2 Self-similar solutions for nonmagnetized plasma

For a Z pinch in a Bennett equilibrium, the characteristic value of $\Omega_e \tau_e$ is

$$
\begin{aligned}
\Omega_e \tau_e &= \frac{3\pi^{1/2} I^4 R}{8 e^3 m_e^{1/2} c^5 \bar{Z}^2 (1 + \bar{Z})^{3/2} \ln \Lambda N^{5/2}} \\
&= 6.6 \times 10^2 \frac{(I/MA)^4 R(100\ \mu m)}{\bar{Z}^2 (1 + \bar{Z})^{3/2} (\ln \Lambda/10)(N/10^{18}\ cm^{-1})^{5/2}}.
\end{aligned}
\tag{3.103}
$$

We see that in the initial stage of the pinch evolution, when $I \ll 1$ MA, the plasma is nonmagnetized. In this case, the ohmic terms dominate both in the induction equation and in the energy equation. In cylindrical coordinates they take the form

$$
\frac{\partial B_\varphi}{\partial t} + \frac{\partial}{\partial r} (u B_\varphi) = \frac{\partial}{\partial r} \left(\frac{v_m}{r} \frac{\partial}{\partial r} (r B_\varphi) \right),
\tag{3.104}
$$

$$
\frac{1 + \bar{Z}}{\gamma - 1} n \left(\frac{\partial T}{\partial t} + u \frac{\partial T}{\partial r} \right) + \frac{(1 + \bar{Z}) n T}{r} \frac{\partial}{\partial r} (r u)
$$

$$
= \frac{v_m}{4\pi r^2} \left(\frac{\partial}{\partial r} (r B_\varphi) \right)^2,
\tag{3.105}
$$

where $v_m = c^2/4\pi\sigma_\parallel$ (see Braginskii, 1963).

Substituting the anzatz (3.63)–(3.65) and (3.67) into (2.4), (3.56), (3.104), and (3.105) and separating the variables ξ and t, we obtain the following two equations for the self-similar profiles (Bud'ko and Liberman, 1989a,b, 1990)

$$\frac{d}{d\xi}\left[\frac{\Theta^{-3/2}}{\xi}\frac{d}{d\xi}(\xi H_\varphi)\right] = sR_M H_\varphi, \tag{3.106}$$

$$\left[\frac{1}{\xi}\frac{d}{d\xi}(\xi H_\varphi)\right] = \left(\frac{R_M}{2}\right)N\Theta^{5/2}. \tag{3.107}$$

Here the magnetic Reynolds number, defined as

$$R_M = \frac{R^2(\dot{I}/I)}{\nu_m}, \tag{3.108}$$

is constant in time for the time dependence of R, I and is given by (3.98)–(3.100). The value of R_M determines the length scale of the problem. If a sharp boundary is present, so that the relevant length scale is and no radial stretching is permitted, then R_M is an eigenvalue of the problem, being determined by the constant s. The boundary conditions for (3.96), (3.106), and (3.107) are $N(0) = \Theta(0) = 1$, $H_\varphi(0) = 0$, $dH_\varphi(\xi = 0)/d\xi = \sqrt{R_M/8}$.

We see that (3.106) is easily solved, if Θ is independent of ξ, and this is exactly the solution obtained by Braginskii and Shafranov (1958). The family of solutions under consideration includes the case of a flat temperature profile, which corresponds to $s = -1/2$. The magnetic field and density profiles are

$$H_\varphi(\xi) = I_1(\xi/j_{0,1})/I_1(1/j_{0,1}), \tag{3.109a}$$

$$N(\xi) = [(I_0(1/j_{0,1}))^2 - (I_0(\xi/j_{0,1})^2] \cdot (I_1(1/j_{0,1}))^{-1}, \tag{3.109b}$$

where $I_0(z)$, $I_1(z)$ are modified Bessel functions, $j_{0,1} = 2.405$ is the first positive root of $I_0(z)$. The eigenvalue is $R_M = 2j_{0,1}^2 = 11.6$. This solution describes expansion of a plasma column, $\alpha(t) \sim t^{5/4}$, with decreasing current, $I(t) \sim 1/\sqrt{t}$.

Another example of a simple analytical solution is found for $s = 0$, i.e., for an expansion of a plasma column with a constant current, $\alpha(t) \sim \sqrt{t}$. The profiles are

$$H_\varphi(\xi) = \frac{\xi\xi_0}{\xi^2 + \xi_0^2}, \quad N(\xi) = \left(\frac{\xi_0^2}{\xi^2 + \xi_0^2}\right)^{2/3}, \quad \Theta(\xi) = \left(\frac{\xi_0^2}{\xi^2 + \xi_0^2}\right)^{4/3}, \tag{3.110}$$

where $\xi_0 = (8/R_M)$ is an arbitrary length scale of the problem; R_M is arbitrary here, too.

The solutions found for an increasing current ($s > 0$) describe a diffuse Z pinch with mass and current densities that are not integrable at $r \to \infty$. They can be interpreted as referring to gas-embedded Z pinches but not to the experiments with deuterium fibers. A sharp boundary is present only if $s < 0$. Figures 3.8 (a–c) show typical self-similar profiles obtained by Bud'ko and Liberman (1989a,b) for $s = 0.11$, -4, and $-2/3$ (the magnetic Reynolds numbers are $R_M = 8, 0.72$, and 5.71, respectively), all of them corresponding to expansion of a plasma column with increasing (a) or decreasing (b, c) current. Figure 3.8(a) corresponds to a diffuse (gas-embedded) Z pinch. The profiles shown in Fig. 3.8(b) represent an

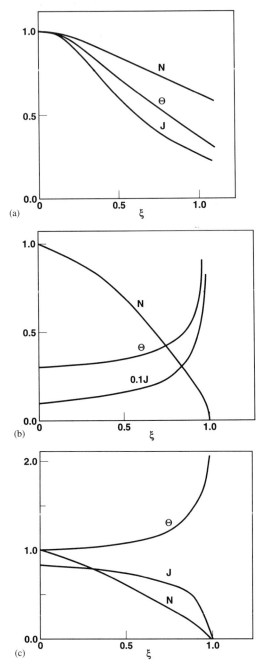

FIGURE 3.8. Self-similar profiles of density N, temperature Θ, and current density J, for a Z pinch in a nonmagnetized plasma: (a) a diffuse pinch, $R_M = 8$; (b) a pinch with a sharp boundary, strong skin effect, $R_M = 0.72$; (c) a case of marked inverse skin effect, $R_M = 5.71$.

example of strong skin effect: the current density diverges near the pinch boundary, the total current being finite. On the other hand, Fig. 3.8(c) demonstrates an interesting feature, a so-called inverse skin effect, with concentration of the current near the axis of the pinch, not at the boundary. The inverse skin effect, known since the 1920s (see Libin, 1927; Tamm, 1956), arises when a current in a conductor is decreased faster than the current can diffuse through the conductor, so that even a region of reversed current can emerge near its external boundary. In our case, the ratio of the magnetic field diffusion time, R^2/v_m, to the time scale of current decrease, $I/\dot{I}$, is given by (3.108). Thus, formation of self-similar profiles is influenced by the inverse skin effect, if the value of R_M is large and the value of s is negative. A numerical study of the inverse skin effect in a dynamic Z pinch is reported by Culverwell et al. (1989a); the profiles obtained using a 1-D code are similar to those presented in Figs. 3.8(a,c). Self-similar solutions for time-dependent Z-pinch equilibria with flow profiles affected by strong inverse skin effect are studied by Culverwell et al. (1989a) for the case of fully magnetized plasma discussed below, $s = -1/3$.

3.5.3 Self-similar solutions for fully magnetized plasmas

The dense high-temperature plasma produced at a developed stage of the fiber-initiated Z pinch is expected to be fully magnetized, with both $\Omega_e \tau_e$ and $\Omega_i \tau_i$ much greater than unity ($\Omega_e \tau_e \sim 3 \times 10^2$, $\Omega_i \tau_i \sim 5$). Then, to describe a time-dependent Z-pinch equilibrium one should use the full system of (3.38)–(3.40), retaining the terms of lowest order in $(\Omega_i \tau_i)^{-1}$, $(\Omega_e \tau_e)^{-1}$. Apart from those ohmic terms describing the transverse heat conductivity, the thermogalvanomagnetic effects called Nernst and Ettingshausen effects, and electron-ion temperature relaxation should be taken into account, all of them being of the same order. The electron and ion temperatures, $T_e(r, t)$ and $T_i(r, t)$, are both presented in the form (3.67). They have similar time dependence given by (3.67), (3.100), and (3.101), but their self-similar profiles $\Theta_e(\xi)$ and $\Theta_i(\xi)$, may differ.

The induction and energy equations for a fully magnetized plasma [below we assume $\bar{Z} = 1$, $P = n(T_e + T_i)$, $n \equiv n_i$, and take the transport coefficients corresponding to this value of $\bar{Z}$] are (Bud'ko and Liberman, 1990a):

$$\frac{\partial B_\varphi}{\partial t} + \frac{\partial}{\partial r}(u B_\varphi) = \frac{\partial}{\partial r}\left[v_m \left(\frac{1}{r} \frac{\partial(r B_\varphi)}{\partial r} + \frac{3\beta_e B_\varphi}{4T_e} \frac{\partial T_e}{\partial r} \right) \right], \qquad (3.111)$$

$$\frac{3}{2} n \left(\frac{\partial T_i}{\partial t} + u \frac{\partial T_i}{\partial r} \right) + \frac{n T_i}{r} \frac{\partial}{\partial r}(ru)$$

$$= v_m \left[\frac{1}{r} \frac{\partial}{\partial r} \left[\frac{n \beta_i}{\delta} (T_e/T_i)^{3/2} r \frac{\partial T_i}{\partial r} \right] + 3\Pi \left(\frac{n}{R^2} \right) (T_e - T_i) \right] \quad (3.112)$$

$$\frac{3}{2} n \left(\frac{\partial}{\partial t} (T_e + T_i) + u \frac{\partial}{\partial r}(T_e + T_i) \right) + \frac{n(T_e + T_i)}{r} \frac{\partial}{\partial r}(ru) \qquad (3.113)$$

$$= v_m \left[\begin{array}{l} \frac{1}{r} \frac{\partial}{\partial r} \left[\frac{n\beta_i}{\delta} (T_e/T_i)^{3/2} r \frac{\partial T_i}{\partial r} + 2.33 n\beta_e r \frac{\partial T_e}{\partial r} + \frac{3\beta_e B_\varphi}{16\pi} \frac{\partial(r B_\varphi)}{\partial r} \right] \\ + \frac{1}{4\pi r^2} \left(\frac{\partial(r B_\varphi)}{\partial r} \right)^2 + \frac{3\beta_e B_\varphi}{16\pi T_e r} \frac{\partial(r B_\varphi)}{\partial r} \frac{\partial T_e}{\partial r} \end{array} \right].$$

Here $\nu_m = c^2/4\pi\sigma_\parallel$, $\beta_{e,i} = 8\pi n T_{e,i}$, $\delta = (2m_e/m_i)^{1/2}$, Π is given by (2.12) (see Braginskii, 1963). The electron heat conductivity term in (3.113) is smaller than the ion heat conductivity term by a factor of order $\delta \ll 1$; nevertheless, it is retained because of the sharper gradient in electron temperature near the boundary.

Equations (3.111)–(3.113) are solved using the self-similarity anzatz together with (3.56) and (3.60), which are not modified (in the latter, $B_z = 0$) (see Bud'ko et al., 1990b). Self-similar solutions of the same equations with fewer terms included (the difference between T_e and T_i and/or the thermogalvanomagnetic effects being neglected) are studied by Coppins et al. (1988), Rosenau et al. (1989), and Glasser (1989). Though the terms in question may have little effect on the self-similar profiles, they are nevertheless of the same order as those retained [see (3.111)–(3.113)], so there is no direct justification for neglecting them from the start.

Typical self-similar profiles calculated for a Z-pinch equilibrium with constant radius [in (3.98), $s = 1/3$] are plotted in Figs. 3.9(a) and (b) for $R_M = 10$ and 0.2, respectively. Solutions corresponding to $R_M > 2.5$ are shown to correspond to a thermally isolated Z pinch, with density and heat flux vanishing at the boundary. The plasma temperature maximum at the boundary is due to the skin effect. Lesser values of $R_M < 2.5$ refer to solutions with nonvanishing density and heat flux, which can be interpreted as describing a gas-embedded Z pinch.

From the inequality $R_M \geq 2.5$, we find for the constant radius of a Z pinch (Coppins et al., 1988; Bud'ko and Liberman, 1990a,b):

$$R \geq R_{\min} = 18.5 \frac{(N(10^{18}\ \text{cm}^{-3}))^{3/4}}{I(\text{MA})\dot{I}(\text{MA}/100\ \text{ns})^{1/2}}\ \mu\text{m}. \qquad (3.114)$$

Therefore, substituting in (3.114) parameters of the NRL experiment (Sethian, 1987) $N = 5 \times 10^{18}\ \text{cm}^{-1}$, $I = 0.5$ MA, $\dot{I} = 0.5$ MA/130 ns, we obtain an estimate for the minimum radius of a stationary Z pinch, $R_{\min} \cong 200\ \mu$m, which is approximately the value observed by Sethian et al. (1987). This scaling was first obtained through analysis of the self-similar solutions by Coppins et al. (1988), where the numerical coefficient was found to be greater than that given in (3.110) by a factor ~ 5. The difference seems to be due to a more accurate account of physical effects, represented by the additional terms in the induction and energy equations (3.111)–(3.113) by Bud'ko et al. (1990b).

3.5.4 Dissipative screw-pinch problem

The ULQ pinch operates at safety factor with values $0 < q(r) < 1$, where $q(r) = rB_z/RB_\varphi$ in the cylindrical approximation, B_z, B_φ are the toroidal (axial) and poloidal (azimuthal) components of the magnetic field, and R is the pinch major radius. Under these conditions, the profiles should be obtained from the full Braginskii system of equations, including all the classical transport processes, e.g., the thermogalvanomagnetic Nernst and Ettingshausen effects. Since the ULQ pinch operates well below the Kruskal–Shafranov limit, $q(a) = 1$, its stability properties against the global $m = 1$ kink modes depend strongly upon the details of the magnetic field and current-density equilibrium profiles. It was shown (Wahlberg, 1991)

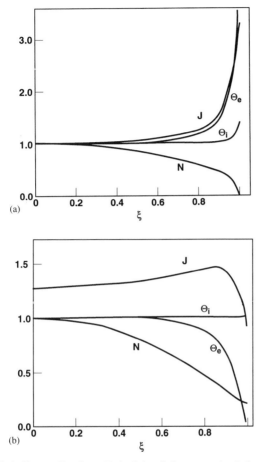

FIGURE 3.9. Self-similar profiles for a Z pinch in a fully magnetized plasma: (a) $R_M = 10$; (b) $R_M = 0.2$.

that hollow current distribution and strongly paramagnetic axial-field profile result in substantially decreased growth rates and cutoff wavenumbers and thus possess optimal stability against kinks. The macroscopic pinch parameter limits imposed by magnetic separatrix and the stability condition against short wavelength kinks (the number of ion Larmor radii) are also sensitive to the profiles (Lehnert and Scheffel, 1992). For these reasons, a detailed knowledge of the current and field distributions provides further insight into the ULQ-pinch stability properties and operational limits. The exact self-similar solutions for the problem were obtained for the full Braginskii set of equations describing dynamics and time-dependent equilibria of a screw pinch in the cylindrical approximation, with a self-consistent account of transport and dissipative processes (Bud'ko et al, 1994). Although the solutions represent a straightforward generalization of those treated in Secs. 3.5.2 and 3.5.3, they incorporate new phenomena associated with magnetic-field component coupling. This makes it possible to reveal the paramagnetic (or diamagnetic) behavior of the ULQ pinch and Z pinch with an applied axial field (Bud'ko et al, 1994). It was found, in particular (Bud'ko et al., 1994), that at the pinch boundary, the axial field is always diamagnetic; whereas, at the axis, the resistivity anisotropy results in a strong paramagnetism, provided the pinch current is strong enough.

Electron temperature profiles established in a screw pinch by classical non-ideal transport phenomena can be found as a solution of the set of one-fluid one-temperature equations of a classical inviscid plasma (3.36)–(3.41). Equations (3.36)–(3.41) together with the corresponding boundary conditions, comprise a boundary value problem, the MHD profiles being the eigenfunctions of this problem. Boundary conditions to be imposed at the pinch boundary, $r = a(t)$, deserve special attention. Under experimental conditions, a completely ionized plasma column is usually surrounded by a cold neutral gas. Consequently, a partially ionized boundary layer at the pinch edge is formed supporting a nonzero particle and heat flux across the boundary. A self-consistent treatment of the edge particle, momentum, and energy transport necessitates an accurate model for plasma-neutral gas collisions, ionization kinetics, etc. The model presented that yields exact solutions of the transport equations involving separation of variables fails to incorporate the plasma-neutral gas interaction. For this reason, the boundary conditions corresponding to a plasma-vacuum interface, i.e., vanishing plasma pressure and total heat flux, $nT(a) = 0$, $\mathbf{q}_e(a) + \mathbf{q}_i(a) = 0$, are imposed. Particle diffusion at the partially ionized boundary layer may result in additional diamagnetism. Self-similar profiles obtained by Bud'ko et al. (1994) always exhibit diamagnetic properties at the boundary. Thus, the edge diamagnetism can be considered as an estimate from below, while the bulk paramagnetic (or diamagnetic) properties are not sensitive to boundary effects.

The problem under consideration represents a straightforward generalization of the problems in Sec. 3.5.3 [see (3.56), and (3.111)–(3.113)]. Taking into account anisotropy of the electric conductivity in a strong magnetic field, and neglecting small terms in the parameter $(\Omega_i \tau_i)^{-1} \ll 1$, we obtain the following set of equations:

$$\frac{\partial B_\varphi}{\partial t} + \frac{\partial}{\partial r}(u B_\varphi) \tag{3.115}$$
$$= \frac{\partial}{\partial r}\left[\nu_m\left(-\left(\frac{\sigma_\parallel}{\sigma_\perp} - 1\right)B_z D(\mathbf{B}) + 2\frac{\partial(r B_\varphi)}{r \partial r} + 6\pi\frac{n B_\varphi}{B^2}\frac{\partial T}{\partial r}\right)\right],$$

$$\frac{\partial B_z}{\partial t} + \frac{1}{r}\frac{\partial}{\partial r}(r u B_z)$$
$$= \frac{1}{r}\frac{\partial}{\partial r}\left[r\nu_m\left(\left(\frac{\sigma_\parallel}{\sigma_\perp} - 1\right)B_\varphi D(\mathbf{B}) + 2\frac{\partial B_z}{\partial r} + 6\pi\frac{n B_z}{B^2}\frac{\partial T}{\partial r}\right)\right] \tag{3.116}$$

$$\frac{\partial}{\partial r}\left(2nT + \frac{B^2}{8\pi}\right) + \frac{B_\varphi^2}{4\pi r} = 0, \tag{3.117}$$

$$3n\left(\frac{\partial T}{\partial t} + u\frac{\partial T}{\partial t}\right) + \frac{2nT}{r}\frac{\partial(ru)}{\partial r}$$

$$= \frac{v_m}{4\pi}\left[\frac{\sigma_{\parallel}}{\sigma_{\perp}}\left(\left(\frac{\partial(rB_{\varphi})}{r\partial r}\right)^2 + \left(\frac{\partial B_z}{\partial r}\right)^2\right) - \left(\frac{\sigma_{\parallel}}{\sigma_{\perp}} - 1\right)B^2 D^2(\mathbf{B})\right.$$

$$\left. + \frac{6\pi n}{B^2}\left(B_{\varphi}\frac{\partial(rB_{\varphi})}{r\partial r} + B_z\frac{\partial B_z}{\partial r}\right)\frac{\partial T}{\partial r}\right]$$

$$+ \frac{1}{r}\frac{\partial}{\partial r}\left[\frac{4\pi rnT v_m}{B^2}\right. \tag{3.118}$$

$$\left. \times\left((4.66 + \delta)n\frac{\partial T}{\partial r} + \frac{3}{8\pi}\left(B_{\varphi}\frac{\partial(rB_{\varphi})}{r\partial r} + B_z\frac{\partial B_z}{\partial r}\right)\right)\right],$$

which, together with (3.56), represent the coupled set of equations. The following notations are used: $\sigma_{\parallel} = 2\sigma_{\perp} = 2ne^2\tau_e/m_e$ are the parallel and perpendicular conductivities; $B^2 = B_{\varphi}^2 + B_z^2$, and

$$D(\mathbf{B}) = B^{-2}\left(B_z\frac{\partial(rB_{\varphi})}{r\partial r} - B_{\varphi}\frac{\partial B_z}{\partial r}\right).$$

Equations (3.115)–(3.118) and (3.56) are solved (Bud'ko et al., 1994) using the self-similarity anzatz (3.63)–(3.67) for a uniform deformation, i.e., with $U(\xi) \equiv 1$. A condition for separation of variables is a power-law time variation of the total current (3.98); the time dependence for the pinch radius is given by (3.100), and the self-similarity exponents in the expression for the self-similarity anzatz (3.64)–(3.67) are given by (3.71) and (3.101). For the self-similar profile functions, $N(\xi)$, $H_{\varphi}(\xi)$, ... the system (3.56), (3.115)–(3.118) can be reduced to the canonical form, and solved by the Runge–Kutta method (Bud'ko et al., 1994). Typical self-similar profiles calculated by Bud'ko et al. (1994) for a screw-pinch equilibrium with constant radius [$s = 1/3$ in (3.98)] and for a constant current ($s = 0$) are shown in Figs. 3.10 and 3.11, respectively. Profiles calculated exhibit plasma dia-magnetism in the vicinity of the boundary, whereas near the axis, they possess strong paramagnetism, provided the applied axial field is weak. The current den-sity profiles are flat ($s = 0$) and hollow ($s = 1/3$), and in the latter case one can observe a considerable skin effect for large values of $b = B_{z0}/B_{\varphi 0}$. One can ob-serve that strong heat conductivity produces a very flat temperature profile, which makes the diamagnetic effect of the Nernst term negligible in the whole plasma bulk, excepting a very thin boundary layer with a sharp temperature gradient. The paramagnetic effect of the anisotropic resistivity tensor is particularly pronounced at a small amplitude of the applied axial field and results in nonmonotonic axial field profiles with strong on-axis paramagnetism. On the contrary, ULQ profiles (large magnitude of the axial field) are generally diamagnetic. In the case of nonmagne-tized ions, the behavior of the axial field profile is quite different: it is generally paramagnetic regardless of the external field value. Nevertheless, the paramagnetic effect of the resistivity anisotropy is insufficient to stabilize the screw pinch against

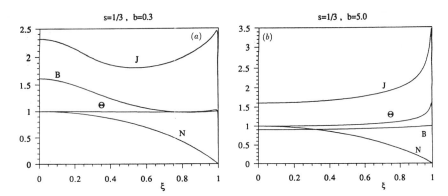

FIGURE 3.10. Self-similar profiles corresponding to the stationary equilibrium, $\alpha = \alpha_0 = cons$, for an increasing current, $I \propto (1 + t/t_0)^{1/2}$, $s = 1/3$.

global kink modes. Calculation of the growth rates demonstrates a relatively weak decrease of the maximal growth rate as the axial field amplitude increases, although the range of unstable wavenumbers becomes substantially narrower.

3.5.5 Self-similar solutions as asymptotic states of subsonic Z-pinch flows

As the pinches under consideration appear to be stable for many Alfvén transit times, the plasma inertia may be ignored, which enables us to separate variables and proceed with an analytical treatment. Nevertheless, whenever dealing with

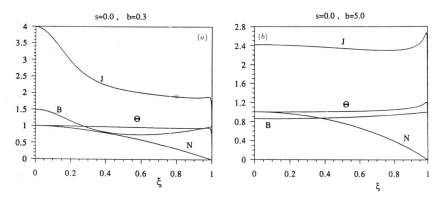

FIGURE 3.11. Self-similar profiles corresponding to dynamic equilibrium at an increasing radius. $\alpha \propto (1 + t/t_0)^{1/2}$, for constant current $I = I_0$, $s = 0$.

self-similar solutions that are exact but still particular solutions of nonstationary equations, one should address the question whether these solutions represent asymptotic states of plasma flow, i.e., are attractors independent of initial conditions. Self-similar solutions are known to be attractors in some cases, representing asymptotic states of plasma flows, which are independent of initial conditions (Zel'dovich and Raizer, 1967; Barenblatt, 1979). This important property should be numerically tested for each particular case. For self-similar solutions describing time-dependent equilibria of fiber-initiated high-density Z pinches, this study is reported by Rosenau et al. (1989), Glasser (1989), Jaitly and Coppins (1994), where very fast convergence (at the time scale of order of $10\tau_A$) to self-similar profiles have been obtained.

Glasser (1989) solved the full system of one-fluid, one-temperature ($T_e = T_i = T$) equations (3.36)–(3.41) with transport coefficients given by Braginskii (1963) and bremsstrahlung radiative loss term. The model of Rosenau et al. (1989) is two-temperature ($T_e \neq T_i$), but inertia and viscosity of the plasma are neglected from the start, and the Braginskii transport coefficients are taken in the fully magnetized limit ($\Omega_e \tau_e \gg \Omega_i \tau_i \gg 1$). Both studies are aimed at simulating the Los Alamos National Laboratory (LANL) and NRL experiments with frozen deuterium fibers, hence the parameter range is much the same. In the paper of Glasser (1989), evolution of an initial self-similar profile with the on-axis number density, $n_i = 6.8 \times 10^{22}$ cm^{-3}, and initial radius, $R_0 = 20$ μm (this simulates an initially solid fiber with the same line density N and R_0, and uniform density profile), is calculated. The current is linearly increased from the initial value of 200 kA, with a rate of 10^{13} A/s, rising to 1.2 MA after 100 ns. Rosenau et al. (1989) programmed the current, using the form (3.98) with various values of s. Typically it is raised from 100 kA to 1.1 MA in 100 ns, but the case of the current exceeding the Pease–Braginskii value, 1.4 kA, is also studied.

Though the initial conditions of Glasser (1989) correspond to an exact self-similar solution of the type considered in Sec. 3.5.3, the numerical solution does not follow it from the very beginning, since both the terms not taken into account, plasma inertia and viscosity, and departure of the transport coefficients from their respective fully magnetized limits are not negligible. However, "late in time, the density profiles take on the appearance of self-similarity, with subsequent profiles unchanged in shape but increasing in amplitude as the radius decreases" (Glasser, 1989). This is illustrated by Figs. 3.12(a) and (b), representing a sequence of 50 profiles of temperature and velocity versus radius, at equal time steps of 2 ns. Figures 3.13(a) and (b) show the profiles of velocity normalized to dR/dt and temperature normalized to the on-axis temperature versus radius, normalized to $R(t)$ during last 2.5 ns of the simulation. A monotonic contraction of the plasma column corresponding to $R(t) \sim (1 + t/\tau_0)^{-1}$, $s = -1$ [see (3.100)], with almost linear velocity distribution is eventually established. The velocity profile is far from linear during the initial stage, because waves propagate in both directions, to and from the axis. This is seen in Fig. 3.14, where the velocity profiles calculated at a time interval of 0.1 ns for the first 2.5 ns are presented. Relaxation of the wave motion in this simulation takes about 3 ns. Other choices of initial conditions, with

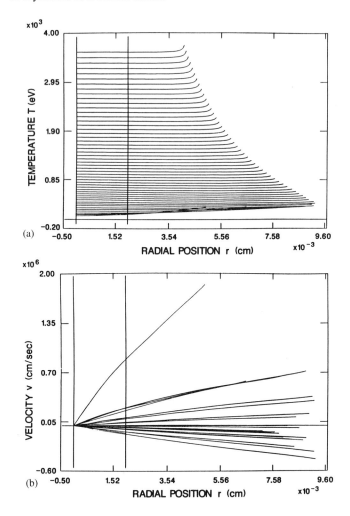

FIGURE 3.12. Evolution of temperature (a) and velocity (b) profiles for the conditions of experiments with fiber-initiated high-density Z pinches [numerical simulation by Glasser (1989)].

initial magnetic pressure decreased by a factor of 2, and initial pressure balance thus eliminated, results in a more violent wave motion and greater amplitude of the oscillations. However, the eventual relaxation of the wave motion produces profiles, which are not only self-similar, but indistinguishable from those presented in Figs. 3.13 and 3.14, for example.

The final self-similar state is not exactly that given by the analytic solutions. This concerns mostly the velocity profile, which deviates from a straight line, particularly near the axis and the boundary [Fig. 3.13(a)]. Glasser (1989) explained

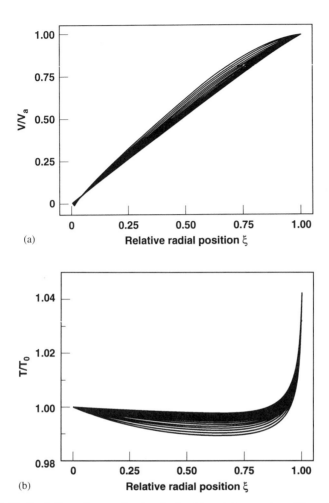

FIGURE 3.13. Evolution of normalized velocity (a) and temparature (b) profile for the conditions of Fig. 3.12.

the difference as being mainly due to the influence of viscosity. The latter is indeed considerable at kilovolt temperatures. Viscosity plays a major role in damping the waves, and of course it affects the final state, too. Thus, an analytical self-similar solution of the simplified equations is not the real attractor for the full system. But the attractor is shown to exist, to be self-similar, and to be very well-approximated by the analytical solutions.

Rosenau et al. (1989) did not take into account relaxation of the wave motion due to neglect of inertia and came to much the same conclusion. Convergence of the numerical solutions to the self-similar ones was studied by two methods.

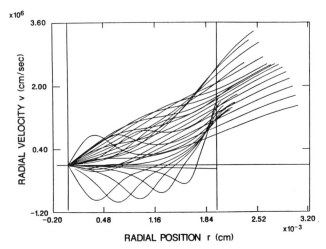

FIGURE 3.14. Evolution of velocity profiles obtained by Glasser (1989) for the initial stage (first 2.5 ns) of a Z-pinch development.

First, evolution of the normalized profiles of flow variables was studied, and their convergence to asymptotic self-similar shapes was demonstrated. For the linear current rise, the convergence is achieved in 60 ns. Second, a "similarity diagnostic" function, $\mathcal{R}(t)$, was introduced to test the convergence of the time-dependence of the variable, $f(t)$, to that given by the self-similar solution. Let $f(t)$ represent a flow variable corresponding to fixed plasma particle (that is, its Lagrangian or self-similar coordinate is constant), for example, the on-axis electron temperature. If a self-similar state is achieved, then the time dependence of any such variable is a power of $(1 + t/\tau_0)$, and thus $\mathcal{R}(t)$ is a linear function of t. As the time dependence characterizing a variable approaches the self-similar one, the corresponding diagnostic function tends to a straight line.

Rosenau et al. (1989) found that the asymptotic self-similar profiles are determined mostly by the parameter s in (3.100) characterizing the current programming regime, and are almost independent of the initial conditions. The radiative energy losses, not taken into account in the equations of Sec. 3.5.3, have no significant effect on the shapes of the asymptotic profiles. Here, again, a self-similar solution of the simplified set of equations provides a good approximation of the real attractor.

The time dependence of the flow variables is more affected by radiative losses. In particular, for $s < 1/3$, (3.100) predicts infinite expansion. Radiative losses decrease the plasma pressure and eventually stop the expansion. Then contraction and radiative collapse of the plasma column follow, as the Pease–Braginskii current $I_{PB} \cong 1.4$ MA is exceeded, with a new self-similar flow regime emerging as an attractor, one similar to that found by Meierovich (1985) and described in Sec. 3.4.3. The behavior of the similarity diagnostic function, $\mathcal{R}(t)$, of the on-axis electron temperature is shown in Fig. 3.15 for a current below and above I_{PB} [curves (a) and (b), respectively]. Curve (a) demonstrates a power-law self-similar behavior. Curve (b) shows a radiative collapse. The self-similar solutions available for this case (Sec. 3.4.3) correspond to constant current and predict a time dependence of $T_e(0, t)$, which differs from a power law. The same appears to be true for the asymptotic self-similar regime of radiative collapse in the case of rising current, which explains the similarity diagnostic tool fails in the case of curve (b).

The self-similar solutions considered by Rosenau et al. (1989) and Glasser (1989) are good approximations to the attractors in the context of a 1-D problem. In the broader context of a 3-D problem, no 1-D attractors exist because of instabilities. Hence, these solutions represent typical 1-D flows realized in the experiments with fiber-initiated high-density Z pinches and remain valid as long as the initial cylindrical shape of the plasma column is not disrupted by instabilities.

3.5.6 Two-dimensional modeling of Z-pinch flows

One-dimensional computations give satisfactory results when the Z pinch implodes uniformly along the axis. This happens when stagnation is late in the current pulse and instabilities do not grow significantly in the course of the Z-pinch implosion. However, sometimes a more detailed description of the dynamic

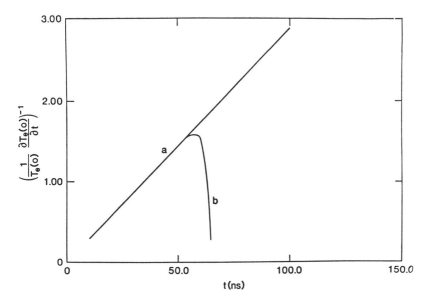

FIGURE 3.15. The similarity diagnostic function $\mathcal{R}(t)$ indicates convergence of time dependence of Z-pinch flow variables to a self-similar power-law form (Rosenau et al., 1989).

behavior of Z pinch is needed than can be obtained using self-similar analytical solutions. Comprehensive modeling of Z-pinch hydrodynamics, which takes into consideration plasma radiation, involves a heavy dose of atomic physics, ionization processes, and radiation transfer (e.g., Clark et al., 1986). Certain aspects of radiative hydrodynamics of Z pinches can be of great importance: collision-radiative equilibrium and the time-dependent approach to equilibrium for optically thick plasmas. Collisional-radiative models for the highly dynamic nonequilibrium pinch plasma were developed by Foord et al. (1990, 1994).

The effects of radiation emission on the development of the pinch compression were examined by Venneri et al. (1990). The emission of radiation results in cooling of the plasma. The effect of this cooling on the dynamics of the plasma compression can be negligible, as in the case of pure bremsstrahlung from unseeded discharges, or dominant, as in the case of heavily seeded discharges. In this latter case, the compression produced by the driving current can continue until very high densities are reached, i.e., it can result in a radiative collapse. These results of radiation-caused changes in the dynamics of compression were investigated in the dense-plasma focus. Venneri et al. (1990) found that radiative cooling of the discharge can be responsible for a more than a tenfold increase in density of the pinch. The sequence of events was found to be different when high-Z seed gas was added the discharge. In this case, the compression phase was more complete as a large fraction of the plasma thermal energy, which would otherwise have opposed compression, leaves the plasma in the form of x-ray radiation. Therefore, stagnation ($\beta = 1$ limit) is reached at a much smaller radius (< 1 mm), and at much higher densities ($> 10^{20}$ cm^{-3}). The measured density (Venneri et al., 1990) in the line-emitting region was as high as 10^{21} cm^{-3}. Also, from interferometer pictures, assuming that all the plasma in front of the collapsing current sheet is actually compressed in the pinch, an upper limit of 10^{21} cm^{-3} was estimated for a final diameter of 1 mm. The temperature of these heavily seeded discharges was found experimentally to be almost a factor 2 larger than the values calculated using the model. One of the possible explanations for this discrepancy is that anomalous ohmic heating becomes important when the pinch is at maximum compression.

Some effects which are not combined into the standard models can be important for an explanation of the experimentally observed behavior of the pinches. The high-voltage, high-current device may result in intense evaporation of plasma from the surface of a Z pinch. The increase in the size of the pinch corona and the MHD instability of a superheated skin layer presumably result in a leakage of part of the pinch current outside an imploding Z pinch. Possible effects and mechanisms of the turbulent spreading of the superheated skin layer of Z pinch were discussed by Sasorov (1991). The leakage currents through the imploding Z-pinch periphery that can have a strong influence on the x-ray yield were discussed by Terry and Pereira (1991). In this case, the appearance of a highly conductive plasma in the pinch periphery can prevent the current from entering the Z pinch.

Hydrodynamic instabilities during the implosion can destroy 1-D radial symmetry of the Z-pinch plasma. In some particular cases, a 2-D self-similar analytical solution of the MHD equations can be obtained (Liberman et al., 1985), but in general, a numerical solution is needed to describe a Z-pinch plasma flow in a proper way. Much is known about 2-D plasma hydrodynamics without radiation, including the growth of the RT instabilities, their nonlinear saturation, etc. (see Chapter 4). Nevertheless, the relevance of the results to radiating Z pinches, formation of hot or bright spots, and radiative collapse remains to be evaluated by extending the considerations to include radiation. Satisfactory analysis of many important aspects of Z-pinch plasmas that are 2-D in nature does not yet exist: development of turbulence, anomalous resistivity due to plasma turbulence, turbulent corona and leakage current outside the pinch core, particle beams along the pinch axis, etc. Comprehensive 2-D MHD analysis of a noncylindrical Z pinch (dense-plasma focus) has been done by D'yachenko and Imshennik (1974). An important conclusion of this modeling was the ejection of an essential amount of plasma along the pinch axis (the ejection involves up to 90% of the mass per unit length of the plasma). Note that such a strong plasma ejection makes doubtful the very high plasma density claimed for hot spots (Koshelev et al., 1989).

Two-dimensional computational models have been presented for a variety of pulsed-power systems, including 2-D modeling of a fiber ablation in the solid-deuterium Z pinches (Lindemuth, 1990; Sheehey et al., 1992), 2-D modeling of gas-puff implosion experiments (Deeney et al., 1993; Cochran et al., 1995), 2-D modeling of magnetically driven RT instabilities in cylindrical Z-pinches (Hammer et al., 1996; Peterson et al., 1996). These calculations are discussed in more detail in Sec. 6.9. 2-D simulations of deuterium-fiber-initiated Z pinches have been done to model low- and high-current deuterium-fiber-initiated experiments on the LANL machines, HDZP-I and HDZP-II (Lindemuth, 1990; Sheehey et al., 1992). Early deuterium fiber experiments, with current peaks up to about half the Pease–Braginskii current, reported very long-lived, compact plasmas showing little indication of disruption by sausage ($m = 0$) and kink ($m = 1$) instabilities (see chapter 4). Results from the 2-D computational models are discussed in Sec. 6.9.

For typical waterline-driven gas-puff Z pinches, the current tends to flow as a skin current on the outside of the gas shell. Consequently, the current follows the divergence of the gas shell, which results in the magnetic pressure being highest near the nozzle. For a constant mass per unit length, the shell will collapse first near the nozzle and then consequently along the remaining pinch length. This is the characteristic zipper effect (Hussey et al., 1986). Two-dimensional MHD modeling of the gas-puff implosion (Deeney et al., 1993) indicates that the collapsing shell reaches a higher density and a smaller diameter when the axial zipper is eliminated. When the zipper effect is eliminated from an argon gas-puff implosion, the K-shell x-ray yields and powers increase significantly. This appears to be due to the fact that zippering produces a preheated plasma jet on axis, onto which the bulk of the shell implodes. The collapsing shell is then inhibited from reaching as small a pinch radius, and, consequently, as high a density. This limits the radiation efficiency.

4

Stability of Z-Pinch Plasmas

The stability of Z-pinch plasmas is a primary research area and has been actively pursued in both theoretical and experimental investigations. Plasma instabilities can be conveniently divided into two classes: hydrodynamic instabilities and kinetic instabilities. Hydrodynamic instabilities are caused by macroscopic motions of the plasma. For example, the electron two-stream instability results when the mean velocity of the electrons relative to the ions is larger than a threshold velocity that is somewhat larger than the electron thermal velocity. This instability is important in either low-density pinches or for a very high pinch current. A kinetic instability is caused by the interaction of a restricted class of particles with the unstable mode. An example of a kinetic instability is the ion acoustic-drift instability, which is caused by the interaction between the electrons with a velocity near the phase velocity of the ion acoustic waves and the ion acoustic waves.

The problem of stability can be stated qualitatively as follows: the existence of an MHD equilibrium state implies a situation where the sum of the forces acting on the plasma is zero. If the plasma is perturbed from this state, the resulting perturbed forces either restore the plasma to its original equilibrium (stability) or cause a further enhancement of the initial disturbance (instability). The simplest type of stability corresponds to exponential stability; that is, if any of the eigenfrequencies ω in the linear stability analysis (any small perturbation of first order is expanded in terms of its Fourier components, and each term can be treated independently in the linearized approximation) corresponds to exponential growth, the system is said to be exponentially unstable. If not, it is considered exponentially stable. Implicit in this definition is the assumption that the modes are discrete with distinguishable eigenfrequencies. This is not true in general. Discussion of these and many related questions, including general reviews of plasma instabilities, may be found in Mikhailovskii (1974, 1977), Bateman (1978), Manheimer et al., (1984), and Freidberg (1987).

So far, the greatest effort has been devoted to linear-stability analysis. This is often adequate for practical situations. Nonlinear analysis of ideal MHD modes provides the details of plasma self-distortion; that is, these modes are so virulent that it is more important, in practice, to avoid them rather than ascertain their nonlinear evolution. There is consensus in the fusion community that ideal MHD

96

stability is a necessary requirement for a fusion reactor. To a large extent, it is the avoidance of ideal MHD instabilities that led to the somewhat unusual, and sometimes technologically complicated, magnetic configurations in the international controlled-thermonuclear-fusion research program.

The problem of MHD stability of a linear pinch is one of the most intensively investigated subjects in plasma physics. To a large extent, it was developed in the 1950s, during the early stage of thermonuclear fusion research, when the basic stability criteria were established (Rosenbluth and Longmire, 1957; Kruskal and Schwarzschild, 1954; Shafranov, 1956, 1958b,c; Suydam, 1958), the Energy Principle was formulated (Rosenbluth and Longmire, 1957; Bernstein et al., 1958; Furth, 1963; Grad, 1966), the equations describing the MHD approximation of the eigenmodes of a Z pinch (including the unstable ones) were derived (Hain et al., 1957; Tayler, 1957; Kruskal and Oberman, 1958), and their solutions were found for some simple models of a Z pinch (Tayler, 1957; Trubnikov, 1958; Shafranov, 1956, 1958b; Furth, 1963; Coppi, 1964a,b, 1965, 1966). The stability criteria were deduced from the Energy Principle, and three important stability conditions were obtained: the Kruskal–Shafranov limit, the Suydam criteria, and the Newcomb theorem. The results of this research were not optimistic for a Z pinch regarded as a magnetic confinement fusion system. Theory demonstrated poor stability of a steady-state Z pinch, which in almost any case exhibited hydromagnetic instabilities with growth rates of order of the inverse Alfvén transit time, τ_A^{-1}. This conclusion was confirmed by many experiments (Anderson et al., 1958; Saywer et al., 1958; Artsimovich et al., 1964; Gorbunov and Rosumova, 1963).

In fact, the usual plasma-stability concept is too strong for a Z pinch, or any other dynamical system. The point is that all confined plasmas are unstable. We are interested in whether or not the growth time for the instability is long enough so that the high-density plasma is confined for a long enough time to attain the goals for the Z-pinch application (i.e., energy gain for fusion applications, x-ray emission for x-ray sources, etc.). In contrast, the usual stability concept requires stability with respect to all growing perturbations, including those that take an extremely long time. We require stability with respect to perturbations that grow in times shorter than some characteristic time needed for the application. This approach to stability is not necessarily restricted to fusion problems, but can be useful for any physical problem where a dynamical system is only studied for a limited period of time.

In recent years there has been renewed interest in the linear Z pinch for several reasons. First, it turns out that linear Z pinches are more stable than predicted by simple ideal MHD theory. Second, power sources have become available that are capable of generating currents up to 20 MA, with rise times of about 100 ns driven by voltages in the MV range, and with peak electric power of about 50 TW. The fast rise time optimizes the power input to the pinch. Such a power source has been used to drive Z pinches in the form of hollow shells and has yielded encouraging results (see Spielman et al., 1992; Spielman et al., 1994; Hammer et al., 1996; Wong et al., 1995). Finally work on dense Z pinches has been stimulated by a number of applications. Imploding Z pinches employing a gas puff, wire arrays, or cylindrical foils are of interest as intense sources of x rays in a broad range of applications, from testing of nuclear weapon effects to the needs of high-energy-density physics.

New interest in Z-pinches, developed mostly during the late 1970s and early 1980s, is due to the enhanced stability of some pinch systems discovered experimentally. One example is the EXTRAP linear plasma confinement system, suggested by Lehnert, which is basically similar to a steady-state Z pinch, but with the plasma surface in it made concave due to magnetic fields of external linear conductors. Plasma in the EXTRAP was found to be stable for $\sim 10^2 \, \tau_A$ (Lehnert, 1983). Good stability properties of compressional Z pinches (Haines, 1982) and gas-embedded Z pinches (Dangor et al., 1983) were observed, the plasma remaining stable for $\sim 10 \, \tau_A$. Stability of the pinch plasma for $\sim 10 \, \tau_A$ was discovered in experiments with frozen deuterium fibers heated by a rapidly rising current (Sethian et al., 1987). Both imploding gas-puff Z pinches (Pereira et al., 1984; Baksht et al., 1997b) and wire arrays (Baksht et al., 1989; Baksht et al., 1997a) demonstrate better stability than that predicted by the simplified calculations.

Apart from the EXTRAP system and, to a lesser extent, compressional and gas-embedded Z pinches, the examples of enhanced stability given above refer to dynamic pinches, whose unperturbed cylindrically symmetrical states are essentially time dependent. Of all the Z-pinch systems listed above, only the EXTRAP system is regarded as a magnetic confinement fusion system in which all kinds of instabilities should be suppressed, the fast hydromagnetic ones as well as the slower instability modes: dissipative, drift, etc. As for the other systems, their present and prospective applications (except the radiation sources), require mainly hydromagnetic stability. This includes the fusion applications of dense Z pinches regarded as inertial (or hybrid: inertial + magnetic) confinement fusion systems. On the other hand, most of the new experimental results concerning the unexpected stability of Z pinches refer to hydromagnetic stability (confinement of a cylindrically symmetric plasma much longer than τ_A). This has caused both a revival of interest in the traditional MHD stability analysis problems and attempts to extend the conventional MHD model by including the factors that influence stability, such as motion of the unperturbed plasma, dissipation, radiative energy losses, etc. An understanding of the basic features involved in stability is clearly crucial to further progress in the field. The goal of Chapter 4 is to develop a basic understanding of the various mechanisms that drive instabilities in Z-pinch plasmas and to discuss possible ways to avoid or to suppress them.

4.1 The stability of steady-state Z pinches

4.1.1 Ideal MHD stability theory: Linearized equations

The ideal MHD theory of stability is presented in detail in a number of reviews and books (e.g., see Kadomtsev, 1963; Bateman, 1978; Freidberg, 1982; Manheimer et al., 1984). Here we give a brief review of the basic concepts of ideal MHD theory. This theory is traditionally oriented to the magnetic confinement fusion concept; hence, its most important results are stability criteria. In most cases when studying Z pinches that are dynamical systems, we deal with unstable systems that live for a limited time of interest, so we are mainly interested in the growth rate of instabilities.

Our starting point is the system of ideal MHD equations (3.41)–(3.44) together with appropriate boundary conditions:

$$\mathbf{J} = \frac{c}{4\pi} \nabla \times \mathbf{B}, \tag{4.1}$$

$$\frac{\partial \mathbf{B}}{\partial t} = \nabla \times (\mathbf{u} \times \mathbf{B}), \tag{4.2}$$

$$\frac{d}{dt}(P\rho^{-\gamma}) = 0 \tag{4.3}$$

$$\rho \frac{d\mathbf{u}}{dt} = -\nabla P + \frac{\mathbf{J} \times \mathbf{B}}{c}. \tag{4.4}$$

We assume that the plasma is in stationary equilibrium specified by conditions of (2.1) and (2.2) together with $\mathbf{u}(r) = 0$ so that there is no equilibrium flow of the plasma and cylindrical symmetry. Instabilities in moving equilibria will be considered below. All equilibrium quantities will be unmarked variables, and all perturbed quantities will be denoted by a superscript 1. All quantities are linearized about the equilibrium state. The linearized, ideal MHD equations are

$$\mathbf{J}^{(1)} = \frac{c}{4\pi} \nabla \times \mathbf{B}^{(1)}, \tag{4.5}$$

$$\frac{\partial \mathbf{B}^{(1)}}{\partial t} = \nabla \times (\mathbf{u}^{(1)} \times \mathbf{B}), \tag{4.6}$$

$$\frac{\partial P^{(1)}}{\partial t} = -\mathbf{u}^{(1)} \cdot \nabla P - \gamma P \nabla \cdot \mathbf{u}^{(1)}, \tag{4.7}$$

$$\rho \frac{\partial \mathbf{u}^{(1)}}{\partial t} = -\nabla P^{(1)} + \frac{1}{c}(\mathbf{J} \times \mathbf{B}^{(1)} + \mathbf{J}^{(1)} \times \mathbf{B}). \tag{4.8}$$

It is convenient to express all perturbed quantities in terms of the displacement vector, $\boldsymbol{\xi}(r, t)$, defined by

$$\mathbf{u}^{(1)} = \frac{\partial \boldsymbol{\xi}}{\partial t}. \tag{4.9}$$

Then, (4.5)–(4.8) can be integrated with respect to time with the aid of (4.9), yielding for perturbations of magnetic field, pressure, and density

$$\mathbf{B}^{(1)} = \nabla \times (\boldsymbol{\xi} \times \mathbf{B}), \tag{4.10}$$

$$P^{(1)} = -(\boldsymbol{\xi} \cdot \nabla)P - \gamma P(\nabla \cdot \boldsymbol{\xi}), \tag{4.11}$$

$$\rho^{(1)} = -\nabla \cdot (\rho \boldsymbol{\xi}). \tag{4.12}$$

The constants of integration in (4.10)–(4.12) were eliminated by specifying appropriate initial data for the problems as follows:

$$\boldsymbol{\xi}(\mathbf{r}, 0) = \rho^{(1)}(\mathbf{r}, 0) = P^{(1)}(\mathbf{r}, 0) = \mathbf{B}^{(1)}(\mathbf{r}, 0) = 0, \tag{4.13}$$

$$\frac{\partial \boldsymbol{\xi}(\mathbf{r}, 0)}{\partial t} \equiv \mathbf{u}^{(1)}(\mathbf{r}, 0) \neq 0. \tag{4.14}$$

Note that the applicability of (4.10)–(4.12) and of the ideal MHD theory do not require that the mechanisms that participate in the formation of the unperturbed plasma state to be dissipationless. In particular, the unperturbed state can represent the balance of ohmic heating and radiative energy losses, with considerable contribution from heat conductivity and other dissipation mechanisms in shaping the unperturbed profiles. Dissipation must be neglected only on the space and time scales characteristic of the perturbations. This condition is expressed by the inequalities (3.49) and (3.51), where the length and velocity scales are, respectively, $1/k$ and γ/k, where $k = 2\pi/\lambda$ is the perturbation wave vector and γ is its growth rate. Similarly, the Hall effect and radiative energy losses can be neglected, if the properly modified conditions (3.47) and (3.55) hold.

The linearized equations (4.10)–(4.12) may then be combined into a single, second-order partial differential equation for the displacement

$$\rho \frac{\partial^2 \boldsymbol{\xi}}{\partial t^2} = \mathbf{F}(\boldsymbol{\xi}), \tag{4.15}$$

where the force operator, $\mathbf{F}(\boldsymbol{\xi})$, is

$$\mathbf{F}(\boldsymbol{\xi}) = \nabla(\boldsymbol{\xi} \cdot \nabla P + \gamma P \nabla \cdot \boldsymbol{\xi}) + \frac{1}{4\pi}(\nabla \times \mathbf{B}) \times [\nabla \times (\boldsymbol{\xi} \times \mathbf{B})]$$

$$+ \frac{1}{4\pi}\{\nabla \times [\nabla \times (\boldsymbol{\xi} \times \mathbf{B})]\} \times \mathbf{B}.$$

Equation (4.15), subject to $\boldsymbol{\xi}(\mathbf{r}, 0) = 0$, $\partial \boldsymbol{\xi}(\mathbf{r}, 0)/\partial t = \mathbf{u}^{(1)}(\mathbf{r}, 0)$, together with appropriate boundary conditions, constitute the formulation of the general linearized stability problem as an initial-value problem. In general, this is a rather complicated sixth-order system of partial differential equations for $\boldsymbol{\xi}$, though the initial value approach is useful, especially in numerical calculations, and has the advantage of directly determining the time evolution of a given initial perturbation.

Since the right-hand sides of (4.10)–(4.12) and (4.15) contain no explicit time derivatives, the problem can be simplified and reformulated as an eigenvalue problem, by assuming that all perturbed quantities vary as $\exp(-\omega t)$. Upon substituting this into (4.15) we obtain

$$-\omega^2 \rho \boldsymbol{\xi} = \mathbf{F}(\boldsymbol{\xi}). \tag{4.16}$$

Such an approach is useful if exponential stability is valid and if the properties of the force operator, $\mathbf{F}(\boldsymbol{\xi})$, are known. In particular, Kadomtsev (1963) showed that the operator, $\mathbf{F}(\boldsymbol{\xi})$, is self-adjoint. The self-adjointness of operator $\mathbf{F}(\boldsymbol{\xi})$ means that the corresponding eigenvalues ω^2 are purely real, and the transition from stability to instability occurs when $\omega^2 = 0$. The spectrum of the force operator, $\mathbf{F}(\boldsymbol{\xi})$, was investigated by Grad (1973) and Goedbloed and co-workers (1974, 1975) (see also Freidberg, 1982). It was shown that a typical ideal MHD spectrum contains discrete modes ($\omega^2 < 0$), with an accumulation point at $\omega^2 = 0$, and continuous eigenvalues for $\omega^2 \geq 0$. This provides further motivation for examining plasma stability by the eigenvalue method, restricting attention only to the question of whether or not exponentially growing modes exist. Still, the situation may become somewhat complicated as $\omega \to 0$, since unstable modes can either accumulate or make a smooth transition from instability to stability.

4.1.2 Ideal MHD stability theory: the energy principle

It is often of primary interest to determine whether a given system is stable or unstable without actually finding the characteristic frequencies. In such cases, the variational formulation determines stability boundaries and an estimation of growth rates. Stability criteria can be formulated with the aid of the Energy Principle (Bernstein et al., 1958). The self-ajointness of operator $\mathbf{F}(\boldsymbol{\xi})$ means that (4.16) can be obtained from a variational principle. Integrating $\boldsymbol{\xi} \cdot \mathbf{F}(\boldsymbol{\xi})$ over the plasma volume, one can show that variation of the plasma potential energy caused by the perturbation $\boldsymbol{\xi}$, in the case when the plasma is either diffuse and unbounded (the perturbations are supposed to vanish at infinity) or bounded by conducting walls but with no free boundary, can be expressed in the form

$$W[\boldsymbol{\xi}] = \frac{1}{2} \int_V L(\boldsymbol{\xi}) \, d\mathbf{r}, \qquad (4.17)$$

where

$$L(\boldsymbol{\xi}) = (\nabla \cdot \boldsymbol{\xi})[(\boldsymbol{\xi}\nabla)P + \gamma P(\nabla \cdot \boldsymbol{\xi})]$$
$$+ \frac{1}{4\pi}[\nabla \times (\boldsymbol{\xi} \times \mathbf{B})][\nabla \times (\boldsymbol{\xi} \times \mathbf{B}) - \boldsymbol{\xi} \times (\nabla \times \mathbf{B})].$$

The unperturbed state is stable, if and only if any displacement increases the potential energy of the plasma. As the displacement field $\boldsymbol{\xi}$ is arbitrary, the energy-functional $W[\boldsymbol{\xi}]$ is positive definite if

$$L(\boldsymbol{\xi}) \geq 0. \qquad (4.18)$$

Alternatively, if all perturbations lead to an increase in the potential energy, then the system is linearly stable to exponentially growing modes (Laval et al., 1965).

Consider the application of the Energy Principle to Z-pinch plasmas. Recall that the equilibrium is now given by $P(r)$, and $\mathbf{B}(\mathbf{r}) = (0, B_\varphi(r), 0)$, where

$$\frac{dP}{dr} + \frac{B_\varphi}{4\pi r} \frac{d}{dr}(rB_\varphi) = 0, \qquad (4.19)$$

and the equilibrium symmetry implies that the perturbations can be Fourier analyzed as follows (the exponential factor will be omitted in most cases below):

$$\boldsymbol{\xi}(\mathbf{r}, t) = \boldsymbol{\xi}(\mathbf{r}) \exp(\Gamma t + im\varphi + ikz), \qquad (4.20)$$

where the eigenvalues Γ^2 are real; hence, the eigenmodes of a steady-state Z pinch represent either oscillations and waves ($\Gamma^2 = -\omega^2$) or instabilities ($\Gamma^2 > 0$); $\mathrm{Re}(\Gamma)$ being the growth rate, m and k are azimuthal and axial wave numbers, respectively.

In some cases, the conclusion concerning stability of a given unperturbed state (i.e., of the presence of positive eigenvalues Γ^2 in the spectrum) can be made using the Energy Principle only, without solving the boundary-value problem. In particular, the necessary and sufficient condition for stability of a pure Z pinch against $m = 0$ (sausage) instability modes is given by the Kadomtsev criterion (Kadomtsev, 1963)

$$-\frac{d \ln P}{d \ln r} < \frac{4\gamma}{2 + \gamma\beta}, \qquad (4.21)$$

where $\beta = 8\pi P/B_\varphi^2$. In a diffuse Z pinch, $\beta \to 0$ at $r \to \infty$, and hence the Kadomtsev criterion can be satisfied for sufficiently gradual pressure profiles; the pressure as $r \to \infty$ should not grow faster than $r^{-2\gamma}$. For a constant entropy profile in the unperturbed state [$S(r) \propto P(r)/\rho(r)^\gamma = \text{const}$], this means that the asymptotic density $\rho(r)$ should decrease slower than r^{-2}, which is not compatible with the condition that the line mass is finite. Thus, the Kadomtsev criterion (4.21) can be satisfied either when the latter condition is not necessary (e.g., in the case of a gas-embedded Z pinch) or if the temperature grows sufficiently fast with increasing distance from the axis, so that the pressure decreases slower than the density (strong-skin effect).

The necessary and sufficient condition for stability against $m \neq 0$ modes is given by (Kadomtsev, 1963; Freidberg, 1982)

$$\frac{r^2}{B_\varphi} \frac{d}{dr}\left(\frac{B_\varphi}{r}\right) < \frac{1}{2}(m^2 - 4). \qquad (4.22)$$

For a typical magnetic field profile in a Z pinch, $B_\varphi \propto r$, and as $r \to 0$, B_φ/r is a decreasing function of r. Thus, for a diffuse Z pinch, (4.22) predicts stability for $m \geq 2$ perturbations at large radii, whereas the $m = 1$ mode is always unstable for the inner core near the axis of a Z pinch. Choosing a power-law profile of current density, $j \sim r^{-s}$, one can present the criterion (4.22) in the form (Tayler, 1957)

$$m^2 < 2s + 1. \qquad (4.23)$$

Thus, the $m = 1$ mode can be stabilized only if the current and magnetic field is singular at the axis, which is hardly practical. Hence, the conclusion that "a Z pinch is always unstable to $m = 1$ perturbations and is likely to be unstable to $m = 0$ perturbations as well" (Freidberg, 1982). This conclusion is valid also for a Z pinch with a sharp boundary.

The physical mechanism of the $m = 0$ sausage and $m = 1$ kink instabilitykink instability modes is illustrated by Fig. 4.1(b) and (a) from Freidberg (1982). The $m = 0$ sausage perturbation [Fig. 4.1(b)] increases the magnetic field in the neck, because the same current flows in a smaller cross section and the increased magnetic pressure tends to deepen the neck further. The $m = 1$ kink perturbation [Fig. 4.1(a)] forces the magnetic lines to concentrate on the concave side of the column, and the increased magnetic pressure acts as a destabilizing force. Figure 4.1 shows that both the sausage and the kink modes develop, if the plasma column has a sharp boundary. (A notable exception is the case of an incompressible plasma

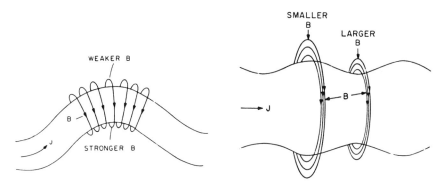

FIGURE 4.1. Physical mechanism of the (a) $m = 1$ and (b) $m = 0$ instability in a pure Z pinch (Freidburg, 1982).

column with distributed current: the sausage perturbation shown in Fig. 4.1(a) is incompatible with incompressibility and magnetic flux conservation; hence, the $m = 0$ mode does not develop.) (See Sec. 4.1.3.) In a diffuse Z pinch, the sausage and kink instabilities, which are driven by the same physical mechanisms, can be partly or even fully suppressed. This makes evident the destabilizing role of a free boundary.

The instability modes of a Z-pinch boundary are known as "flute modes" because the perturbations are fluted along the magnetic lines, or as "interchange modes" because the magnetic field and the plasma interchange: the field penetrates into the column and the plasma is forced out of it. The development of interchange instability at the plasma surface is determined by the geometry of the magnetic field near this surface. If $dB^2/d\mathbf{n} < 0$ ($\mathbf{n}$ is the normal vector to the plasma boundary), then the ejected plasma finds itself in the region of smaller magnetic pressure, and the perturbation grows. Otherwise, if $dB^2/d\mathbf{n} > 0$, which is characteristic of a concave plasma surface (e.g., in an EXTRAP system), the interchange perturbations are stabilized. In a cylindrical Z pinch, $dB^2/d\mathbf{n} = B^2(R)/R$ (R is the pinch radius), and the development of the sausage instability mode results in an interchange between a "heavy fluid" (plasma) and "light fluid" (magnetic field).

Thus, the interchange mode is to some extent similar to the known RT instability of a heavy fluid, supported by a light fluid against gravity. To elucidate the similarity, let us compare the expressions for drift velocity of a particle in a curved magnetic field and in crossed magnetic and gravitational fields:

$$\mathbf{V}_D = \frac{1}{\Omega}[\mathbf{b} \times \nabla(\ln B)](v_\parallel^2 + 0.5v_\perp^2),$$

$$\mathbf{V}_D = -\frac{1}{\Omega}[\mathbf{b} \times \mathbf{g}], \tag{4.24}$$

where Ω is the gyrofrequency, $\mathbf{b} = \mathbf{B}/B$; indices $\parallel$ and $\perp$ correspond to parallel and transverse directions with respect to the magnetic field. Assuming the magnetic field to be curl-free (this means that we consider a zero-pressure plasma without axial current), we find that the transverse component of the magnetic field gradient is $\nabla_\parallel (\ln B) = \mathbf{n}/R$, where R is the radius of curvature of a magnetic line and $\mathbf{n}$ is the main normal vector to it. We see that the gravitational drift is similar to that caused by the curvature of the magnetic lines (in particular, charges of different signs drift in opposite directions) so that the latter can be regarded as a drift due to an effective gravity

$$\mathbf{g}_{\text{eff}} = \frac{v_{thi}^2}{R} \mathbf{n}, \tag{4.25}$$

because it is the drift of ions that matters here, the electrons being glued to the ions by the quasineutrality condition. For a magnetic line passing near the surface of a cylindrical Z pinch, R in (4.25) is just the pinch radius. Using the classical expression for the RT instability growth rate, $\Gamma = \sqrt{gk}$, we can estimate the growth rate of the sausage instability mode as

$$\Gamma \cong \frac{v_{thi}}{R} \sqrt{kR}. \tag{4.26}$$

An estimate of Γ, which differs from (4.26) by a factor of order unity, has been obtained in early work on stability of Z pinches (Trubnikov, 1958). However, the similarity between interchange and RT instability modes is limited. In particular, the RT instability is not only a surface mode, it can develop in the volume of a fluid whenever a heavy fluid is supported by a light one, whereas the $m = 0$ instability mode of a Z pinch is absent when the condition in (4.21) holds (see the more detailed discussion in Chap. 5).

Stabilization of the hydromagnetic instabilities of a Z pinch by an external axial magnetic field was also studied. The most dangerous of the internal instability modes are the interchange and/or quasi-interchange ones, those for which there is no, or almost no, bending of magnetic lines by perturbations. This means that the value of

$$\mathbf{k} \cdot \mathbf{B} = k B_z + \frac{m}{r} B_\varphi \tag{4.27}$$

is zero or small, where $\mathbf{k} = \{0, m/r, k\}$ is the wave vector of perturbation. Suydam (1958) obtained the following criterion of stabilization of the internal modes localized near the point where the right-hand side of (4.27) vanishes:

$$r \frac{B_z^2}{4\pi} \left(\frac{d \ln \mu}{dr} \right)^2 + 8 \frac{dP}{dr} > 0, \tag{4.28}$$

where $\mu(r) = B_\varphi(r)/r B_z(r)$ is the pitch of the magnetic field and $d\mu/dr$ is a parameter characteristic of the magnetic shear. Suydam's criterion, (4.28), is a necessary, but not sufficient condition for stability, because an unstable mode may be not localized in the radial direction.

If a cylindrical Z pinch in an axial magnetic field has a free boundary, then $dB^2/d\mathbf{n} < 0$ near it, and the interchange instability can develop. The surface perturbations are helical flutes, the screw pitch being $h = 2\pi R B_z(R)/B_\varphi(R)$, so that $k/m = 2\pi/h$. The only way to avoid the interchange instability is to abandon cylindrical geometry. In particular, rolling the cylindrical Z pinch with an axial magnetic field into a torus with major radius $R_m \gg R$, and thus producing a simple model of the tokamak configuration, we can get rid of the perturbations with wavelengths $\lambda = 2\pi/k$ greater than $2\pi R_m$ due to periodicity with respect to the toroidal angle. The corresponding condition for hydromagnetic stability, $2\pi/h < 1/R$, is thus reduced to the well-known Kruskal–Shafranov's condition

$$q(r) = \frac{r B_z(r)}{R B_\varphi(r)} > 1 \qquad (4.29)$$

written for $r = R$. The quantity $q(a)$ (a being the minor radius of the torus) is called the safety factor. If the condition in (4.29) is satisfied in the whole plasma volume, then the interchange perturbation would not develop there. This advantage of toroidal systems of magnetic confinement over the cylindrical ones in stability (which became evident in the 1950s), and the absence of the electrode problems, stimulated the shift of interest of the fusion community to toroidal configurations and the abandonment of linear Z-pinch research.

4.1.3 Ideal MHD stability: The instability eigenmodes

The growth rates of the instability modes of a Z pinch can be found as eigenvalues of a boundary-value problem for (4.16). Substituting (4.20) into (4.16), one can reduce the system (4.16) of three equations for components of the displacement vector, $\boldsymbol{\xi}(r)$, to the following second-order differential equation for the radial displacement, $\xi_r(r)$:

$$\frac{d}{dr}\left(\frac{K}{r}\frac{d}{dr}(r\xi_r)\right) - L\xi_r = 0, \qquad (4.30)$$

where

$$K = \frac{1}{D}(\rho\Gamma^2 + F^2)\left[\rho\Gamma^2\left(\frac{B^2}{4\pi} + \gamma P\right) + \gamma P F^2\right], \qquad (4.31)$$

$$L = \rho\Gamma^2 + F^2 + \frac{B_\varphi}{2\pi}\frac{d}{dr}\left(\frac{B_\varphi}{r}\right) - \frac{k^2 B_\varphi^2}{\pi r^2 D}\left(\rho\Gamma^2\frac{B^2}{4\pi} + \gamma P F^2\right) \qquad (4.32)$$

$$- r\frac{d}{dr}\left\{\frac{k B_\varphi}{2\pi r^2 D}\left(k B_\varphi - \frac{m}{r}B_z\right)\left[\rho\Gamma^2\left(\frac{B^2}{4\pi} + \gamma P\right) + \gamma P F^2\right]\right\};$$

$$D = \rho\Gamma^4 + \left(k^2 + \frac{m^2}{r^2}\right)\left[\rho\Gamma^2\left(\frac{B^2}{4\pi} + \gamma P\right) + \gamma P F^2\right],$$

$$B^2 = B_\varphi^2 + B_z^2, \qquad F^2 = \frac{1}{4\pi}\left(k B_z + \frac{m}{r}B_\varphi\right)^2,$$

and $P(r)$, $B_\varphi(r)$ and $B_z(r)$ are related by the equilibrium condition (2.1, 2).

The boundary conditions for (4.30) are determined by the problem: for a diffuse Z pinch bounded by conducting walls of inner radius R they are

$$r\xi_r = 0, \quad \text{for } r = 0, R. \tag{4.33}$$

The first boundary condition (4.33) means the solution is regular at the axis ($r = 0$). The second one reflects rigidity of the walls. If the outer boundary of the plasma is free, then the corresponding boundary condition stems from the requirements of continuity of tangential magnetic field and total pressure (magnetic + thermal) at the perturbed surface. For a Z pinch in a vacuum, where $B_\varphi(r) = B_\varphi(R)(R/r)$, and when a constant external axial magnetic field, B_z, is present and there is no discontinuity of current and pressure at the pinch surface, the boundary condition at $r = R$ is

$$r\frac{d\xi_r}{dr} = \left(\frac{k^2 r^2 + m^2}{kr} \frac{K_m(kr)}{K_m'(kr)} + \frac{mB_\varphi - krB_z}{mB_\varphi + krB_z} \right) \xi_r, \tag{4.34}$$

where $K_m(x)$ is the modified Bessel function, $B_\varphi = B_\varphi(R)$. If (4.30) has a singular point at the boundary (i.e., the coefficient K vanishes at $r = R$), which is the case, in particular for $m = 0$ perturbations in the absence of an external axial magnetic field when $\rho(R) = 0$, then the corresponding boundary condition is replaced by the requirement of regularity of ξ_r at $r = R$. Finally, if a diffuse Z pinch has no definite external boundary, one boundary condition is the requirement that perturbations of density, pressure, and magnetic field vanish in the limit $r \to \infty$. Note that the displacement $\xi_r(r)$ does not necessarily vanish at infinity; in some cases it can tend to infinity itself.

The specific features of the second-order ordinary differential equation (4.30), making this boundary-value problem differ from the standard Sturm–Liouville problem, are the following (Goedbloed and Sakanaka, 1974; Goedbloed, 1984). First, the coefficient K before the second derivative has no definite sign. It can be positive or negative, depending on the sign of the eigenvalue, Γ^2. Second, the eigenvalue, Γ^2, enters the expressions for K and L in a complicated nonlinear way, which is why this boundary-value problem is characterized by rather complicated spectra. The oscillatory part of the spectrum ($\Gamma^2 < 0$, standing or running waves in the plasma column) generally includes five discrete subspectra and two continua, which correspond to the Alfvén and slow magnetosonic waves in a uniform plasma (for details, see Goedbloed and Sakanaka, 1974; Goedbloed, 1984). We are mainly interested in the unstable part of the spectrum, $\Gamma^2 > 0$. It turns out that this part consists of discrete eigenvalues and is bounded from above. The maximum growth rate corresponds to an eigenfunction $\xi_r(r)$, which has no zeros between $r = 0$ and $r = R$, the next growth rate corresponds to an eigenfunction having one zero in this interval, etc. In other words, the well-known Sturmian properties are valid for the spectra of growth rates. This fundamental result, first obtained by Goedbloed and Sakanaka (1974) for the boundary conditions (4.33) at $r = 0, R$, is known as *the oscillation theorem* and appears to be valid for any physically meaningful formulation of the boundary value problem.

Thus, a spectrum of instability growth rates of a Z pinch for any value of m and k is qualitatively similar to an energy spectrum of a quantum particle, described by the Schroedinger's equation, in a potential well. In both cases, the eigenfunction corresponding to the ground state (number of the mode, $n = 0$) has no nodes, whereas an eigenfunction with n nodes corresponds to the $(n + 1)th$ eigenvalue. The spectrum can contain a finite number of discrete levels or an infinite series of them, accumulating from the unstable side of the spectrum to $\Gamma = 0$. An example is illustrated in Fig. 4.2 from Kanellopoulos et al. (1988), where the growth rates of the first six $m = 0$ instability eigenmodes, calculated for a Z pinch with a free boundary and uniform current density. The profiles (2.5) are plotted versus kR. The eigenfunctions $\xi_r(r)$ corresponding to the first six $m = 1$ instability modes found for the same conditions and $kR = 10$, are shown in Fig. 4.3.

The contribution of the plasma volume to the energy integral (4.17) can be reduced also to the one-dimensional form:

$$W_{ID}[\xi_r] = \pi \int r \, dr \left(K(r) \left| \frac{d\xi_r}{dr} \right|^2 + L(r)|\xi_r|^2 \right). \qquad (4.35)$$

For unstable modes, the first term is always positive: $K > 0$ [see (4.31)]. It can be shown that if the plasma column is surrounded by conducting walls, so that the boundary conditions are given by (4.33), then $L \geq 0$ is a sufficient condition for stability. If instead of the eigenvalue Γ we substitute an arbitrary value σ into (4.32) and find that it makes $L(r) \geq 0$ for all r, this means that there is no instability eigenmode growing faster than $\exp(\sigma t)$. This method (the σ-stability method of

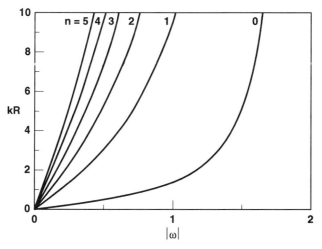

FIGURE 4.2. kR versus growth rate of the first six $m = 0$ radial unstable eigenmodes (Kanellopoulos et al., 1988).

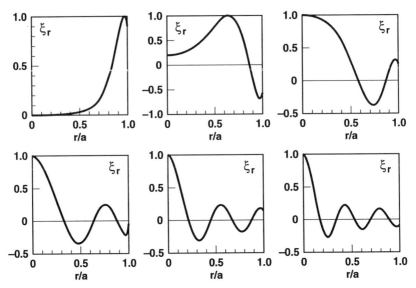

FIGURE 4.3. Eigenfunctions $\xi_r(r)$ corresponding to the first six unstable eigenmodes, $m = 1$, $kR = 10$ (Kanellopoulos et al., 1988).

Goedbloed and Sakanaka, 1974) provides a convenient way to estimate the growth rate of the dominate mode.

We see that the ideal MHD model yields a well-defined dominant instability mode for any wave numbers m and k. Its growth rate, being an isolated eigenvalue, is readily found with the aid of standard numerical techniques (shooting, Galerkin's, Rayleigh-Ritz's, etc.). If Suydam's criterion (4.28) is violated, then instability modes that are localized in the radial direction exist, with the number of nodes $n \to \infty$; hence, a dominant mode with $n = 0$ exists, too. Thus, violation of (4.28) is a sufficient condition of instability. Below, the term "growth rate of the hydromagnetic instability, Γ" always means the growth rate corresponding to the dominant mode. We are interested in the dependence of Γ on the wave numbers m and k, on the boundary conditions, and on the unperturbed profiles of density, pressure, and current density in the Z pinch.

4.1.4 Ideal MHD instability: The spectra of the growth rates

The equations of Sec. 4.1.3 can be solved analytically in some important cases for instability modes $m = 0$ and $m = 1$. For a current concentrated in a skin layer of a uniform plasma column, (4.30) is reduced to the Bessel equation, and we obtain the well-known dispersion relation for Γ (Tayler, 1957; Shafranov, 1958a):

$$\frac{\Gamma^2 I_m(\Delta)}{\Delta I_m(\Delta)} = 1 + \frac{m^2 K_m(kR)}{kR K_m(kr)}, \qquad (4.36)$$

where $\Delta = (k^2 R^2 + 2\Gamma^2/\gamma)^{1/2}$, $I_m(x)$ and $K_m(x)$ are modified Bessel functions. In the limit of an incompressible fluid $\gamma \to \infty$, the transcendental equation (4.36) is simplified, and we obtain

$$\Gamma = \frac{V_A}{R} \sqrt{\left(1 + \frac{m^2 K_m(kR)}{kR K_m(kR)}\right) \frac{kR I_m'(kR)}{I_m(kR)}}. \qquad (4.37)$$

The growth rate Γ as a function of wave numbers m and k from (4.37) is shown in Fig. 4.4. The sausage ($m = 0$) and kink ($m = 1$) instability modes are seen to be dominant for any k; the modes with $m \gg 1$ arising only for $k \sim m^2$. It can be shown (Pereira et al., 1984) that in the case when the whole current is concentrated in an infinitely thin skin layer, the dependence of growth rates on γ is very weak, except for small values of kR.

The $m = 0$ sausage mode can be stabilized, if the plasma pressure decreases slowly enough (see Sec. 4.1.2). But nothing within the ideal MHD model can stabilize the long-wavelength kink mode, the latter thus being either the only long-wavelength instability mode or co-existing only with the sausage mode. The growth rate of the kink mode in the limit $kR \to 0$ for an arbitrary current profile in a Z pinch is given by the asymptotic expression (Shafranov, 1956, 1958a; Nycander and Wahlberg, 1984)

$$\Gamma \cong (kV_A) \left(\ln\left(\frac{2}{kR}\right) + \frac{c^2}{4I^2} \int_0^R B_\varphi^2(r)\, r\, dr - \gamma_E\right)^{1/2}, \qquad (4.38)$$

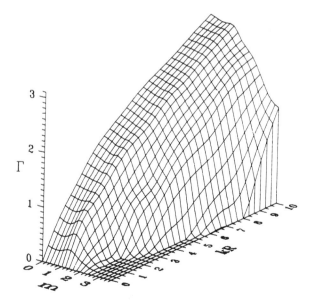

FIGURE 4.4. Instability growth rate Γ versus wavenumbers m and kR for a Z-pinch with a skin current in an incompressible fluid.

where $\gamma_E = 0.5722\ldots$ is Euler's constant, and the integral in the parentheses is proportional to the inductance of the plasma column. The growth rate of the sausage mode in the same limit is, to a greater extent, dependent on the unperturbed profiles. Typical examples are

$$\Gamma \cong \begin{cases} \left(\dfrac{\gamma}{2(\gamma-1)}\right)^{1/2} (kV_A), & \text{skin current}, \\ (\sqrt{3}kR)^{1/2} \cdot (V_A/R), & \text{uniform current}, \gamma = 1, \end{cases} \tag{4.39}$$

where the asymptotic expression for the skin current case can be obtained from (4.36), and the other one is given by Trubnikov (1958).

Consider the influence of the skin effect on the spectra of $m = 0$ growth rates. Suppose that the plasma is incompressible and a part, I_{vol}, of the total current, $I_{\text{tot}} = I_{\text{vol}} + I_{\text{skin}}$, is uniformly distributed in the volume, whereas the other part, I_{skin}, is concentrated in the skin layer. The ratio, $\eta = I_{\text{vol}}/I_{\text{tot}}$, characterizes the uniformity of the total current. Then (4.30) remains a Bessel equation, and the growth rate of the $m = 0$ mode is given by

$$\Gamma = \frac{V_A}{R}\sqrt{1 - \eta^2}(kR)\left(1 + \frac{kRI_1(kR)}{I_1(kr)}\right)^{-1/2} \tag{4.40}$$

In particular, in the limit $k \to \infty$ we obtain

$$\Gamma \cong \frac{V_A}{R}\sqrt{kR(1 - \eta^2)} \tag{4.41}$$

[cf. the estimate (4.26)]. We see that the growth rate decreases with increasing η, vanishing in the limit of uniformly distributed current $\eta = 1$. In other words, the $m = 0$ instability mode cannot develop in a column of incompressible plasma with uniformly distributed current, because the sausage perturbations are found to be incompatible with flux conservation and incompressibility. The same conclusion remains valid if the unperturbed density profile is not uniform.

To test whether the key condition for suppressing the $m = 0$ mode in the case of distributed current is incompressibility, one can solve the same problem analytically for an isothermal, compressible plasma, i.e., for the case of $\gamma = 1$. Then the solutions of (4.30) are expressed via the Coulomb wave functions (Abramowitz and Stegun, 1964), and we find that in the limiting case of a strong skin effect (i.e., for $\sqrt{kR(1 - \eta^2)} \gg 1$), the expression for Γ is given by the right-hand side of (4.41), whereas in the opposite extreme for $kR \to \infty$, saturation of the growth rate is found in contrast with the known result of Trubnikov (1958):

$$\Gamma \cong \sqrt{5}\frac{V_A}{R} = \sqrt{10}\frac{v_{thi}}{R}, \tag{4.42}$$

where v_{thi} is given by (2.14). Note that the growth rate (4.42) does not vanish in the limit $\eta \to 1$.

Thus, the $m = 0$ instability mode (and the $m = 1$ mode to a lesser extent) is suppressed by uniform current distribution, if the plasma can be treated as incompressible. Pereira et al. (1984) used a finite width of the surface current layer to explain why the observed growth rates of $m = 0$ and $m = 1$ modes were considerably smaller than those given by (4.36) and (4.37). The experiments were performed on the BLACKJACK-3 ($I = 0.8$ MA) and the BLACKJACK-5 ($I = 3$ MA) facilities. For each time, the structure of the optical emission was Fourier analyzed, and the observed values of Γ were found for perturbation modes characterized by the wave numbers m and k. The growth time, $1/\Gamma$, of the $m = 0$ mode in the BLACKJACK-3 experiment was found to be 13 ns, whereas (4.37) predicts 3.7 ns. In the BLACKJACK-5 experiment, the $m = 0$ mode dominated, the observed and calculated growth times being 14 and 5.7 ns, respectively.

The model of a Z pinch used by Pereira et al. (1984) is an ideal MHD incompressible plasma, with the current density equal to zero for $r < AR$ and the uniform current density for $AR < r < R$ (here $0 \leq A \leq 1$), so that the limits $A = 0$ and $A = 1$ correspond to uniform current and to a strong skin effect, respectively (i.e., $A \to 0$ means the same as $\eta \to 0$ above, and vice versa). Figures 4.5(a) and (b) present the growth rates calculated by Pereira et al. (1984) for the sausage $m = 0$ (a) and kink $m = 1$ (b) instability modes. Indeed, a factor of 3 reduction of the growth rates that is needed for agreement between theory and experiment was obtained for $A = 0.25$ (around $kR \cong 4$). The estimated depth of the current channel due to diffusion, $(v_m \tau)^{1/2} \cong 0.3R$, is consistent with the assumption

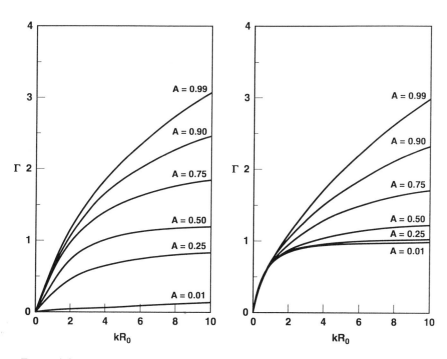

FIGURE 4.5. Instability growth rates Γ versus kR_0 for instability modes (a) $m = 0$ sausage and (b) $m = 1$ kink (Pereira et al., 1984).

that the current is homogeneously distributed over the outer 75% of the pinch plasma. As this result relies heavily on the assumption of incompressibility, the only question that requires further explanation is why the plasma in the experimental conditions under consideration behaves as an incompressible fluid.

One can attempt to explain this by noting the influence of the low-pressure plasma surrounding the main plasma column and increasing its inertia. The stabilizing effect of a surrounding gas was studied in the long-wavelength limit by Manheimer et al. (1973). It is known to be important, even in cases when its pressure is small compared with the pinch pressure. The Z pinch behaves like a vibrating string in air, its motion exciting waves in the surrounding gas. It is possible that the additional inertia in the case when the pressure is high only in the plasma column makes the perturbation modes dominant. This issue certainly deserves further consideration.

Another important aspect of a low-density plasma corona surrounding the main plasma column has been discussed by Sasorov (1991), Neudachin and Sasorov (1991), and Terry and Pereira (1991). They suggested some mechanisms that can be responsible for leakage currents outside a pinch. Note that the models discussed above assume that the current density in the pinch is constant and that all the current that goes into the diode flows through the pinch. In fact, even a small current leakage through the pinch periphery (for example, due to a superheated skin layer, turbulence, etc.) can essentially affect the estimate for the perturbations growth.

The agreement between the estimates (4.26) and (4.41) should not lead to the conclusion that the square-root behavior of the growth rate, Γ, in the short-wavelength limit is due to the physical similarity between sausage and RT instabilities. In fact, this behavior in (4.41) is determined by the discontinuity in the magnetic pressure, due to the strong skin effect. Choosing instead a continuous unperturbed density profile, $\rho = \rho_0(1 - r/R)^s$, with density vanishing at the boundary [$0 < s < 2$; s smaller than 2 for convergence of the energy integral, $\int T(r)r\,dr$]; and a uniform profile of current density, we find that in the short-wavelength limit the growth rate is (Bud'ko, 1990)

$$\Gamma \cong 2\frac{V_A}{R} \left(kR \frac{2\Gamma\left(\dfrac{2}{s-1}\right)}{\sqrt{\pi}\,\Gamma\left(\dfrac{1}{s}\right)} \right)^{s/2} \left(\frac{2}{(s+1)(s+2)} \right)^{1/2}, \qquad (4.43)$$

here $\Gamma(x)$ is Euler's gamma function. In particular, for $s = 1$ we obtain on the right-hand side of (4.43) the value $(V_A/R)\sqrt{(2kR/3)}$, which coincides with the result of Trubnikov (1958). Other values of s give other behavior of the density near the boundary; hence, other asymptotic shapes of the spectrum, $\Gamma(k)$, in the short-wavelength limit. Thus, the coincidence of the asymptotic law (4.26) and that given by Trubnikov (1958) and reproduced by (4.43) for $s = 1$ does not stem from some deep physical similarity between the flute and RT instability modes. It is determined by this particular choice of s and hence is fortuitous.

Now let us consider the case of a diffuse Z pinch, when there is no free plasma boundary. A diffuse Z pinch with a standard (flat or Gaussian) current-density profile exhibits only $m = 0$ and $m = 1$ instability modes [see (4.22)]. Figure 4.6 presents the dependence $\Gamma(k)$, calculated by Felber (1982) and Bud'ko (1989) for a pure Z pinch with a Gaussian density profile [Fig. 4.6(a)], and the same for the case when an axial magnetic field is present [Fig. 4.6(b)]. Here, the sausage mode, $m = 0$, is the fastest growing mode. For a Gaussian pure Z pinch, the growth rate of this mode saturates at $k \to \infty$ at the level (Felber, 1982; Bud'ko, 1989)

$$\Gamma = \frac{1}{\tau} = \frac{V_A}{R} = \sqrt{2}\,\frac{v_{thi}}{R}, \tag{4.44}$$

where R is the characteristic pinch radius $[\rho \propto \exp(-r^2/R^2)]$. Note that the sausage mode is not necessarily the dominant one: the kink $m = 1$ mode can dominate with other choices of the unperturbed profiles. A sufficiently strong

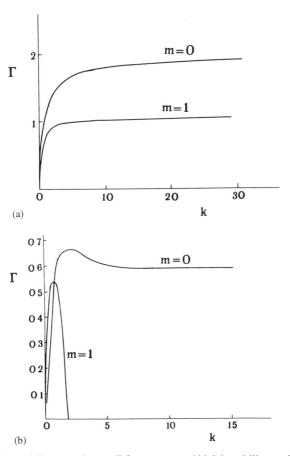

(a)

(b)

FIGURE 4.6. (a) Instability growth rates Γ for sausage and kink instability modes of a diffuse pure Z pinch with a Gaussian density profile; (b) the same with an axial magnetic field.

external axial magnetic field [for a Gaussian Z pinch, $B_z > 0.7(I/Rc)$] can fully suppress the sausage mode, and then other instability modes (the flute modes) would dominate (see Sec. 4.1.3).

Coppins (1988) used the ideal MHD model of a Z pinch with a free boundary to explain the unexpected stability of compressional and gas-embedded Z pinches observed in the experiments by Dangor et al. (1983) and Haines et al. (1986). The compressional Z pinch was produced by a current discharge of amplitude $I_{max} = 130\,\text{kA}$ in hydrogen in a quartz tube of radius $R_0 = 1$ cm and length $L = 5$ cm. The Z pinch of radius $R = 7$ with temperature $T \cong 130$ eV (the Bennett estimate) remained stable for a time of about 300 ns. "The stability of the pinch is particularly striking since large (10–20%) nonuniformities in the number-density profile are seen at early times after pinch formation, probably due to an asymmetric collapse caused by preferential current flow on one side" (Coppins, 1988).

Substituting the above values of R and T into (4.44), we obtain an estimate of the maximum growth rate of the sausage mode: $\Gamma^{-1} = 32$ ns. Calculations made by Coppins (1988) for a flat current-density profile yield for the growth times, Γ^{-1}, 15 ns and 27 ns, for the $m = 0$ and $m = 1$ modes, respectively, when $kR = 4$ (i.e., for the perturbation wavelength $\lambda = 1.57R$). For a Z pinch with a free boundary, the growth rates are not limited in the limit $k \to \infty$, and we can obtain much smaller estimates of the growth times. However, the influence of the extreme short-wavelength perturbations is limited both by dissipation and nonlinearities. As we are not able to take all these factors into account, it is natural to suppose that the wavelength of the dominate instability mode is of order of R. Nevertheless, there is no appreciable perturbation growth during 10–20 growth times, though the initial nonuniformities are by no means small!

Coppins (1988) explained the stability of the compressional Z pinch by assuming rapid cooling of the pinch plasma after the peak current. This results in a decreased value of v_{thi} and of the growth rates, which scale as v_{thi}/R [see (4.44)]. The initial nonuniformities, being produced by the asymmetric initial discharge, correspond to the $m = 1$ perturbation mode with a wavelength, $\lambda \cong 21$ cm, $kR \cong 4$. These long-wavelength perturbations interact with the outer layers of the plasma near the walls of the tube. If the plasma near the wall can be regarded as a conducting wall situated at $r = R_0$, then the calculations of Coppins (1988) demonstrate the stability of the pinch with respect to such perturbations. Thus, it appears that, in principle, the enhanced stability of compressional Z pinches can be explained by the ideal MHD model. A more definite conclusion can be only based on using MHD stability analysis performed for a realistic unperturbed state and calculated taking account of all the physical mechanisms that take part in shaping the unperturbed profiles. Of particular importance is a correct calculation of the profiles for the peripheral plasma layers, including those adjacent to the walls.

Stability of gas-embedded Z pinches can be explained in a simpler way. The absence of the $m = 0$ instability mode, confirmed by many experiments with gas-embedded Z pinches, can be naturally explained by the criterion (4.21), because the slow decrease of the unperturbed pressure is consistent with the way such pinches are produced. For a Kadomtsev marginally stable profile [that is, the left- and right-hand sides of (4.21) are supposed to be equal], which provides stability of a diffuse Z pinch with respect to all $m = 0$ and $m \geq 0$ perturbations, Coppins (1988) found the growth time of 7.9 ns for the kink mode ($kR = 4$). Thus, the observed gas-embedded Z pinches can remain stable during 20–40 ns simply because the perturbations have not enough time to develop, the initial state being highly symmetrical. Of course in this case, a realistic simulation of the unperturbed state is desirable.

4.2 Effect of ohmic heating and radiative losses: Overheating instability and filamentation

Filamentation or azimuthal instability produces stratification of an initially cylin-drically symmetric current channel into separate current filaments, thus giving the plasma "the appearance of a multistranded cable" (Coppins, 1988). Current filamentation has been observed in many experiments (Filippov, 1980; Nardi et al., 1980; Sadowski et al., 1984; Herold et al., 1989). Still, the physical mech-anisms responsible for current filamentation are not well known. Presumably, filaments observed in different experimental conditions are caused by different mechanisms. For instance, in some experiments the observed filaments may be just the nonuniformities produced by the initial breakdown. But this simple ex-planation is definitely inadequate for many recent experiments (Sadowski et al., 1984; Herold et al., 1989).

We have seen that the ideal MHD model predicts no instability with $m > 2$ and $k = 0$ for a stationary diffuse Z pinch. This is why the observed filaments con-ventionally have been explained by taking into account some factors not described by the ideal MHD equations. The first theoretical analysis of current filamentation in high-energy plasma-focus devices (Kaeppeler, 1983) was based on a two-fluid MHD model. Some other mechanisms were discussed, including a current–current interaction according to Biot–Savart's law, initiated by microinstabilities due to electron cyclotron activity. Plasma overheating (Imshennik and Neudachin, 1987, 1988) and effects specific to two-fluid hydrodynamics (Meierovich, 1983) were also suggested as possible mechanisms of filamentation.

It is generally believed that filamentation occurs due to the overheating instabil-ity, that is, the local loss of energy balance. This instability is well known, both in laboratory and space plasma physics. A simple estimate of its growth rate can be obtained (Imshennik and Neudachin, 1987) by linearization of the energy equation

$$\frac{\partial}{\partial t}\left(\frac{\rho u^2}{2} + \frac{P}{\gamma - 1}\right) + \nabla \cdot \left[\left(\frac{\rho u^2}{2} + \frac{\gamma P}{\gamma - 1}\right)\mathbf{u}\right] = Q_j - Q_r, \quad (4.45)$$

where the first term in the right-hand side represents ohmic heating and the second one radiative energy losses. Neglecting the perturbations of Q_j, the temperature dependence of Q_r, and the nonuniformity of the unperturbed variables (the perturbation is thus supposed to be local), one obtains for plane geometry:

$$\frac{\Gamma P^{(1)}}{\gamma - 1} + ik\gamma \frac{Pu^{(1)}}{\gamma - 1} = \left(\frac{dQ_r}{d\rho^{(1)}} \right). \tag{4.46}$$

It is natural to suppose that isobaric perturbations, i.e., those with $P^{(1)} = 0$, which do not generate running sonic waves, are the fastest ones. Expressing $\rho^{(1)}$ via $u^{(1)}$ with the aid of (4.46) and substituting the result into the linearized continuity equation $\Gamma\rho^{(1)} = -ik\rho u^{(1)}$, we obtain for the isobaric perturbations the following estimate of Γ:

$$\Gamma \cong \frac{\rho}{\gamma E_T} \left(\frac{dQ_r}{d\rho} \right) = \frac{Q_r}{E_T}, \tag{4.47}$$

where $E_T = P/(\gamma - 1)$ is the plasma thermal energy density. The physical mechanism of this instability is quite clear. If a given plasma volume radiates more thermal energy than is dissipated in it, then the adjacent plasma layer is attracted to it, thereby maintaining the pressure on the same level, and thus increasing the density and further increasing the radiative losses. Thus, the radiating plasma can be regarded as a fluid with a negative value of the adiabatic exponent γ. One can advance this metaphor (Torricelli-Ciamponi et al., 1987) by presenting (4.45) in Lagrangian variables, that is, for a fixed plasma particle:

$$\rho^\gamma \frac{d}{dt} (P\rho^{-\gamma}) = (\gamma - 1)(Q_j - Q_r). \tag{4.48}$$

Linearizing (4.48) with the same assumptions as before and using (4.11) and (4.12), (4.20), we obtain:

$$\gamma_{eff} = \frac{P_L^{(1)}/P}{\rho_L^{(1)}/\rho} = \gamma - \frac{\rho}{\Gamma E_T} \left(\frac{dQr}{d\rho} \right), \tag{4.49}$$

where the Lagrangian perturbation $f_L^{(1)}$ refers to the quantity f in a fixed plasma particle. Taking into account that it is related to the Eulerian perturbation $f^{(1)}$ by $f_L^{(1)} = f^{(1)} + (\xi\Delta)f$, one can immediately find that (4.10)–(4.12) correspond to $\gamma_{eff} = \gamma$. Thus, the growth rate can be obtained from the boundary-value problem formulated in Sec. 4.1.3, where $\gamma_{eff}(r; \Gamma)$ should be taken instead of γ. Negative values of γ_{eff}, obtained when Γ exceeds the estimate (4.47), mean that initial perturbations do not propagate as running waves (this behavior corresponds to $\gamma_{eff} > 0$) but keep growing, and this indeed corresponds to a strong thermal instability (Torricelli-Ciamponi et al., 1987). Calculations made in this paper for cylindrical plasma equilibria in force-free magnetic fields demonstrated the possibility of achieving $\gamma_{eff} < 0$. This type of instability can develop faster than those given by the ideal MHD model, if the inequality (3.55) is violated. Radiative losses are important in determining both the dynamics and stability of a Z pinch.

Of course, the conventional mechanism of overheating is also acting (Kadomtsev, 1963): a local increase in the plasma conductivity due to increased temperature produces greater current density, increased ohmic heating, and thus a further rise of the conductivity and temperature. The effect of ohmic heating on perturbations of wavelength $\lambda \approx R$ is characterized by the Lundquist number (3.50). For typical Z-pinch experiments $L_u \gg 1$ (see Table 2), so that the effect of finite conductivity is important only for shorter wavelength

$$\lambda \ll RL_u^{-s}, \tag{4.50}$$

where $s = 2/3$ and $1/2$ for a free boundary Z pinch and diffuse Z pinch, respectively. In particular, for the experiments with frozen deuterium fibers (Scudder, 1985) the right-hand side of (4.50) is about 10 μm, which is consistent with numerical results (Lindemuth, 1989a,b).

Numerical calculations of the growth rates for azimuthal instability modes were performed by Imshennik and Neudachin (1988), and Belova and Brushlinskii (1988). They used a one-fluid MHD model that accounted for ohmic heating and radiative losses. The dominant instability mode was found numerically by imposing an arbitrary perturbation on the steady state and integrating the linearized equations until an exponentially growing dominating eigenmode was singled out. They have shown that finite conductivity influences Z-pinch stability in two ways. Magnetic field diffusion, as with any process that smoothes out nonuniformities, can be stabilizing; on the other hand, the ohmic overheating mechanism can drive instabilities absent in the ideal MHD model.

Belova and Brushlinskii (1988) studied the dependence of the growth rate on the current density profile and the type of boundary conditions for a model dependence, $Q_r(r, t) = \text{const} \cdot T^{3/2}$ and $m = 2$. Their calculations show a stabilizing role of finite magnetic viscosity. They found that the $m = 2$ instability mode can develop only for large (but finite) conductivity and is switched off with decreased conductivity. A flat current density $j_z(r)$ profile was found to be the most unstable: a minimum of j_z near the wall or the axis increases locally the value of magnetic viscosity, thus stabilizing the instability. A typical growth rate obtained for a flat profile is about 5% of the characteristic value, v_{thi}/R. For a free-boundary Z pinch, the instability is much stronger, and the stabilizing role of magnetic field diffusion less effective. Increasing the magnetic viscosity results in increasing the growth rate by up to 20% of v_{thi}/R. This is interpreted by Belova and Brushlinskii (1988) as a manifestation of the overheating instability. It is interesting that pinches with current and temperature increasing in the periphery turned out to be stable with respect to $m = 2$ modes. However, it should be noted that these results, like those of Brushlinskii and Shatanov (1980), essentially depend on the model adopted for radiative energy losses (in particular, they are changed by choosing $Q_r \propto T^4$), so that the above conclusions should be regarded as preliminary ones.

The growth rates of the azimuthal instability were calculated by Imshennik and Neudachin (1988) for conditions typical of plasma-focus experiments with deuterium, where radiative losses are due to radiation from heavy impurity ions (Xe). The unperturbed state was represented by the nonhomogeneous equilibrium solutions obtained by Bobrova and Razinkova (1984, 1985), and the plasma conductivity σ and radiative losses term Q_r were described by realistic approximations. For a wide range of parameters, the calculated growth rates were found to be in good agreement with the estimate (4.47) and almost independent of the wave number, m, for $m > 2$, as it should be for local instability modes. For the experimental condition when $C_r < 1$ [see (3.55)], the azimuthal modes driven by the radiative mechanism can dominate. The estimates obtained for the growth rates, $\Gamma \cong 10^8 - 10^9 \, s^{-1}$, are in agreement with filamentation growth rates observed in plasma focus experiments (Filippov, 1980; Sadowski et al., 1984; Herold et al., 1989).

Thus, two hydrodynamic mechanisms of filamentation in steady-state Z pinches are available: ohmic heating and radiative losses. The latter appears to be more suitable for explaining current filamentation observed in high-energy plasma focus experiments. An alternative explanation is based on the RT instability of the plasma accelerated in the plasma-focus device.

4.3 Resistive and viscous effects on Z-pinch stability: Heat conductivity

The theoretical description of the Z pinch in terms of ideal MHD suggests that it is violently unstable due to instabilities growing on the Alfvén time scale. Ideal MHD stability theory, on which the early decision to abandon the linear Z pinch as a fusion device was made, neglects resistivity, viscosity, thermal conduction, pressure anisotropy, finite Larmor radius effects, and the Hall effect. However, recent experiments with dense pinches (Haines et al., 1988) and with deuterium-fiber-initiated Z pinches at the Naval Research Laboratory (NRL) (Sethian et al., 1987) and LANL (Hammel and Scudder, 1987) exhibit apparent stability over a very large number of Alfvén growth times. These pinches are formed from frozen deuterium fibers, in which the current rises very rapidly due to the application of a very high voltage. Unlike earlier pinches, there is no dynamic phase. Only a slow expansion is observed at much less than the Alfvén speed, so it appears that the assumption of pressure balance is a good approximation in this case. Stability is observed as long as the current is rising, but the pinch succumbs to a rapidly growing $m = 0$ instability at current maximum. The instability invariably occurs when $dI/dt = 0$, independent of the current amplitude and other pinch parameters, at least within the range of the experiment, and so it is tempting (but misleading) to conclude that stability is a consequence of the rapid current rise. Cochran and Robson (1990) have concluded that the most likely explanation of this anomalous behavior lies in the resistive nature of these pinches, which makes ideal MHD models invalid. These observations have prompted theoretical efforts to reexamine stability with emphasis on the sausage mode, since the kink mode is not seen in recent experiments.

One possible explanation for this behavior is that these pinches remain cold for a large fraction of the current rise time and that resistivity could, therefore, have an important effect on their dynamics. Theoretically, one expects ideal behavior if the Lundquist number is greater than, or of the order of one, resistive effects only being significant if L_u is smaller than this. The importance of resistivity is determined by the Lundquist number, (3.50). For ideal MHD theory to be applicable, R_M or L_u should be much greater than unity. For frozen pinches, L_u starts out much less than unity and rarely exceeds 200. Thus, realistic attempts to describe these pinches should include the effects of resistivity.

Note that a compressible pinch with finite resistivity and carrying a constant current will expand continuously, owing to ohmic dissipation. For stability studies, it is desirable to have a stationary equilibrium, and this may be obtained by making the current rise at such a rate that the magnetic pressure always balances the plasma pressure. The pinch radius will then remain constant, while important quantities, such as temperature and Lundquist number, change in time. The stability of the pinch may be tested at any time by applying a perturbation and watching it grow or decay.

Culverwell and Coppins (1990) have studied the linear stability of an ohmically heated Z pinch under pressure balance, with fixed radius, and using a 1-D linearized initial-value code. If the current rises according to a power law, the current-density profile (as well as other profiles) will be self-similar (see Sec. 3.5). It can be shown (Culverwell and Coppins, 1990) that for an isothermal pinch with the current rising as $t^{1/3}$, the current-density profile and the density profile are unstable to $m = 0$ perturbations in the ideal, compressible MHD model, according to Kadomtsev's criterion (Sec. 4.1.2). This is an equilibrium, since the increase in thermal pressure is balanced by the increasing magnetic pressure caused by the rising current. Culverwell and Coppins (1990) have found the critical Lundquist number L_u^*, below which the pinch is stable. They have found an approximate relation $L_u^* = 50(kR)^{-0.86}$, which indicates that for long-wavelength modes, the stability threshold may be more than two orders of magnitude greater than might be expected intuitively. In general their results agree with results of the numerical simulations of the NRL groups (Cochran and Robson, 1990; Glasser and Nebel, 1989). Cochran and Robson (1990) employed a 2-D resistive MHD simulation. They found that the time dependence of the equilibrium can have a stabilizing influence on global $m = 0$ and $m = 1$ modes for sufficiently small Lundquist numbers. Typically, below a Lundquist number of about 10^2, the pinch is found to be stable. They have shown that a linear Z pinch, in which the current is rising in such a way as to keep the radius constant, and which has a current profile that is unequivocally unstable in the ideal MHD model, may be stable to $m = 0$ perturbations, if it is sufficiently resistive, that is, if the Lundquist number is sufficiently small. A simple physical picture is that the distortion of the pinch radius by an $m = 0$ perturbation is opposed by the tendency of a resistive pinch with this form of current rise to assume a constant current radius, as evidenced by 1-D calculations. The reason why the stability of fiber pinches appears to be connected with the current rise may be because the rising current prevents them from expanding significantly from their initially very small radius, and so the Lundquist number, which is proportional to the radius, is kept small. When the current stops rising, the pinch expands rapidly and the assumption of constant radius, which is the central feature of the model, no longer holds. Whether this can explain the observed onset of instability at this time will be examined in future work. Applicability of various theoretical models have been discussed by Haines and Coppins (1991). The results are significant because they have shown that a Z pinch may be stable to $m = 0$ perturbations, even though its pressure profile fails the Kadomtsev criterion for stability in the ideal MHD picture. One-dimensional simulations of fiber ablation (Lindemuth et al., 1989) suggested that the presence of a residual core of frozen deuterium might explain the enhanced stability of fiber pinches.

A solvable model has been developed by Lampe (1991) for the linearized sausage mode within the context of resistive MHD. The model is based on the assumption that the plasma motion is self-similar, but the perturbations are not assumed to be self-similar. He considered the current rising as t^α, so that ohmic heating and the time variation of the pinch radius have been included in the analysis. Results by Lampe disagree with those of Culverwell and Coppins (1990) and Cochran and Robson (1990). First, Lampe found that if $\alpha \leq 1$, the growth rate is reduced, as it also is for the case $I = I_0 (t/\tau)^{1/3}$ where the radius is time independent, but not by more than about a factor of 2, as compared with the known ideal MHD value with an assumed time-independent equilibrium. In contrast, if $\alpha \geq 1$, the net result is that the growth rate is larger than the ideal MHD value with a time-independent equilibrium! Thus, Lampe concluded that within the context of a linearized fluid treatment, resistivity alone cannot account for the long period of stability observed in the deuterium-fiber experiments.

The resistive stability of a static equilibrium, including the combined effect of resistivity and viscosity, has been examined by Cox (1991) with a linear initial value code. He found that resistivity alone has only a minor effect. However, a scalar viscosity leads to some suppression of sausage and kink modes. Recently, detailed 2-D "cold-start" resistive MHD simulations of deuterium-fiber Z pinches (Sheehey and Lindemuth, 1994) have shown that when the fiber becomes fully ionized, $m = 0$ instabilities develop rapidly and drive intense nonuniform heating and rapid expansion of the plasma column.

Still, there is no consensus in understanding the anomalous stability of the fiber-initiated pinches. An understanding of the basic features involved in the stability problem of fiber-initiated pinches is clearly crucial to further progress in the field and deserves further investigation.

The influence of viscosity is determined by the Reynolds number, (3.51) and (3.52). Typically for Z-pinch plasmas, $R_e \gg 1$, and the stabilizing effect of viscosity is weak for most experimental conditions with ion temperature below 1 keV. For higher temperatures, the effect of viscosity, which scales as $T^{5/2}$, can be considerable (Glasser, 1989). But, the high-temperature plasma is not collision-dominated in most cases, so that the validity of the hydromagnetic models becomes questionable. The effect of ion viscosity in the weakly collisional regime deserves further study. At present, attempts to use ion viscosity to explain the observed stability of Z pinches have not been successful. For instance, even the variation of the viscosity coefficient within six orders of magnitude from its classical value (Braginskii, 1963) is not enough to explain the stability of the long-wavelength $m = 1$ kink mode in the EXTRAP 1987. The stabilizing effect of ion viscosity can be significant only for RFPs, where the plasma density is comparatively low, and plasma rotation and wall effects related to it are important. Gimblett (1988) derived a second-order equation similar to (4.30), taking into account parallel ion viscosity. The example of a Z pinch with skin current, separated by a vacuum region from conducting walls, was used by Gimblett (1988) to demonstrate that both the growth rate of the $m = 1$ mode and the value of wavenumber k, above which this mode is stabilized by conducting walls, are decreased due to viscosity.

The electron heat conductivity in the direction parallel to the magnetic field is the strongest mechanism of heat conductivity in the plasma [the estimate (3.54) of the Peclet number refers just to this mechanism]. The parallel heat conductivity does not contribute to shaping the unperturbed cylindrically symmetrical state, because the magnetic lines lie on the constant r surfaces and all the variables depend only on r. But, the perturbations imposed on this state break the uniformity in azimuthal and axial directions. The parallel heat conductivity may be a factor smoothing out such perturbations.

It turns out that the effect of parallel heat conductivity is equivalent to renor-malization of the adiabatic exponent γ, and can be exactly taken into account by substituting $\gamma_{eff}(r; \Gamma)$ instead of γ into the boundary-value problem of Sec. 4.1.3 (Torricelli-Ciamponi et al., 1987). To show this, we take the plasma-energy equation in Lagrangian variables, the parallel heat conductivity being the only dissipative term included:

$$\rho^{\gamma} \frac{d}{dt}(P\rho^{-\gamma}) = (\gamma - 1)\nabla \cdot (\kappa_{\parallel}^{e} \nabla_{\parallel} T), \tag{4.51}$$

where $\kappa_{\parallel}^{e}$ is the coefficient of parallel electron heat conductivity. Linearizing (4.51), we obtain, after simple calculations, an expression similar to (4.49):

$$\gamma_{eff} = \gamma \frac{P_e + 1/\gamma}{P_e + 1}, \tag{4.52}$$

where

$$P_e = \Gamma/(K_{\parallel}^2 \chi_{\parallel}), \quad k_{\parallel}^2 \equiv (kB_z + \frac{m}{r}B_{\varphi})^2/B^2, \quad \chi_{\parallel}^e \equiv \kappa_{\parallel}^e/E_T.$$

The value of P_e depends on plasma density and temperature; therefore, it is a function of Γ and r. Note that when $\gamma_{\text{eff}}(r; \Gamma)$ given by (4.52) is substituted into (4.31) and (4.32), the boundary-value problem of Sec. 4.1.3 for (4.30) is no longer self-adjoint now that Γ enters the equations together with Γ^2. Hence, the con-clusions of Sec. 4.1.3 concerning the spectral properties are not valid, so that the shooting method is not appropriate for calculating the eigenvalues; other methods, like Galerkin's or Rayleigh–Ritz's, suitable for finding complex eigenvalues, are to be used.

If $|P_e| \ll 1$, then $\gamma_{\text{eff}} = 1$ and the results of Secs. 4.1.3 and 4.1.4 are reproduced. In the opposite extreme, when for any r we have $|P_e| \gg 1$, then (4.52) yields

$$\gamma_{\text{eff}} = \gamma. \tag{4.53}$$

Since this limiting value of γ_{eff} is constant, real, and independent of Γ, we again return to the results of Secs. 4.1.3 and 4.1.4. If one also cannot neglect the transverse heat conductivity, then in the limiting case of large P_e, we again return to the boundary-value problem of Sec. 4.1.3, with γ given by (4.53) and the only additional condition of temperature uniformity in the unperturbed state, i.e., $T(r) = \text{const.}$

The calculations of ideal MHD growth rates done by many authors (Goedbloed and Hagebeuk, 1972; Pereira et al., 1984; Culverwell and Coppins, 1989) have shown that, for a given unperturbed state, the eigenvalue Γ very weakly depended on the parameter γ (variation of γ from 1 to 5/3 changed Γ by no more than several percent). Thus, we conclude that the stabilizing effect of heat conductivity on the perturbations of a Z pinch is small, though readily taken into account. The same conclusion applies to the radiative energy transfer-stabilizing mechanism (see Davis and Cochran, 1990).

4.4 Effects of finite and large ion Larmor radius: The Hall effect

The influence of finite ion Larmor radius (FLR) effects on plasma instabilities was first considered by Rosenbluth et al. (1962) and then studied by many authors (see Mikhailovskii 1975, 1977). In a traditional formulation of the problem, the parameters $k\rho_i$ and ω/Ω_i (ρ_i and Ω_i are the ion Larmor radius and gyrofrequency, k and ω are the wavenumber and frequency of interest, respectively), which are zero in the ideal MHD approximation, are supposed to be small but finite. The contribution due to FLR effects was shown to be stabilizing, with the main stabilizing mechanism being the ion gyroviscosity. The coefficient of gyroviscosity is estimated as nT_i/Ω_i. Note that it is about $\Omega_i\tau_i$ times smaller than the collisional viscosity coefficient (Braginskii, 1963). Since the stabilizing effect of the latter is not very important for dense Z pinches (see Sec. 4.1.6), and the effect of gyroviscosity being considered for $\Omega_i\tau_i > 1$ is even weaker, one can hardly expect the FLR effects to be major stabilizing mechanisms for dense Z pinches. The possibility of Hall term destabilization of MHD stable equilibria has been suggested (Coppins et al., 1984). An interesting issue of the fast penetration of magnetic fields into plasma due to the Hall field was examined (Fruchtman and Maron, 1991; Fruchtman and Gomberoff, 1993; Sarfaty et al., 1995a, 1995b; Swanekamp et al., 1996).

In recent years, there has been renewed interest in this issue for several reasons. First, it was noted that for a Z-pinch fusion plasma, the parameter ρ_i/R is of order unity (Haines, 1978a). Second, an unexpectedly fast growth of perturbations was observed (see also Sec. 4.2.6) in a number of experiments in space (Bernhardt et al., 1987) and laboratory (Ripin et al., 1987) plasmas. Finally, attempts to explain the stability of some plasma configurations like EXTRAP stimulated new studies of stabilizing FLR effects.

Let us consider the effect that dominates in the opposite limit of large ion Larmor radius (LLR): $k\rho_i \gg 1, \omega/\Omega_i \gg 1$. Hassam and Lee (1984) (see also Hassam and Huba, 1988) pointed out that the plasma dynamics in the LLR limit is described by the equations of the one-fluid Hall MHD. They differ from those of the conventional isotropic/isothermal MHD equations by a single term in the induction equation

$$\frac{\partial \mathbf{B}}{\partial t} = \nabla \times \left(\mathbf{u} \times \mathbf{B} - \frac{m_i c}{e} \frac{D\mathbf{u}}{Dt} \right) = \nabla \times \left(\mathbf{u} \times \mathbf{U} - \frac{1}{en_e} \mathbf{J} \times \mathbf{B} \right), \quad (4.54)$$

where $D/Dt = \partial/\partial t + \mathbf{u} \cdot \nabla$, and the plasma is supposed to be quasineutral and singly ionized. The term in question is due to the Hall effect. It constitutes a vorticity-driven source of the magnetic field (Hassam and Huba, 1988). Comparing the two terms in brackets in (4.54), we see that for perturbations of a steady state ($D/Dt = \Gamma$), the second term is of the same order as the first one, retained in (3.42), if $|\Gamma| \geq \Omega_i$. Derivation of (4.54) is very simple in the collision-dominated limit (Braginskii, 1963) that is the most important for us. [It should be noted that the one-fluid Hall MHD used here is not a universally accepted model of collisionless LLR plasmas (see the discussion between Lehnert and Scheffel, 1988, and Huba et al., 1988.] The Hall term is large compared to the ohmic term, not included in (4.54), if the electrons are magnetized: $\Omega_e \tau_e \gg 1$. In general, one cannot tell from the form of the equations if the Hall term is stabilizing or not. By linearizing the equations and Fourier analyzing in space and time, a non-Hermitian eigenvalue equation can be derived, the eigenvalues of which are complex, i.e., in general, the unstable solutions are both growing and oscillating.

If a Z pinch is cylindrically symmetric, plasma rotation and nonuniform shear flow along the z axis are absent, then both vectors $\mathbf{J}$ and $\mathbf{B}$ are normal to the radial direction. Thus, their vector product is radial, depends only on r, and hence does not contribute to the right-hand side of (4.54). Therefore, all the cylindrically symmetric solutions of the ideal MHD equations, described in Chap. 3, are also exact solutions of the Hall MHD equations. However, for perturbations breaking the uniformity in the azimuthal and axial directions, the additional term in (4.54) can be important, just as is the parallel heat conductivity term in the energy equations in Sec. 4.1.6. Equations (4.10) and (4.12) evidently remain valid [the Hall model of collisionless LLR plasma, Hassam and Lee (1984), corresponds to $\gamma = 1$ in (4.11)], and instead of (4.10), we obtain from (4.54):

$$\mathbf{B}^{(1)} = -(\boldsymbol{\xi}\nabla)\mathbf{B} + (\mathbf{B}\nabla)\boldsymbol{\xi} - (\nabla \cdot \boldsymbol{\xi})\mathbf{B} - \frac{m_i c}{e} \Gamma \nabla \times \boldsymbol{\xi}. \quad (4.55)$$

Substituting (4.11), (4.12), and (4.55) into the equation of motion (3.44), one obtains a boundary-value problem for the vector function $\boldsymbol{\xi}(r)$. The instabilities of a Z pinch in the Hall MHD model were studied by Schaper (1983) and Coppins et al. (1984) for $m = 0$ modes, and by Spies and Faghihi (1987), Scheffel and Faghihi (1989) and Ågren (1988, 1989) for $m = 1$ modes.

In the limit of incompressible plasma ($\gamma \to \infty$), the stabilizing effect of the Hall term on the $m = 0$ sausage instability mode was confirmed by Schaper (1983), at least for the case of a uniform density profile. Coppins et al. (1984) studied the development of an arbitrary initial perturbation numerically, so that the maximum growth rate, $\Gamma = \operatorname{Re}(\widehat{\Gamma})$, was found (when a Hall term was present, the eigenvalue, $\widehat{\Gamma}$, could be complex) as a function of parameter

$$\varepsilon = \frac{\rho_i}{R} = \frac{c}{\omega_{pi} R} = C_H^{-1} \approx \Pi^{-1/2}, \qquad (4.56)$$

where C_H and Π are given by (3.47) and (2.12), respectively. The effect of the Hall term is important for ε of order unity, that is, for C_H and Π not too large. The results obtained by Coppins et al. (1984) for a Z pinch bounded by a conducting wall at $r = R$, and for a constant value of $kR = 1$, are shown in Fig. 4.7 [the growth rate $\Gamma \equiv \operatorname{Im}(\omega)$]. For small values of ε, the growth rate, Γ, is indeed decreasing with increasing ε, in agreement with Schaper (1983). But when ε exceeds 0.42, another radial mode becomes the dominant one, and the growth rate increases appreciably; then, after $\varepsilon \cong .087$, the third one arises. Will it continue to increase further? If we can rely upon the similarity between the flute and RT instabilities (see Sec. 4.1.2), then the answer is affirmative. In the short-wavelength limit, the growth rate of the dominant RT mode in a LLR plasma is estimated as $\Gamma \cong gk/\Omega_i$ (see Sec. 4.2.6). Estimating the effective acceleration due to magnetic field geometry by (4.25), we can obtain a similar estimate of the $m = 0$ mode growth rate in the LLR limit, when the effect of the Hall term is strong:

$$\Gamma \cong \frac{v_{thi}}{R} (k\rho_i). \qquad (4.57)$$

In Fig. 4.7 the value of Γ is presented in units of v_{thi}/R, $k = 1/R$, so that the estimate (4.57) yields $\Gamma \cong \varepsilon$. The dashed straight line corresponding to this

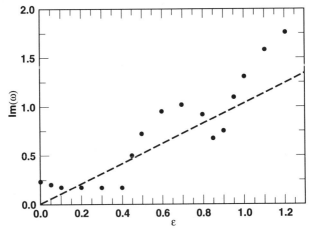

FIGURE 4.7. The points show the dominating $m = 0$ instability growth rate, $\Gamma = \operatorname{Im}(\omega)$, calculated by Coppins et al. (1984) for a pure Z pinch, bounded by conducting walls in the Hall MHD model versus the Hall parameter; points; the dashed line is the estimate (4.57).

dependence is plotted in Fig. 4.7. We see that (4.57) correctly estimates both the order of magnitude of Γ and an average linear increase of Γ with increased ε. The same linear dependence is predicted by (4.57) for Γ as a function of k, ε being constant (see below Fig. 4.13). Thus, the results of Coppins et al. (1984) can be interpreted as an indication that the effect of the Hall term on the short-wavelength, $m = 0$ instability modes of a Z pinch is essentially destabilizing, providing a linear (not a square-root) increase of Γ with increasing k, just as in the case of RT instability (see Sec. 4.2.6). Further study of this issue would be appropriate.

The $m = 1$ instability mode was studied by Spies and Faghihi (1987), Ågren (1988), and Scheffel and Faghihi (1989) with the aid of the Hall MHD model, and by Ågren (1989) and Åkerstedt (1989) with the aid of the Vlasov equations, which describe the drift-kinetic effects and the influence of resonant particles. These results refer to the traditional FLR range (smallness of the parameter $k\rho_i \ll 1$ is explicitly in the asymptotic expansions). In particular, marginally unstable (ideal) modes are stabilized by gyroviscosity (Scheffel and Faghihi, 1989); the Hall term has a damping, but not absolutely stabilizing, effect. Stabilization of the dominant radial mode due to FLR effects was studied for short-perturbation wavelengths. The stabilizing effect was found to be essentially dependent on the unperturbed profiles. In particular, for a flat current profile the decrease of the growth rates in comparison with those given by the perpendicular MHD does not exceed 30%, whereas for the Bennett equilibrium profile $[B_\varphi(r) = 4Ir/c(r^2 + R^2), T(r) = \text{const}]$, the corresponding reduction in growth rates calculated for $kR = 4$–8 was about five times. The eigenfunction profiles of the dominating instability modes are not similar to the ideal MHD profiles: in the short-wavelength limit the perturbations are localized not at the axis, as the MHD predicts, but at $r = 0.4R$. As for the long-wavelength modes, it is natural to expect, that in the limit $k \to 0$, all the effects characterized by finite length scales, be it dissipative mechanisms, FLR, or other kinetic effects, are negligible if the corresponding length scales are much smaller than $\lambda = 2\pi/k$. For the Hall model, this was confirmed by a direct calculation of Γ by Ågren (1988), where the result of (4.38) of Nycander and Wahlberg (1984) was reproduced.

4.5 Kinetic effects

For a dense Z-pinch collision-dominated plasma, we are interested in the instabilities that are hydrodynamic by nature, like the sausage and kink instability modes. A vast amount of literature exists on kinetic-beam, instabilities drift, etc. [e.g., microinstabilities of plasma equilibria, Z pinches included (see Mikhailovskii, 1975, 1977)], and most of it is certainly beyond the scope of the present review. Still, it should be noted that the kinetic-beam instabilities may become important for a new generation of very high-power machines. Kinetic-stability analysis in some cases provides important corrections to the estimates given by the fluid models.

The preceding refers mainly to the short-wavelength instability modes, because in the long-wavelength limit the fluid models are justified (see Sec. 4.1.7). Kinetic theory, in most cases, predicts a slower growth of short-wavelength perturbations than MHD theory. This was demonstrated for both linear (Haines et al., 1984) and nonlinear (Imshennik et al., 1984) stages of the $m = 0$ mode development, and for the $m = 1$ mode (Åkerstedt, 1988, 1989) (see Sec. 4.1.6). Physically, this mechanism of stabilization is due to the mixing effect specific to the kinetic theory: some particles leave the area where the perturbation develops, while some other particles with arbitrary phases enter it. This effect is evidently absent in hydrodynamics, where each of the perturbed fluid particles moves as a whole, remaining close to its unperturbed position.

The first analytic kinetic treatment of the $m = 0$ instability development is given independently by Isichenko et al. (1989) and Arber and Coppins (1989). According to both papers, the current is supposed to be concentrated in a skin layer, so that no magnetic field is present in the plasma column. Kinetic theory is used to describe the collisionless ions, the cold electrons representing the background that neutralizes the ion space charge. Thus, the skin layer is a potential barrier confining the ions, which are otherwise free and interact only with the barrier, reflecting from it. A sausage perturbation of the cylindrical shape of the barrier [see Fig. 4.1(a)] causes perturbations of the magnetic pressure outside it and of the ion motion inside it, so that its deformation develops further.

In the long-wavelength limit, a motion of an ion successively reflected by a slowly moving, deformed pinch boundary is characterized by an adiabatic invariant. This fact was used by Isichenko et al. (1989) to calculate explicitly the perturbed ion distribution function, and hence, the growth rate:

$$\Gamma \cong 0.433 k v_{thi}, \quad kR \ll 1. \tag{4.58}$$

A linear dependence, $\Gamma(k)$, known from hydrodynamics [see (4.39)] is obtained here; however, the numerical coefficient in (4.58) is noticeably smaller [the corresponding value is 1.12 in (4.39) for $\gamma = 5/3$] than that given by (4.39) for the case of strong-skin effect and cold electrons. Even in the incompressible plasma limit $\gamma \to \infty$, the corresponding coefficient is $1/\sqrt{2} = 0.707$. Since for long-wavelength perturbations, the fluid and kinetic approaches should yield the same results, it would be interesting to find which of the fluid models provides the same estimate (4.58).

In the short-wavelength limit, for $1/R \ll k \ll 1/\delta$ (δ is the thickness of the skin layer), a strong mixing takes place: the phase of the perturbed boundary profile where an ion strikes depends mostly on the initial conditions determined by the previous collision. Then, one can neglect perturbations of the initial ion distribution function (taken as a Maxwellian distribution function) and determine the perturbed pressure, considering the motion of the deformed boundary in the ion gas. The result of Isichenko et al. (1989) and Arber and Coppins (1989) for a Maxwellian ion distribution is

$$\Gamma \cong \frac{v_{thi}\sqrt{\pi}}{2R}. \tag{4.59}$$

The growth rate given by kinetic theory is shown to saturate for large values of k, thus being approximately $\sqrt{kR}$ times smaller than the MHD estimate predicts [see (4.43)]. Isichenko et al. (1989) pointed out that if the ratio ρ_i/R is not small, the above results are not likely to be modified considerably, because this ratio means only that ions are reflected not from a vertical potential barrier, but rather from a smooth potential profile. This does not prevent kinetic mixing in the short-wavelength limit, and hence, saturation of the growth rate for large k. The linear stability of a Z pinch confined by a skin current has been studied in the collisionless regime, using the Vlasov fluid model for the $m = 0$ mode (Arber and Coppins, 1994) and for an arbitrary m number (Arber, 1990). For the case $T_e = 0$, an analytical solution is found for the growth rate in the short-wavelength limit, which saturates according to (4.59). Kinetic effects of finite Larmor radius have been discussed by Åkerstedt (1990).

4.6 Nonlinear evolution of the $m = 0$ mode

As a rule, the instabilities of Z pinches and other plasma systems become noticeable when amplitudes of perturbations are not small, so that their nonlinear interaction cannot be neglected. The problem of their nonlinear evolution is not limited by analysis of evolution for any particular motion. Since now we cannot construct a general solution as a linear combination of particular solutions, the main problem always is the degree of generality of the obtained analytical or numerical solutions. Can we be sure that they indeed describe characteristic properties of the nonlinear flows generated by sufficiently broad classes of initial conditions?

These problems were studied in most detail for the nonlinear evolution of the sausage ($m = 0$) instability mode of a Z pinch with skin current. The attention to this particular case is due not only to some convenient simplifications of the general problem related to it, but also to the widespread interest in the bright spots in Z pinches, which are conventionally identified with the necks representing the final stages of development of the $m = 0$ perturbations. In the long-wavelength limit, i.e., for $kR_0 \ll 1$, where R_0 is the initial radius of the pinch, the following equations describing the nonlinear evolution of the sausage mode were obtained by Book et al. (1976) and, in a more general form, by Trubnikov and Zhdanov (1985):

$$\frac{\partial A^2}{\partial t} + \frac{\partial}{\partial z}(vA^2) = 0, \tag{4.60}$$

$$\frac{\partial v}{\partial t} + v\frac{\partial v}{\partial z} = v_0^2 \frac{\partial}{\partial z}(A^{-2}). \tag{4.61}$$

Here $A(z, t) = (R(z, t)/R_0)^\zeta$; $R(z, t)$ is the radius of the plasma column, depending both on time and on the axial coordinate z; $\zeta \equiv (\gamma - 1)/\gamma$; v is the axial velocity of the plasma; $v_0^2 \equiv \gamma V_A^2/2(\gamma - 1)$, γ being the adiabatic exponent. In the first approximation in $kR_0 \ll 1$ plasma density ρ, pressure P, axial velocity

v, and the ratio u/r (u is the radial velocity) are uniform over the cross section, that is, independent of r. The boundary condition is determined by the variation of the pressure due to the azimuthal magnetic field $B_\varphi = 2I/cR(z, t)$ in the given cross section due to variation of $R(z, t)$, so that $P(z, t) = P_0(R(z, t)/R_0)^{-2}$. In the limit of incompressible plasma, first studied by Book et al., we have $\gamma \rightarrow \infty$, $\zeta = 1$, $v_0 = V_A/\sqrt{2}$.

Equations (4.60) and (4.61) are identical to the equations of 1-D compressible gas dynamics, with A^2 representing density, v velocity, and $-v_0^2 \ln(A^2)$ pressure. The effective adiabatic exponent is

$$\gamma_{\text{eff}} = [\ln(A^2)]^{-1}. \tag{4.62}$$

Since γ_{eff} is negative for $A < 1$, the flow is unstable (see Sec. 4.1.5).

A less-evident fact, pointed out by Trubnikov and Zhdanov (1985), is that the system of (4.60) and (4.61), being quasilinear and homogeneous, can be reduced to one linear, second-order equation for $t(A^2, v)$ by the hodograph transform (dependent and independent variables switch roles; e.g., see Whitham, 1974). This made it possible for Trubnikov and Zhdanov (1985) to find, for some cases, exact analytic solutions of (4.60) and (4.61), which previously had been studied numerically by Book et al. (1976) for $\zeta = 1$. One of them, obtained for a perturbation periodic in z, is presented in Fig. 4.8. The nonlinear evolution thus produces a

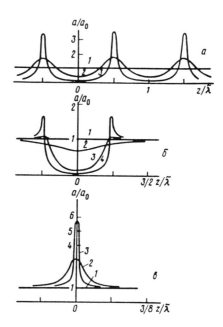

FIGURE 4.8. The analytic solution obtained by Zhdanov and Trubnikov (1985) represents the nonlinear evolution of the sausage instability mode.

"spindle" structure of the plasma column: near the axis, a cylinder of dense plasma (the neck) is formed, its minimal radius being $R_{min}(t)$, whereas a part of the plasma mass is ejected to infinity by sharp disk-like jets, whose radius, $R_{max}(t)$, increases with time. At some finite moment, $t = 0$, the current channel breaks, R_{min} vanishes, and the jets reach infinity. Near this moment

$$R_{min}(t) \cong R_0[4v_0(-t)/\lambda]^{\frac{\gamma}{(\gamma-1)}},$$

$$R_{max}(t) \cong R_0[4v_0(-t)/\lambda]^{-\frac{\gamma}{2(\gamma-1)}},$$

(4.63)

where λ is the wavelength of the initial perturbation ($t < 0$); in the limit $t \to -\infty$ both $R_{min}(t) \to R_0$ and $R_{max}(t) \to R_0$. When the compression of the plasma in the neck is high, we obtain from (4.63)

$$\rho(t) \propto (-t)^{-\frac{2\gamma}{\gamma-1}}, \; P(t) \propto (-t)^{-\frac{2}{\gamma-1}}, \; T(t) \propto (-t)^{-2}.$$

(4.64)

The length of the neck, $L < \lambda$, can be regarded as constant near the instant $t = 0$. Its value is established when transition occurs from linear to the nonlinear stage of evolution. The asymptotic expressions (4.64) and constancy of L were confirmed by the numerical study by Garanin and Chernyshev (1987). They have also shown that the plasma motion in the neck near $t = 0$ becomes self-similar. The time dependence of the radius of the neck, density, velocity, and temperature are given by (4.63) and (4.64), $v \propto L/(-t)$, $u \propto \dot{R}_{min}(t)$, and their coordinate dependence is given by some universal functions of a self-similar variable, $\eta = z/L$. If $\Psi(\eta)$ represents the profile of the variable $(\rho/\rho_m)^{\gamma-1}$, where ρ_m is the maximum density in the center of the neck, the self-similar solution found by Garanin and Chernyshev (1987) yields a universal implicit expression for $\rho(\eta)$:

$$\eta = \frac{1}{4}(2 - \psi^{1/3})(1 - \psi^{1/3})^{1/2},$$

(4.65)

since the profile of $R/R_{min}(t)$ is given by $\Psi^{-\gamma/2(\gamma-1)}$, (4.65) determines the shape of the neck as well. The transition to the self-similar asymptotic behavior is seen in Fig. 4.8. After the self-similar stage of motion, the neck breaks. The time of its development is estimated by (4.63) as λ/V_A. The breaking of the neck was first demonstrated numerically for an ideal MHD model by D'yachenko and Imshennik (1974) (see also Imshennik et al., 1973). It also has been shown that dissipation does not prevent the breaking of the current channel. Thus, the described picture represents a quite general nonlinear behavior of the long-wavelength perturbations.

The equations obtained by Book et al. (1976) for the short-wavelength limit $kR_0 \gg 1$ are identical to those describing an RT instability of a plane fluid layer supported against gravity by a massless incompressible fluid. Therefore, it is natural to expect that the nonlinear stage of evolution will be similar to that characteristic of the RT instability (see below, Sec. 4.2.7). For a periodic perturbation whose wavelength is λ, the magnetic field should penetrate into the plasma as broad bubbles moving at constant velocity of order $\sqrt{\lambda g_{eff}}$, and the plasma should be ejected from the axis as sharp spikes moving at constant acceleration g_{eff} (g_{eff} being the effective gravity). When the bubbles approach a distance comparable with λ from the axis, the flow pattern changes: the growth of magnetic pressure due to decreased radius of the plasma becomes noticeable, and the self-similar regime described above is established (Garanin and Chernyshev, 1987).

This type of nonlinear evolution was studied numerically by Vikhrev et al. (1989, 1993). The ideal MHD model was used, with the unperturbed state represented by a Z pinch with skin current and uniform density profile $\rho = \rho_0$, surrounded by a uniform low-density plasma ($\rho = 10^{-3}\rho$). A small ring cavity of depth and height of about $0.03 R_0$ (that is, the characteristic value of $k R_0 \approx 10^2$) was chosen as the initial perturbation. Its evolution is shown in Fig. 4.9. The deviation from the simple scenario described above is due to the fact that greater magnetic pressure in the cavity causes its expansion in both the radial and the axial directions. The magnetic field filling the growing cavity drags the surrounding gas into it, due to the $\mathbf{J} \times \mathbf{B}$ force. This low-density gas is accelerated to very high velocities, exceeding the characteristic value of v_{thi} in the unperturbed pinch by factor of about 30. At the pinch axis, a cylinder of dense, high-temperature plasma is formed, its behavior being close to that described by the above self-similar solution. Then it, in turn, is destroyed by the MHD instabilities.

Most of the numerical studies start from an initial perturbation with a fixed axial wavelength λ. Interaction of perturbations having different wavelengths and

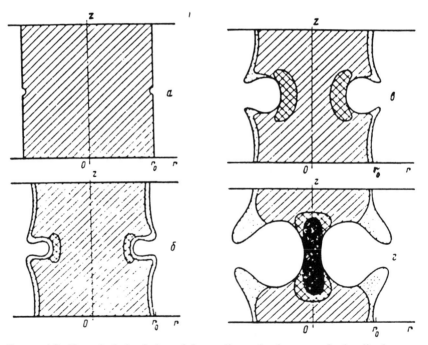

FIGURE 4.9. Numerical simulation of the nonlinear development of a localized sausage perturbation (Vikhrev et al., 1989); the lines are contours of constant density.

formation of nonlinear flow from the chaotic small perturbations were not studied in the context of Z-pinch applications. Analysis of a model problem related to the RT instability (Sharp, 1984) indicates that the energy of short-wavelength perturbations, which at first develop faster than the long-wavelength ones, is subsequently pumped to the large-scale perturbations. Indeed, the wavelength, which corresponds to the instability mode that is dominant during the nonlinear stage, is known to emerge gradually from the chaos. This has been observed in many of the early Z-pinch experiments (e.g., see Alexandrov et al., 1973) and found also in some of the numerical calculations (see also Hussey et al., 1980; and Sec. 4.2.6).

The behavior of currents and fields near the moment of breaking the current channel has attracted considerable attention since the early 1950s, when neutron radiation from high-current Z pinches in deuterium was first observed by Artsimovich et al. (1956). It was soon understood that the observed neutron yield was due to fast ions with energies of order 10^2 keV, much greater than the plasma temperature. A number of physical mechanisms that could provide acceleration of a relatively small number of fast ions in the necks of Z pinches were discussed, from the conventional ones like acceleration of ions by the axial electric field (the overvoltage was supposed to be produced by current channel breakdown, see Imshennik et al., 1973) to some recently suggested; e.g., the radial acceleration of the low-density plasma by the $\mathbf{J} \times \mathbf{B}$ force (Vikhrev et al., 1989, 1993) described above. There is no generally accepted model yet; readers are referred to the reviews by D'yachenko and Imshennik (1974), Trubnikov (1986), and Vikhrev (1986), where the problem of ion acceleration is discussed from different points of view. We note only that the model of thermal, that is, thermonuclear mechanism of the neutron production in high-current Z-pinch facilities encounters difficulties in obtaining agreement with the time of existence of a dense Z pinch. For typical conditions produced at the final stage of high-energy plasma-focus experiments with deuterium (Filippov, 1983), e.g., $I \cong 1$ MA, $T_e \cong 5$ keV, $n_i \cong 10^{19-20}$ cm^{-3}, the characteristic Alfvén transit time, τ_A, does not exceed 3 ns. The observed neutron yield is two orders of magnitude higher than thermonuclear reactions can produce in this time interval; in addition, the measured duration of the observed neutron pulse (e.g., Filippov, 1983) is about 50 ns.

An interesting explanation of the stability of a dense Z pinch emitting neutrons was proposed (Sasorov, 1985, 1991; Neudachin and Sasorov, 1991). Stabilization is supposedly due to the effect of the low-density plasma surrounding the neck. They have shown that the presence of the low-density conducting plasma can essentially influence the dynamics of the neck development in the late stages of its evolution. Indeed, even if the plasma corona carries a small part of the total current, when a plasma particle is attracted to the axis, the magnetic flux in it is conserved: $B_\varphi / nr = $ const. Since at the periphery $B_\varphi \propto r^{-1}$, we obtain $n \propto r^{-2}$, so that the pressure in the particle is $P \propto r^{-10/3}$ because the flow is adiabatic ($\gamma = 5/3$ is taken for the fully ionized deuterium plasma). Hence, the thermal pressure increases with decreasing distance, r, from the axis faster than the magnetic pressure ($B_\varphi \propto r^{-2}$) attracting the particle to the axis. Thus, collapse of the current does not take place: after the breakdown of the main current channel all the current is switched to the surrounding plasma, a plasma corona of the Z pinch formed by it. This conclusion is consistent with the self-similar solution describing a collapse of a converging current channel (see Sec. 3.4.4). At the moment of collapse, the current flowing along the axis vanishes, and then all the current flows in the volume. Substituting the conditions of flux and entropy conservation into the equilibrium equation (2.4), one obtains the solution describing the equilibrium state of the plasma corona:

$$(n/n_c)^{-1} - (n/n_c)^{-1/3} = \frac{r^2}{R^2}, \tag{4.66}$$

$$B_\varphi = \frac{2I}{Rc} \frac{nr}{n_c R}, \tag{4.67}$$

$$T = T_c(n/n_c)^{2/3}, \tag{4.68}$$

where n_c and T_c are characteristic values of number density and temperature in the corona, and R is its characteristic radius determined from the condition $I^2 = 5\pi c^2 R^2 n_c T_c$. An interesting feature of this equilibrium is its stability with respect to sausage perturbations. It is easily verified that substitution of the solution (4.66)–(4.68) into (4.21) makes its left- and right-hand sides equal, so that the profiles given by (4.66)–(4.68) are Kadomtsev marginally stable. Thus, the long stability of the neutron source may be explained by the hypothesis that the neutrons are generated in the corona, which represents a high temperature, stable with respect to the sausage-mode cylindrical Z-pinch configuration, formed by the surrounding plasma after breaking the current channel of the initial Z pinch. For the conditions of the experiment of Filippov (1983), this model yields $T_c = 8$ keV, $n_c = 3 \times 10^{19}$ cm^{-3}, $R = 0.4$ mm, allowing one to explain the neutron yield observed during 50 ns. The density of the corona near the outer walls of the device positioned at $r = R_w \cong 30$ cm is estimated as 5×10^{13} cm^{-3}, i.e., about three orders of magnitude less than the initial plasma density. The line mass, μ_c, of the corona can be estimated as $\pi R^2 m_i n_c \ln(R_w/R)$; it is large compared to the mass of the plasma in the neck, due to the logarithmic factor.

This mechanism stabilizing the sausage instability mode can be effective not only in the plasma focus experiments. The important factor is that a sufficient plasma mass is required to form the corona capable of stabilizing the main pinch. Indeed, some indications are that Z pinches with larger mass-line density (including dynamic ones, the gas-puff Z pinches and exploded wires) are less likely to exhibit the sausage instability, e.g., see Pereira et al. (1984). In a recent paper by Filippov and Yan'kov (1988) the authors analyzed this mechanism of stabilization on the basis of the CGL hydrodynamics. They obtained the following inequality for the line mass of the plasma corona:

$$\frac{\mu_c}{\mu_p} < \frac{2m_i c}{e\sqrt{\mu_p}} = \overline{A}(4 \times 10^{-2} \ \mu\text{g} \cdot \text{cm}^{-1}/\mu_p)^{1/2}, \tag{4.69}$$

where μ_p is the line mass of the neck, the dense pinch core. If the inequality (4.69) holds, then the mass of the corona is insufficient to stabilize the sausage modes.

5

Rayleigh–Taylor Instability of a Plasma Accelerated by Magnetic Pressure

5.1 Rayleigh–Taylor instabilities of dynamic plasmas

There are two important specific features in the stability analysis of dynamic Z pinches and plasma liners. First, the unperturbed state is itself time dependent, so that the time dependence of the perturbations cannot be taken in the form $\exp(\Gamma \cdot t)$, and it is not obvious which parameters should be used to characterize the growth instead of the conventional exponential growth rate. Second, the RT instability (see review by Kull, 1991) now comes into play. Indeed, plasma is imploded (accelerated) by the pressure due to the azimuthal magnetic field in a dynamic Z pinch or liner. Thus, the plasma, in its own frame of reference, represents a "heavy fluid," which is supported by a "massless fluid" (i.e., by the magnetic field) in the gravitational field $\mathbf{g} = -\mathbf{a}$, where $\mathbf{a}$ is its acceleration, and this configuration inevitably exhibits the RT instability. In this respect, Z pinches are quite similar to other systems in which plasma or solid objects are electromagnetically accelerated, including rail guns and other electromagnetic guns, and also to laser fusion systems of plasma acceleration, where the high-temperature, low-density plasma corona acts as a "light fluid."

The evolution of perturbations imposed on a time-dependent unperturbed state is a separate problem to be studied, its most important aspect being the invariance of the results with respect to the arbitrarily chosen boundary conditions (see Chap. 6). The following method is widely used. Let τ be the characteristic time of the Z-pinch implosion. Then only the instability modes whose growth times are much smaller than τ, i.e., those with characteristic growth rates Γ satisfying $\Gamma\tau \gg 1$, are important. This allows us to represent the time dependence of perturbations, $F(t)$, sought for in the conventional form of a Wentzel–Kramer–Brillouin (WKB) integral:

$$F(t) = \exp\left[\int_0^t \Gamma(t')dt'\right] \equiv \exp\{n_{\mathrm{eff}}(t)\}, \qquad (5.1)$$

where $\Gamma(t)$ is the instantaneous growth rate of the corresponding instability mode, depending on the state of the system at the instant t. The integral in the exponent represents simply the number of e-foldings of the perturbation during the time t, and $n_{\text{eff}}(t)$ is the same as $\Gamma \cdot t$ in the case of a steady equilibrium state. Calculating the time derivative of $F(t)$, we find that the value of $\Gamma^2(t) F(t)$ is a good approximation to $\ddot{F}(t)$ if

$$\left| \frac{d \ln \Gamma(t)}{dt} \right| \ll \Gamma(t). \tag{5.2}$$

This is the usual condition of validity of the WKB approximation (see Landau and Lifshitz, 1977). Its applicability to the class of problems in question will be discussed in more detail in Chap. 6. Here, $\Gamma(t)$ is the instantaneous growth rate of the perturbation, regarded in the frame of reference of the accelerated plasma, the evolution of the unperturbed state being neglected.

A certain similarity between the RT instability and the flute instability modes of a Z pinch was discussed in Sec. 4.1.2. However, in the experiments with dynamic Z pinches and liners, when the plasma is rapidly accelerated, the corresponding acceleration, g, is, in most cases, much greater than the effective acceleration, g_{eff}, due to the curvature of the magnetic lines. Indeed, consider an annular cylindrical liner of radius R, thickness $d \ll R$, and average density $\bar{\rho}$, accelerated inwards by the azimuthal magnetic field B_φ. Its acceleration may be estimated as $g = B_\varphi^2 / 8\pi \bar{\rho} d$. Comparing this acceleration with the effective gravity, g_{eff}, given by (4.25), we find

$$g_{\text{eff}}/g = (16\pi n_i T_i / B_\varphi^2)(d/R). \tag{5.3}$$

The right-hand side of (5.3) is much smaller than unity, due not only to the smallness of d/R, but also because the implosion is effective only if the thermal pressure is much less than magnetic pressure during the acceleration phase. For instance, in the experiment of Felber et al. (1988a) the ratio g_{eff}/g is estimated as 0.01, for the observed value of $d/R = 0.2$.

Therefore, the RT instability modes develop together with other hydromagnetic instabilities of a Z pinch (Secs. 4.1.2-4), but their growth can be much faster when $g \gg g_{\text{eff}}$. In this case, they are the most dangerous instabilities that destroy the cylindrical symmetry of the pinch, which explains the importance of their consideration.

The growth rate of the RT instability of an interface separating two layers of uniform incompressible fluids of densities ρ_1 and ρ_2 is given by the classical expression

$$\Gamma^2 = Agk, \tag{5.4}$$

where k is the wavenumber of the perturbation that is assumed to be periodic in the plane of the interface and $A = (\rho_2 - \rho_1)/(\rho_2 + \rho_1)$ is the Atwood number. The gravitational acceleration g is directed from the layer of density ρ_2 (the upper) to the layer of density ρ_1 (the lower). If the right-hand side of (5.4) is negative ($\rho_2 < \rho_1$), then the dispersion relation (5.4) describes surface Rayleigh's waves propagating along the interface. If the density of the light fluid ρ_1 can be neglected in comparison with ρ_2 (evidently this is so if the light fluid imitates the magnetic field), we obtain the well-known expression of the growth rate of perturbations for a liquid supported by gravity:

$$\Gamma = \sqrt{|gk|}. \tag{5.5}$$

To what extent is this model applicable to the RT instability of a plasma accelerated by magnetic pressure? What is the effect of the factors not included in the simple analysis: cylindrical geometry of plasma acceleration, plasma compressibility, the magnetic field penetration into the plasma, finite and large Larmor radius effects, etc.? An understanding of these factors is crucial to any application of Z pinches.

5.2 Ideal MHD model: The Rayleigh–Taylor instability modes

Consider a plasma layer of width $2a$ being accelerated in $+z$ direction by the pressure of uniform magnetic field, which is in the (x, y) plane: $\mathbf{B} = B_x(z)\widehat{\mathbf{e}}_x + B_y(z)\widehat{\mathbf{e}}_y$. We will transform to a frame of reference moving with the plasma layer. In this reference frame there will be an effective gravitational field

$$\mathbf{g}(z, t) = g\widehat{\mathbf{e}}_z = -\frac{du_z}{dt}. \tag{5.6}$$

We assume that the plasma is in a stationary equilibrium in the co-moving frame of reference specified by conditions of (2.1) and (2.2) in a gravitational field, and that there is no equilibrium flow of the plasma in this frame of reference. Then, the equilibrium force equation will be

$$\frac{dP}{dz} + \frac{1}{8\pi}\frac{d}{dz}(B_x^2 + B_y^2) = \rho g, \tag{5.7}$$

and the equation of motion now differs from (3.44) by a gravity term on the right-hand side:

$$\rho\frac{d\mathbf{u}}{dt} = -\nabla P + \frac{1}{4\pi}(\nabla \times \mathbf{B}) \times \mathbf{B} + \rho g. \tag{5.8}$$

This configuration of a heavy fluid that is supported by a massless fluid is evidently RT unstable. The growth rates of the instability modes can be found as eigenvalues of a boundary-value problem in a way similar to (4.16). We assume a small perturbation of the layer and that all quantities are linearized about the equilibrium state. Introducing the displacement vector $\boldsymbol{\xi}(\mathbf{r}, t)$ defined by (4.9), and taking into account that for a plane geometry, equilibrium symmetry implies that the perturbations can be Fourier analyzed as

$$\boldsymbol{\xi}(\mathbf{r}, t) = \boldsymbol{\xi}(z)\exp(\Gamma t + ik_x x + ik_y y), \tag{5.9}$$

we can derive an equation for the z component of the displacement vector similar to (4.30), with (5.6) and (5.7) taken into account:

$$\frac{d}{dz}\left(K\frac{d\xi_z}{dz}\right) - L\xi_z = 0. \tag{5.10}$$

Here K is given by the same expression (4.31), and

$$L = \rho\Gamma^2 + F^2 + g\frac{d\rho}{dz} - (g^2 k^2 \rho^2/D)(\rho\Gamma^2 + F^2)$$

$$- \frac{d}{dz}\left[\frac{\rho^2\Gamma^2 g}{D}(\rho\Gamma^2 + F^2)\right], \tag{5.11}$$

$$D = \rho^2\Gamma^2 + k^2\left[\rho\Gamma^2\left(\gamma P + \frac{B^2}{4\pi}\right) + \gamma P F^2\right], \tag{5.12}$$

$$F = \frac{1}{4\pi}(k_x B_x + k_y B_y), \tag{5.13}$$

$$k^2 = k_x^2 + k_y^2, \quad B^2 = B_x^2 + B_y^2. \tag{5.14}$$

In the general case, the boundary conditions for the planar problem, if there is no skin current at the layer surfaces, are

$$\frac{d\xi_z}{dz} \pm k\xi_z = 0, \tag{5.15}$$

where $k > 0$, and the signs $+$ and $-$ refer to the upper and lower boundary, respectively. If the plasma occupies only the upper or only lower half-space, only one of the boundary conditions (5.15) is left, the other being replaced by the requirement that the perturbations vanish at infinity. There is an important particular case of flute perturbations, when (5.10) has a singular point at the boundary ($\rho = 0$, $F^2 = 0$, hence $K = 0$). Then, the corresponding boundary condition is replaced by the requirement that $\xi_z(z)$ is regular at the boundary surface.

Goedbloed (1984) pointed out that though (4.30) and (5.10) are similar, the difference between them is no less important: neither of them is reduced to the other by a simple transformation. In other words, there is no direct way to represent the curvature of magnetic lines as an effective gravity, and one should be careful when using the similarity between them for estimates. Thus, we cannot make use of the results of Sec. 4.1 here, and the main properties of the RT instability modes should be discussed separately. However, the spectral properties of the boundary-value problem under consideration are the same as discussed in Sec. 4.1.3. Here again, the instability growth rates for any given wavenumbers, k_x and k_y, form a discrete spectrum bounded from above. Below in this section, Γ denotes the maximum growth rate corresponding to the given wavenumbers, if not stated otherwise.

In widespread opinion, the RT spectrum (5.5) is characteristic of a discontinuity, a density jump at the boundary between a heavy incompressible fluid and the magnetic field supporting it against gravity, whereas a smooth boundary characterized by a finite-length scale, d, exhibits smaller growth rates, at least for $kd > 1$. To prove that this opinion is erroneous, Bychkov et al. (1990) consider pure flute perturbations. Supposing $\mathbf{B} = B_y(z)\widehat{\mathbf{e}}_y$, $\mathbf{k} = k\widehat{\mathbf{e}}$, $g < 0$, the plasma is supported by magnetic pressure from below. Then we have $F^2 \equiv 0$, and a direct substitution shows that

$$\xi_z(z) = \exp(-kz) \tag{5.16}$$

is an eigenfunction that identically satisfies both (5.10) and the boundary conditions. In particular, it is regular everywhere, since the plasma is bounded from below and corresponds to the eigenvalue (5.5). This remarkable solution exists for any unperturbed profiles $\rho(z)$, $P(z)$, $B_y(z)$, and even when some of the physical effects beyond the ideal MHD model are taken into account, like the presence of several ion species (see Rahman et al., 1985) or the Hall effect (see Sec. 4.4). This solution is independent of the value of the adiabatic exponent γ, which can also depend on z, due to the following property of this mode:

$$\nabla \cdot \boldsymbol{\xi} = 0, \, \nabla \times \boldsymbol{\xi} = 0. \tag{5.17}$$

The first of these relations indicates that a volume of any plasma particle remains unchanged by perturbation, so that the perturbation of pressure in it, which could be dependent on γ, is identically zero. Only two conditions are really important for the existence of this mode: uniformity of the gravitational field, $g(z) = $ constant, and the presence of the plasma–vacuum boundary at finite z.

Since the eigenfunction (5.16) does not vanish anywhere (see Bud'ko et al., 1989b, 1990a), the corresponding eigenvalue (5.5) corresponds to the fastest instability mode for any wavenumber k. This mode is called the global RT mode. Its remarkable properties were discovered in hydrodynamics by Mikaelian (1982) for an incompressible fluid, and by Inogamov (1985) for an arbitrary equation of state of the fluid.

In the short-wavelength limit, $k \rightarrow \infty$, the perturbation corresponding to the global RT mode is localized at a distance of about $1/k$ from the lower boundary of the plasma. The inverse is also true. Let δ be the scale length of the variation of the magnetic field near the lower boundary (that is, the skin depth). Then for $k\delta \gg 1$ on length scales of order of k^{-1}, the conditions of the existence of the global RT mode with growth rate (5.5) are satisfied (if the effect of magnetic shear can be neglected, see below), and this mode is the fastest one for a given k, as is the global mode. All the properties of the global RT modes, such as independence of the eigenvalue (5.5) and the eigenfunction (5.16) from the unperturbed profiles and γ displacement satisfying the relations (5.17), remain valid on this length scale. The short-wave perturbation modes with growth rates given by (5.5), which develop near the lower plasma boundary, are called local RT modes. Note that (5.16), in contrast with the other solutions of the same boundary-value problem, does not depend on the unperturbed profiles and has zero divergence. Therefore, the local RT modes reproduce locally the properties of the global RT modes. To illustrate this, we consider a simple example of a power-law profile near the boundary, i. e., near the boundary the density profile will be $r \propto z^s$, with $s = 1, 2, \ldots$. In this case, (5.10) can be transformed into the equation for a degenerate hypergeometric function. The corresponding eigenfunctions can be expressed via generalized Laguerre polynomials:

$$\xi_{zn}(z) = \exp(-kz)L_n^{(s-1)}(2kz), \quad n = 1, 2, \dots . \tag{5.18}$$

The corresponding spectrum of the boundary-value problem is

$$\Gamma_n^2 = gks/(s + 2n), \quad n = 1, 2, \dots . \tag{5.19}$$

All the eigenvalues and eigenfunctions, except the ones corresponding to the local RT mode ($n = 0$), explicitly depend on the parameter s characterizing the density profile near the boundary. Thus, the plasma boundary behaves just like a sharp boundary, as a density "step" producing the RT spectrum of growth rates of (5.5), no matter how smooth the density profile near the boundary.

In the same short-wavelength limit, we can use the WKB approximation to study the spectrum of local RT instability modes asymptotically for $n \gg 1$. Introducing the new variable, ζ, defined by

$$\frac{d}{d\zeta} = (\rho \Gamma^2 + F^2)\frac{d}{dz}, \tag{5.20}$$

(5.9) can be transformed to the following form

$$\frac{d^2\xi_z}{d\zeta^2} + q^2(\zeta)\xi_z = 0, \tag{5.21}$$

where

$$q^2(\zeta) = k^2\rho^2\left(-1 + \frac{g}{\Gamma^2\rho^2}\frac{d\rho}{d\zeta}\right). \tag{5.22}$$

The eigenvalues Γ^2 for (5.21) with the boundary conditions given above can be found using the well-known Bohr–Sommerfeld quantization rule (Landau and Lifshitz, 1977)

$$\int q\,d\zeta = k \int dz\sqrt{-1 + \frac{g}{\Gamma^2}\frac{d\ln\rho}{dz}} = \pi\left(n + \frac{1}{2}\right), \tag{5.23}$$

which is justified asymptotically for $n \gg 1$. In particular, for the power-law density profile considered above, we find from (5.23) the spectrum

$$\Gamma_n^2 = gks/(2n + 1), \tag{5.24}$$

which for $n \gg s$ coincides with that given by (5.19), as it should.

The RT instability belongs to the class of convective instabilities. Convection can take place locally in the plasma volume, if an interchange of adjacent plasma particles, the upper and lower, produces energy gain. The general condition of stability with respect to convection is

$$g\frac{dS}{dz} < 0, \tag{5.25}$$

where S is the effective entropy of the plasma in the magnetic field. To obtain a more convenient form of this stability criterion, it should be noted that a sufficient condition of local stability derived from the Energy Principle is

$$\lim_{\Gamma^2 \to 0} \{L\} > 0. \tag{5.26}$$

Since we are interested in short-wavelength modes, localized in the z direction, there are two ways of making the double limiting transition $k \to \infty$, $\Gamma^2 \to 0$. First, having assumed $F \equiv 0$ (that is, considering the interchange modes with $k \perp B$), we can pass to the limit $k \to \infty$, $\Gamma^2 \to 0$ and obtain the stability criterion in the form

$$\Gamma_{BV}^2 \equiv -g \frac{d\rho}{dz} + \frac{g^2 \rho^2}{\gamma P + B^2/4\pi} < 0, \tag{5.27}$$

where Γ_{BV} is the modified Brunt–Vaisala growth rate (Chen and Lykoudis, 1972). Second, assuming $\Gamma^2 \equiv 0$, we can let the angle between the vectors k and B approach $\pi/2$, passing in the limit $\Gamma^2 \to 0$. We obtain, then, the so-called quasi-interchange modes (Newcomb, 1961), for which the value of $(k \cdot B)$ is small but finite in the limit $k \to \infty$. The corresponding stability criterion is

$$\Gamma_B^2 \equiv -g \frac{d\rho}{dz} + \frac{g^2 \rho^2}{\gamma P} < 0, \tag{5.28}$$

where Γ_B is the Brunt–Vaisala growth rate, the quantity used in the physics of neutral atmospheres (Eckart, 1960). We see that the condition (5.28) is more demanding than (5.27), the former being the true condition of stability with respect to convection (Newcomb, 1961).

Note that both conditions (5.27) and (5.28) are particular forms of (5.25). Indeed, both denominators, $\gamma P + B^2/4\pi$ and γP, can be presented as $\rho(dP^*/d\rho)_{S'}$, where P^* is the pressure of the plasma with a frozen-in magnetic field. For the interchange modes, the plasma is compressed in the direction transverse to the magnetic field. The magnetic pressure contributes to the plasma elasticity $\gamma P + B^2/4\pi = \rho C_f^2$, where C_f is the fast magnetosonic velocity at which perturbations propagate in the transverse direction. On the contrary, displacements along the magnetic lines dominate for quasi-interchange modes, and the magnetic pressure is not perturbed: $P^* = P$, $\gamma P = \rho C_0^2$, where C_0 is the conventional speed of sound, which explains why the quasi-interchange modes are the most unstable. The corresponding perturbations, on the one hand, do not feel the stabilizing influence of the term F^2, like the interchange modes do, and on the other hand, they have an additional advantage of not perturbing the magnetic pressure. Thus, the similarity between the RT and hydromagnetic instability modes of a steady-state Z pinch corresponds to the interchange and the quasi-interchange modes, which are represented by the sausage and kink instability modes, respectively. In the long-wavelength limit, when the stabilizing effect of the F^2 term on the kink mode can be neglected, this mode may be the dominate one [cf. (4.38) and (4.39)].

If the condition (5.28) is not satisfied, then internal RT instability modes can develop in the volume of the plasma. Their growth rates are estimated as Γ_{BV} and Γ_B for interchange and quasi-interchange modes, respectively. If a density gradient is so high that the first terms on the right-hand sides of (5.27) and (5.28) dominate, i.e., if $h \equiv |d(\ln \rho)/dz|^{-1} \ll \{C_f^2/g, C_0^2/g\}$, then an estimate of the growth rates of the internal instability modes in the limit $kh \gg 1$ will be

$$\Gamma \equiv \sqrt{g/h}. \qquad (5.29)$$

This does not mean that an internal instability eigenmode is localized near a given value of z, like the local RT mode necessarily is. When the condition (5.28) is violated, internal eigenmodes exist and are localized in the plasma volume, not at the surface. For large wavenumbers, their localization depends on the shape of the unperturbed profiles. It should be noted that internal instability modes can develop, depending on the sign of Γ_B^2, whereas the local RT modes, whose growth rate is given by (5.5), always exist, and they dominate in the short-wavelength limit for the direction of k, satisfying the condition $k \cdot \mathbf{B} = 0$ at the plasma surface. If the direction of the magnetic field is not changed so that flute perturbations with $k \cdot \mathbf{B} = 0$ can develop in the whole plasma volume, then the global RT mode dominates for any value of k. In the ideal MHD model, the internal instability modes developing together with the RT modes correspond to higher values of the radial wave number (greater number of nodes), and hence to lower growth rates.

5.3 Ideal MHD model: Effects of plasma compressibility and magnetic shear

Consider the situation when neither global nor local RT instability modes described in the previous section can develop. In particular, this is the case when there is no lower plasma bound or if there is a skin current concentrated at the plasma–vacuum boundary. At the same time, other instability modes can exist that do not have the universal growth rate $\sqrt{gk}$. A simple example is the case of two half-spaces filled by incompressible fluids of densities ρ_1 and ρ_2 and separated by an interface, where a skin current flows so that the magnetic field is discontinuous. The dispersion relation can be easily derived from (5.10) or it can be obtained directly by considering the stability of the plasma–vacuum boundary or the interface between two plasma layers, with the boundary conditions at the interface resulting from the conditions of continuity of the total pressure, $P = nT + \mathbf{B}^2/8\pi$, and the vanishing of the normal field component on the perturbed boundary. Then, instead of (5.4) we obtain the dispersion relation in the form

$$\Gamma^2 = Agk - \frac{1}{4\pi}((k \cdot \mathbf{B_1})^2 + (k \cdot \mathbf{B_2})^2)(\rho_1 + \rho_2)^{-1}. \qquad (5.30)$$

For the case of a plasma–vacuum interface $\rho_1 = 0$, $A = 1$, (5.30) differs from (5.5) by the second term in the right-hand side, whose influence is stabilizing. The jump in the magnetic field at the interface, like a surface tension, stabilizes the short-wavelength unstable modes. When the magnetic fields $\mathbf{B}_1$ and $\mathbf{B}_2$ outside and inside are parallel, then the negative stabilizing term in brackets on the right-hand side (5.30) reduces to zero at $k \perp \mathbf{B}$, otherwise it is a stabilizing factor for any direction of the wave vector, k. For flute perturbation modes with respect to the vacuum magnetic field, B_1, supporting the plasma from below, the instability develops in the limited range of wave numbers

$$0 < k < k_m = |g|/V_A^2 \cos^2 \psi, \tag{5.31}$$

where $\cos^2 \psi = (k \cdot B_2)^2/k^2 B_2^2$. For a given direction of the magnetic field ($\cos \Psi = $ const) the maximum growth rate corresponds to $k = k_m/2$ and is equal to $|g|/2V_A \cos \Psi$.

When both global and local RT modes, which are independent of the equation of state, cannot develop, another factor capable of affecting the growth rates becomes important—the plasma compressibility. In contrast to magnetic shear, plasma compressibility is a destabilizing factor. The destabilizing role of finite compressibility is evident from (5.28). With finite g, the convective instability can develop even for $g(d\rho/dz) > 0$, when an incompressible fluid ($\gamma \to \infty$) would be stable.

Newcomb (1983) has used the MHD energy principle in a very elegant way to show that for an arbitrary magnetostatic equilibrium, instability growth rates are decreasing functions of γ, i.e., $d\Gamma/d\gamma \leq 0$, or, equivalently, a decreasing function of compressibility. Indeed, using the MHD energy principle, let us assume a purely discrete spectrum for simplicity and arrange the eigenvalues resulting from the solution of the Sturm–Liouville problem determining ξ to be ordered by magnitude according to $\Gamma_0^2 \leq \Gamma_1^2 \leq \Gamma_2^2 \ldots$. Note that γ is involved only in the perturbation equations of motion $[\Gamma_n^2(\gamma)]$, not in the equilibrium conditions. The potential energy associated with a small perturbation $\xi(\mathbf{r})$ can be expressed in the form

$$W = W_0 + \gamma \int P(\nabla \cdot \xi) d\mathbf{r}, \tag{5.32}$$

where W_0 is a quadratic functional of ξ and the trial function ξ are independent of γ. The lowest (most unstable) eigenvalue, Γ_0^2, is determined by the energy principle (in fact, by a variational principle) as $\min_\xi (W/K)$, where K is a second quadratic functional of ξ, and the minimum is taken over all ξ satisfying the boundary conditions. Thus, the ratio W/K is an increasing function of γ for any fixed ξ, and therefore, Γ_0^2, the minimum, also is an increasing function of γ. Newcomb (1983) has shown that the same result can be extended for the higher eigenvalues, also.

Note that the degree of compressibility plays no role in determining the threshold for stability. The result obtained implies only that the more compressible of two otherwise identical unstable systems has the larger growth rate.

The influence of compressibility on the RT instability mode in the case when no magnetic field was present has been studied by Bernstein and Book (1983) for a discontinuity between two exponentially stratified fluids; also, several analytical solutions have been presented by Book (1986). Let the interface separating the two fluids be located at $z = 0$, i.e., coinciding with the x-y plane, and the system be in a uniform gravitational acceleration directed in the negative z direction.

The simplest solvable configuration is the basic isothermal state, i.e., $T(z) \propto P(z)/\rho(z) \equiv C^2 = $ const, and hence, the sound speed, C, is independent of z. In this case the pressure and density profiles can be presented as

$$P(z) = P_0 \exp(-gz/C^2), \quad \text{or} \quad P, \rho \propto \exp(-z/h), \tag{5.33}$$

where the vertical characteristic length scales, h, above and below the discontinuity can differ. Substituting (5.33) into (5.28), we find that $\Gamma_B^2 \propto (1 - \gamma)$, so that the stability condition (5.28) is satisfied both above and below the discontinuity, the instability thus being localized near its surface. Equation (5.10) is simplified for the exponential profiles (5.33), being reduced to a linear differential equation with constant coefficients, and its solutions are exponential functions of z, too. Bernstein and Book (1983) have shown that the instability growth rate, Γ, increases with decreasing γ, i.e., that compressible fluids are more unstable.

The stability of a plane, stratified, perfectly conducting compressible plasma with exponential profiles (5.33) supported from below by a uniform magnetic field, $\mathbf{B} = B_0 \widehat{\mathbf{e}}_y$, in the lower half-space has been considered by Parks (1983). In this case, the current is concentrated in the skin layer, so that the field does not penetrate into plasma. For $k \perp \mathbf{B}$, the solution reduces to the familiar relation for interchange modes where the plasma moves through the field lines without distorting them. Thus the global RT mode (5.5, 5.16) dominates. But for $k \| \mathbf{B}$, the instability develops only in the range $kh \leq 0.5$, and

$$\frac{\Gamma^2}{gk} = \begin{cases} (\sqrt{2} - 1)(1 - 2kh), & \gamma = 1, \\ ((1 - 2kh)^2 + 1)^{1/2} - 1, & \gamma \to \infty. \end{cases} \tag{5.34}$$

For any k in this range $\Gamma(k, \gamma = 1) > \Gamma(k, \gamma \to \infty)$, so that the growth rates are decreasing functions of γ, or decreasing functions of compressibility. In this particular case, the increase of the growth rate due to plasma compressibility is about 25%.

In a sense, the RT instability can be compared to the instability of an inverted pendulum. If the pendulum is rigid, then all its parts move at the same angular velocity. However, if it consists of parts capable of falling separately, like a brick smokestack, it is destroyed more rapidly by a small perturbation. Note that this analogy (Bernstein and Book, 1983) is adequate only when the global RT mode cannot develop; otherwise, it is the rigid pendulum that falls faster.

The influence of compressibility and of a magnetic field on the localized internal modes and on global internal modes for isothermal plasma have been studied by Gratton et al. (1988). In the scope of ideal MHD, the magnetic field lines are frozen in the plasma. In addition to the exponential profiles (5.33), we have

$$B_x, B_y \propto \exp(-z/2h), \tag{5.35}$$

so that all the characteristic velocities, those of Alfvén velocity (V_A), fast (C_f) and slow (C_s) magnetosonic velocities, and the plasma temperature and the value of the parameter $\beta = 8\pi P/B^2$ are constant, i.e., independent of z. Substituting (5.33) and (5.35) into (5.28), and making use of the equilibrium condition (5.7), we find that $\Gamma_B^2 \propto (1 + 1/\beta - \gamma)$. In contrast to the case of pure gas dynamics, the presence of a magnetic field makes possible the development of convective instability in the exponentially stratified plasma, if

$$\beta < \beta_c = 1/(\gamma - 1). \tag{5.36}$$

The maximum growth rate corresponding to the quasi-interchange internal modes, i.e., to the short-wavelength limit $k \to \infty$, $(\mathbf{k} \cdot \mathbf{B}) = \text{const}$, is estimated as

$$\Gamma \cong \frac{|g|}{V_A} \left[1 - \left(\frac{\beta \gamma}{\beta + 1} \right)^{1/2} \right]. \tag{5.37}$$

Surface instability modes that are localized at the plasma-vacuum interface have been studied by Gonzalez and Gratton (1990). A finite compressibility of the plasma was shown both to expand the range of wavenumbers, where the unstable modes exist, and to increase the growth rates of unstable modes, (5.37) being an example. When the surface instability modes exist, their growth rates are higher than those of the internal modes with the same wavenumber, just as in the case of local RT modes in the absence of a discontinuity representing the skin layer.

Consider now the influence of magnetic shear on the RT instability modes. It is evidently stabilizing. For instance, in the limiting case of magnetic-field discontinuity at the plasma surface, no short-wavelength instability modes are localized at this surface; all perturbations are stabilized. Stabilization of internal modes by magnetic shear is also possible in the case when the direction of the magnetic field does not jump but changes smoothly, just as in the case of internal instability modes of a steady-state Z pinch. The modified Suydam's criterion obtained for this case by Gratton et al. (1988) takes the form

$$\frac{B_x^2 B_y^2}{16\pi\rho B^2} \left(\frac{d \ln \mu}{dz} \right)^2 - \Gamma_B^2 > 0, \tag{5.38}$$

where the pitch number in the 1-D geometry under consideration is $\mu \equiv B_y/B_x$, and Γ_B^2 is determined by (5.28).

Consider a further generalization of (5.38) for cylindrical geometry of an imploding Z pinch. The stability analysis can be done in the same way as in Sec. 4.1.2. To minimize the energy functional, we need to solve a Euler–Lagrange equation similar to (4.30) with the boundary conditions: ξ_r is finite at $r = 0$ and $\xi_r \to 0$ at the outer boundary of the pinch or at infinity. The equilibrium equation including the inertial force at each instant of time is

$$\frac{dP}{dr} + \frac{1}{4\pi} \left[B_z \frac{dB_z}{dr} + \frac{B_\varphi}{r} \frac{d}{dr} (r B_\varphi) \right] = \rho g, \tag{5.39}$$

where

$$g(r, t) = - \left(\frac{\partial u_r}{\partial t} + u_r \frac{\partial u_r}{\partial r} \right) \tag{5.40}$$

is the local plasma acceleration. The growth rates are eigenvalues of the same boundary-value problem for (4.30), where K is given by (4.31), and

$$
\begin{aligned}
L = \rho \Gamma^2 + F^2 &+ \frac{B_\varphi}{2\pi} \frac{d}{dr} \left(\frac{B_\varphi}{r} \right) + g \frac{d\rho}{dr} \\
&- \frac{1}{D} \left[\rho^2 g^2 \left(k^2 + \frac{m^2}{r^2} \right) (\rho \Gamma^2 + F^2) - \frac{\rho^2 \Gamma^2 g}{\pi r} k B_\varphi \left(k B_\varphi - \frac{m}{r} B_z \right) \right] \\
&+ r \frac{d}{dr} \left\{ \frac{k B_\varphi}{2\pi r^2 D} \left(k B_\varphi = \frac{m}{r} B_z \right) \left[\rho \Gamma^2 \left(\frac{B^2}{4\pi} + \gamma P \right) + \gamma P F^2 \right] \right. \\
&\left. + \frac{\rho^2 \Gamma^2 g F^2}{r D} \right\},
\end{aligned}
\tag{5.41}
$$

where F^2, D, and B^2 are determined by (4.32). For a steady-state equilibrium ($g = 0$), (5.41) is reduced to (4.31) and the case of plane geometry (5.10) and (5.11) is recovered by the substitution $r \to z$, $\varphi \to x$, $1/r \to 0$.

If there is a singular point where $F^2 = \frac{1}{4\pi} \left(k B_z + \frac{m}{r} B_\varphi \right)^2$ vanishes, the modified Suydam's criterion takes the following form

$$
\frac{B_\varphi^2}{4B^2} \left[\frac{r B_z^2}{4\pi} \left(\frac{d \ln \mu}{dr} \right)^2 + \frac{dP}{dr} \right] - \rho g \frac{2 B_\varphi^2}{B^2} - r \left(\frac{\rho^2 g^2}{\gamma P} + g \frac{d\rho}{dr} \right) > 0.
\tag{5.42}
$$

Consider the implosion of a thin shell whose thickness is δ, which is much less than its radius, R. Let us suppose that there is an axial magnetic field inside and azimuthal field outside the shell, but the magnetic field does not penetrate the plasma. Then the solutions of the Euler–Lagrange equation are $\xi_r = I_m'(kr)$ in the inner region, and $\xi_r = r^2 K_m'(kr)$ in the outer region, respectively. Substituting these solutions into (5.42) the stability condition can be obtained in the following form:

$$
\begin{aligned}
k^2 R^2 B_z^2 \frac{I_m(kR)}{I_m'(kR)} &- m^2 B_\varphi^2 \frac{K_m(kR)}{K_m'(kR)} + \frac{1}{2} k R (B_z^2 - 3 B_\varphi^2) \\
&- k R \frac{R}{2\gamma \delta} \left| \ln(B_z^2 / B_\varphi^2)(B_z^2 - B_\varphi^2) \right| > 0.
\end{aligned}
\tag{5.43}
$$

The first two terms on the left-hand side of (5.43) are always positive. The third term is negative, if $B_z < \sqrt{3} B_\varphi$; in particular, the third term is negative in the equilibrium for $B_z = B_\varphi$. This term is responsible for the kink instability that is always present for sufficiently small k, but can be stabilized if k is limited from below. The last term represents the contribution from the RT instability (it vanishes in equilibrium, when $g = 0$). However, for the imploding shell, when $g \neq 0$, its contribution depends most of all on the value of parameter R/δ. The thinner the shell (the smaller the value of δ/R), the more unstable is the implosion in the vicinity of an equilibrium point. We see that a sufficiently high magnetic shear can suppress all the internal modes, i.e., stabilize convection in the plasma. However, the surface instability modes discussed below are more dangerous.

5.4 Effect of magnetic shear

We shall show that in the presence of magnetic shear, the local RT modescan be suppressed for any density profile near the plasma boundary, demonstrating the existence of a "window of stability" for an accelerated plasma-pinch column (Bud'ko et al., 1989b, 1990a). Let us choose $z = 0$ to be the lower boundary of the plasma, $\mathbf{B} = B_1 \widehat{\mathbf{e}}_y$ in the lower half-space, $\mathbf{k} = k \widehat{\mathbf{e}}_x$, $\rho(0) = 0$, and the magnetic field to be continuous at $z = 0$; then we find $F^2 = k^2 B_x^2 / 4\pi$. If near $z = 0$, one can neglect the influence of magnetic shear represented by the terms F^2 in expressions (4.31) and (4.32) for K and L, respectively. Then, for $\mathbf{k} \perp \mathbf{B}$ in the limit $k \to \infty$, the local RT modes dominate, the growth rates being given by (5.5). Integrating the equilibrium equation (5.7), we find

$$\frac{B_x^2}{8\pi} + \frac{1}{8\pi}(B_y^2 - B_1^2) + P = \int_0^z dz' g\rho(z'). \tag{5.44}$$

In the absence of density and current discontinuities at the surface, all the terms on the left-hand side of (5.44) are of the same order of magnitude, so that B_x^2 decreases faster than ρ with decreasing distance from the boundary. A similar behavior is expected from $F^2 \propto k^2 B_x^2$. But the expression for F^2 contains a large parameter, k^2, so that the limiting transition $k \to \infty$ is not trivial. Let us, for simplicity, confine ourselves to the case of power-law density profiles, $\rho = \rho_0(z/h)^s$, near the boundary (Bud'ko et al., 1989b, 1990a). Then, according to (5.44) we obtain

$$F^2 = (k^2 B_{\parallel}^2 / 4\pi)(z/h)^{s+1}, \tag{5.45}$$

where h is the characteristic length of the nonuniformity in the z direction, and ρ_0 and $B_{\parallel}^2$ are the characteristic values of the plasma density and $(\mathbf{k} \cdot \mathbf{B})^2 / k^2$, respectively. The influence of magnetic shear is characterized by the dimensionless parameter

$$\chi^2 = B_{\parallel}^2 / 4\pi \rho_0 sh|g|. \tag{5.46}$$

In the limit $\chi \to 0$, the growth rate of local RT modes is given by (5.5), as it should be. In the opposite limiting case $\chi \gg 1$, the surface RT instability modes are stabilized by the magnetic shear. A rough estimate of the growth rate of the dominant eigenmode

$$\Gamma \cong |gk|^{1/2} 2^{7/4} (s\chi)^{1/2} \exp(-\pi\chi/2) \tag{5.47}$$

shows that the value of Γ can be made arbitrarily small, if the value of χ is sufficiently high. For instance, consider a thin shell confining an axial magnetic field, B_z, inside and being accelerated by the pressure of an azimuthal magnetic field, B_φ, outside. If the magnetic fields below (B_1) and above (B_2) a plasma layer of finite thickness, d, are perpendicular to each other, then the layer acceleration and parameter χ can be estimated as

$$g \cong (B_1^2 - B_2^2)/8\pi\rho_0 d, \quad \chi^2 \cong \min(B_1^2, B_2^2)/|B_1^2 - B_2^2|. \tag{5.48}$$

Note the difference between the cases of an infinitely thin shell and plasma layer of finite thickness in the dependence of the growth rates on the magnetic shear parameter. In the former case

$$\Gamma \cong \frac{|gk|^{1/2}}{\sqrt{2}\chi} = \left(\frac{|g|}{2g_{\max}}\right)^{1/2}, \tag{5.49}$$

where χ^2 is given by (5.48), and $g_{\max} = B_1^2/8\pi\rho_0 d$ is the maximum acceleration that could by achieved with the magnetic pressure $B_1^2/8\pi$ and the given mass per unit area of the shell, provided that there is no magnetic field from the other side of the shell. Thus, the growth rate in this approximation decreases with increasing χ, almost to the same extent as the acceleration, g. A given increment of the shell velocity $\Delta u = |g|\Delta t$ corresponds to the number of e-foldings of the perturbations, $n_{\mathrm{eff}} = \Gamma\Delta t \cong \Delta u\sqrt{k/2g_{\max}}$, which is only $\sqrt{2}$ times smaller than the corresponding value calculated for the case $B_2 = 0$, $|g| = g_{\max}$, when there is no stabilization by magnetic shear. On the contrary, (5.47) demonstrates that for a given increment, Δu, the respective value of $\Gamma\Delta t$ can be made arbitrarily small. Thus, the stabilization of a plasma accelerated by magnetic pressure due to magnetic shear allows one to suppress the dominating RT instability modes, and hence, to increase the degree of stable radial compression, in particular for a Z pinch or a liner stabilized by an axial magnetic field. Note that the mechanism of stabilization by magnetic shear is effective only in a plasma layer of finite thickness. The thinner is the layer, the shorter the wavelength of the perturbations that cannot be stabilized, and the faster the local RT instability modes develop.

5.5 Dissipative effects

The classical expression (5.5) for the growth rate Γ of the RT instability predicts its unlimited increase with increasing wavenumbers. However, for sufficiently short wavelengths, the conditions of applicability of the ideal MHD model are violated, and dissipative effects become important. It is natural to expect that dissipation tends to smooth out the short-wavelength perturbations; therefore, their net effect is stabilizing. Dissipation should be essential for $\Gamma \cong k^2 D$, where D is a corresponding coefficient whose dimensions are the diffusion coefficients (Sec. 3.3). Thus, we have the following estimates for the maximum wavenumber and the corresponding growth rate:

$$k_{\max} = |g|^{1/3}D^{-2/3}, \quad \Gamma_{\max} = |g|^{2/3}D^{-1/3}. \tag{5.50}$$

This estimate is basically quite realistic. Let us suppose viscosity to be the dominating instability mechanism, whose effect on the RT instability modes has been studied in the most detail (Gonzalez et al., 1989). The growth rates versus wave number presented in Fig. 5.1 (where the growth rate is designated by W) are calculated by Gonzalez et al. (1989) for the RT modes, which are flute with respect to the vacuum magnetic field in the lower half-space, i.e., for the conditions corresponding to $\rho_1 = 0$, $(k \cdot B_1) = 0$, and $A = 1$ in (5.30). The plasma is supposed to be incompressible and viscous. The wavenumber is expressed in units of k_m defined by (5.31); viscosity is characterized by a dimensionless parameter $S = k_0/k_m$. The unstable modes thus exist for $0 < k < 1/S$, the maximum growth rate corresponds to $k = 1/2S$ and is equal to $1/2\sqrt{S}$. Figure 5.1(a) is plotted for isotropic collisional ion viscosity $D = 2\eta_0^i/\rho$, where the viscosity coefficient is $\eta_0^i = 0.96 n_i T_i \tau_i$ (see Braginskii, 1963). We see that the effect of viscosity consists in a shift of the value of k, which corresponds to the maximum growth rate shifting to longer wavelengths and a decrease in the maximum growth rate. For the global RT mode, $\cos \psi = 0$, $S = 0$, and the growth rate is not limited from above, if viscosity is not taken into account, whereas, viscosity makes the maximum growth rate $0.58\Gamma_0$. The range of wavelengths where the unstable modes exist is not changed.

The stabilizing effect of parallel ion viscosity in a fully magnetized plasma ($k \| B$, $\Omega_i \tau_i \gg 1$) is qualitatively the same, though a little weaker [Fig. 5.1(b)]. In the case $k \perp B_2$, i.e., when $B_1 \| B_2$ and the ideal MHD model predicts the existence of the global RT mode, the dominant role is played by the ion gyroviscosity, i.e., by a finite Larmor radius effect. The corresponding value of D is $2\eta_4^i/\rho$, where the viscosity coefficient $\eta_4^i = n_i T_i / \Omega_i$ (see Braginskii, 1963), is independent of the collision frequency and can be obtained for a collisionless plasma with the aid of a guiding center motion treatment. FLR effects are known to provide stabilization

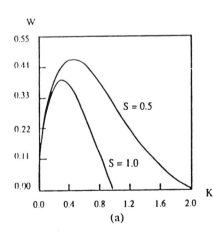
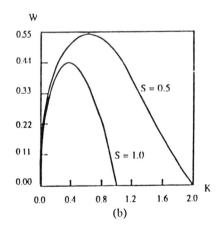

FIGURE 5.1. Normalized RT instability growth rate (W) versus the normalized wavenumber in a viscous plasma for the cases of (a) isotropic collisional ion viscosity and (b) gyroviscosity (Gonzalez et al., 1989).

for sufficiently short wavelengths (Rosenbluth, 1965; Gonzalez et al., 1989):

$$\Gamma = \mathrm{Re}(\widehat{\Gamma}) = \sqrt{|gk|}\left[\left(1 - \frac{1}{4}(k/k_0)^3\right)\right]^{1/2}, \qquad (5.51)$$

Therefore, the RT instability develops for $0 < k < 2^{2/3}k_0$, where the maximum growth rate is achieved for $k = k_0$ and is equal to $(\sqrt{3}/2) \cdot \sqrt{gk_0}$. It is interesting that if one takes into account the corrections due to collisional viscosity [they are small as $(\Omega_i \tau_i)^{-1}$] in addition to the FLR effects, their influence turns out to be destabilizing. The instability is then found for all wavenumbers, just as in the case of isotropic viscosity, though its growth rate is small (Gonzalez et al., 1989).

Hence, the effect of viscosity on the RT instability modes is mainly stabilizing, though sometimes viscous dissipative mechanisms can play a destabilizing role. The same seems to be true for other dissipative mechanisms. In particular, parallel electron heat conductivity produces an effective decrease of the adiabatic exponent γ, making γ_{eff} closer to unity, if the value of $k_{\parallel}^2$ is not too small [see (4.52)]. As was discussed in Sec. 5.3, the increase of plasma compressibility results in increased growth rates. This applies also to the resistive dissipation mechanism which can itself drive instabilities (see Sec. 4.1.5). The so-called resistive g-modes of instability have been studied mostly in the context of magnetic confinement fusion research (Mikhailovskii, 1975, 1977).

5.6 Large Larmor-radius effects

Standard ideal MHD theory is applicable when $k\rho_i \ll 1$ (where k is the wavenumber and ρ_i is the ion gyroradius), while FLR MHD theory is used when $k\rho_i \lesssim 1$. However the RT instability in the parameter regime $\rho_i \gg L$ and $\omega \gg \Omega_i$, but $\rho_e/L \ll 1$ and $\omega/\Omega_e \ll 1$, where Ω_i and Ω_e are the ion and electron gyrofrequency, respectively, and ω and L are the frequency and length scales of interest, respectively, has recently attracted the attention of many researchers. A number of interesting situations have been observed, in both the laboratory and space plasmas, where a plasma motion and/or its expansion is caused by the deceleration of a plasma expanding in a strong magnetic field (e.g., see Bernhardt et al., 1987; Ripin et al., 1987). Finite- or large-Larmor-radius (LLR) effects under certain conditions can significantly influence the stability of fiber-initiated HDZPs, imploding gas-puff Z pinches, and θ pinches.

The stabilizing effect of FLR, mostly due to gyroviscosity, was discussed above (Secs. 4.4 and 5.5). The development of the RT instability is different in two regimes: magnetized and unmagnetized ion systems. The distinction between these two regimes is the time scale of the instability (Hassam and Huba, 1987, 1988; Huba et al., 1987, 1989): for $\Gamma \ll \Omega_i$, the ions can be considered to be magnetized, and for $\Gamma \gg \Omega_i$, the ions can be considered to be unmagnetized. It was shown that a new physical picture of the development of RT instability arises. In particular, when its growth rate is $\Gamma \gg \Omega_i$, the RT instability develops much faster and has a dramatically different nonlinear behavior than the conventional RT instability. In this case, the major difference between the modified-fluid model and a collisionless LLR plasma is the inclusion of the Hall term. A fluid model of a collisionless LLR plasma with conventional isotropic/isothermal Hall MHD, and induction equation in the form (4.54), has been used by Hassam and Huba (1987, 1988, and 1989), and Huba et al. (1989).

For an interface separating two plasma layers of different density in a gravitational field in the limit of cold plasma, when the variation of magnetic field near the boundary can be neglected, the Hall MHD model yields

$$\widehat{\Gamma}^2 = |gk|(A + i\widehat{\Gamma}/\Omega_i) \cdot (1 + iA\widehat{\Gamma}_i)^{-1}, \tag{5.52}$$

where $\widehat{\Gamma}$ is a complex eigenvalue, $\Gamma = \mathrm{Re}(\widehat{\Gamma})$ is the growth rate, A is the Atwood number, and Ω_i is a characteristic value of ion gyrofrequency near the interface (Huba et al., 1989). In the MHD limit, for $|\widehat{\Gamma}| \ll \Omega_i$, (5.52) is reduced to (5.4), but in the opposite limiting case, $|\widehat{\Gamma}| \gg \Omega_i$, we obtain

$$\Gamma^2 = |gk|/A, \tag{5.53}$$

which is quite different from the growth rate in the magnetized-ion limit, given by (5.5). For short wavelengths, i.e., for k^{-1} small compared to the characteristic length scale $|g|/\Omega_i^2$, the result is surprising: the smaller the Atwood number, the greater is the growth rate. On the contrary, in the limit, a plasma–vacuum boundary $(A \to 1)$ (5.5) is reproduced for any value of k. Of course, any real interface has a finite width, h, and it can be regarded as a discontinuity only for sufficiently long perturbation wavelengths, i.e., $kh \ll 1$. In the opposite extreme of local RT instability modes, $kh \gg 1$, Huba et al. (1987) obtained the following estimate of the growth rate, instead of expression (5.29):

$$\Gamma \cong k|gh|^{1/2}. \tag{5.54}$$

Thus, the growth rate of the local RT modes in the LLR limit is found to be $kh \gg 1$ times greater than the MHD estimate (5.29). The smoother is the density profile, the greater the growth rate, in contrast with the conventional MHD result (5.29). A qualitative explanation of this, given by Huba et al. (1989), is based on an almost curl-free behavior of dominant instability modes, the main displacement of plasma particles being along the wave vector, k. For such perturbations, the higher the density gradient, the greater plasma mass is involved in the motion, hence the lower is the growth rate. Of course, (5.54) does not predict an unlimited growth of Γ with increasing h. When h is of order of $|g|/\Omega_i^2$, LLR effects will become insignificant.

The growth rates of RT instability modes were calculated numerically by Huba et al. (1989) to compare the effects corresponding to the length scales $|g|/\Omega_i^2$ and h. The density profiles were chosen of the form

$$\rho(z) = \rho_1\{[1 + A\tanh(z/H)]/(1 - A)\}. \tag{5.55}$$

It was shown that in the limit when the parameter $g_0 \equiv |g|/h\Omega_i^2 \ll 1$ LLR, the effects are negligible; in the limits $kh \ll 1$ and $kh \gg 1$, the growth rates are given by (5.5) and (5.29), respectively. In the opposite limiting case, $g_0 \gg 1$, the estimates (5.53) and (5.54) apply respectively for $kh \ll 1$ and $kh \gg 1$. Figure 5.2(a) presents the spectra of Γ versus kh, calculated by Huba et al. (1989) for $n = 0$ radial RT instability modes of smooth density profiles (5.55), with $A = 0.818$, $g_0 = 10^{-2}$, and $g_0 = 10^2$. The dependence of Γ on the number n of radial eigenmode for the same conditions is shown in Fig. 5.2(b). For $g_0 = 10^{-2}$, we see the conventional behavior characteristic of the Sturm–Liouville problem: the $n = 0$ mode dominates, the growth rates of the other modes are smaller; whereas for $g_0 = 10^2$, the mode with $n = 2$ dominates, its growth rate being greater than that shown in Fig. 5.2(a). The number of the dominating instability modes grows with increasing k.

It was shown above that the global RT instability mode of a plasma–vacuum boundary is specific for the ideal MHD model. Comparing (4.54) and (5.29), we see that the Hall term on the right-hand side of (4.54) vanishes for perturbations corresponding to the global RT mode. Thus, this mode with growth rate (5.5) develops in a LLR plasma, too. However, it is no longer the dominant instability mode for all values of k. To illustrate this (Velikovich, 1991), let us consider smooth density profiles of a plasma that occupies the upper half-space, with $|g|/\Omega_i^2$ as the characteristic length scale. We assume that the unperturbed magnetic pressure is large, compared to both the plasma thermal pressure and to the hydrostatic pressure $\bar{n}m_i gh$. Then, the magnetic field is essentially constant, and the perturbation equation can be reduced to the following second-order equation:

$$\frac{d}{dz}\left(n_0 \frac{dV}{dz}\right) - k^2 \left(n_0 + \frac{1}{k} D \frac{dn_0}{dz}\right) V = 0, \qquad (5.56)$$

where

$$V = u_z - i(\omega/\Omega_i)u_x \quad \text{and} \quad D = \left(\frac{\omega}{\Omega} + \frac{gk}{\omega}\right)\left(\omega + \frac{gk}{\Omega_i}\right)^{-1}. \qquad (5.57)$$

The solution of the boundary-value problem (5.56), corresponding to a power-law density profile $\rho = \bar{\rho}(z/L)^s$, $s > 0$, is

$$V(z) = \exp(-kz)L_n^{(s-1)}(2kz), \qquad n = 0, 1, 2, \ldots \qquad (5.58)$$

where $L_n^\alpha(x)$ is the generalized Laguerre polynomial. The spectra $\Gamma(k) = \text{Im}(\omega)$ for $n = 1$–10 and $s = 1$ (Velikovich, 1991) are shown in Fig. 5.3. The number $n = 0$ corresponds to the global RT mode. For modes with $n \geq 2$, the dispersion curves $\Gamma(k)$ consist of two branches separated by a "band of stability." Within this band the eigenmodes are stable, running gravitational drift waves. There is no effective stabilization compared to the MHD case because the global RT mode always exists where some other modes are stabilized [in contrast to the case studied by Coppins et al. (1984) where the plasma was bounded by conducting walls]. Thus, the net effect of the Hall term is always destabilizing. A finite number of instability modes exists for any $k > \Omega_i^2/|g|$, with the maximum growth rate corresponding

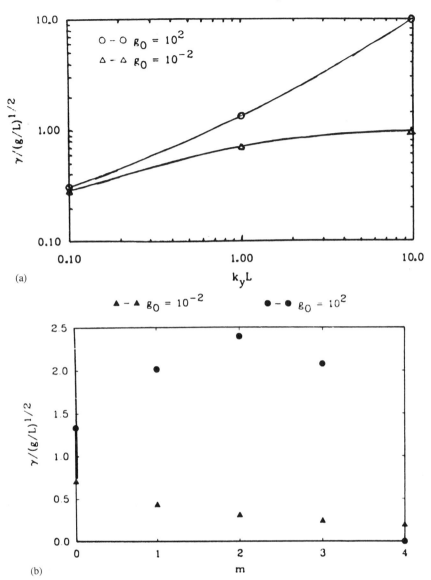

FIGURE 5.2. Normalized RT instability growth rates versus (a) the wavenumber calculated by Huba et al. (1989) for two values of the acceleration parameter, g_0; (b) the same versus the number m of the radial eigenmode.

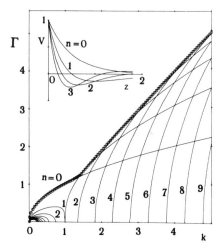

FIGURE 5.3. Normalized RT instability growth rates of the first 10 unstable eigenmodes versus the wavenumber for a plasma–vacuum interface in the Hall MHD model. Inset: profiles of the first four displacement eigenfunctions.

to a mode number $n > 1$ (cf. Huba et al., 1989). A mutual upper boundary of the dispersion curves $\Gamma(k)$, corresponding to various instability modes is shown in Fig. 5.3:

$$\Gamma_{\max}(k) = \begin{cases} |gk|^{1/2}, & k \leq \Omega_i^2/g, \\ \Phi(k), & k \leq \Omega_i^2/g, \end{cases} \tag{5.59}$$

where $\Phi(k)$ is the envelope curve of the family of dispersion curves, branching off the curve (5.5) at $k = \Omega_i^2/|g|$, the second derivative of the combined curve having a discontinuity at this point. The shape of $\Gamma(k)$ does not depend on the value of s in a power-law density profile. The smaller the value of s, the more rarely the successive dispersion curves lie in the plane (k, Γ). In particular, in the sharp boundary limit $s \to 0$, only the global RT mode remains for any finite k, the next $n = 1$ mode appearing only for $k \cong \Omega_i^2/2|g|s \to \infty$. On the other hand, the larger the value of s, the more densely the successive dispersion curves cover the same area of the plane (k, Γ). The smoother the density profile is near the boundary, the better is the growth rate of the dominant instability mode estimated by (5.59). In the limit $k \ll \Omega_i^2/|g|$, we have

$$\Gamma_{\max}(k) \cong |gk/\Omega_i|. \tag{5.60}$$

This result follows from (5.54) by substituting there the length scale $h = |g|/\Omega_i^2$. Hence, the growth rate of the dominant flute-instability mode of a plasma–vacuum boundary in the LLR limit (i.e., when the destabilizing effect of the Hall term is taken into account) can be estimated as

$$\Gamma \cong \max\{|gk|^2, |gk/\Omega_i|\}, \tag{5.61}$$

which is correct for the whole range of wavenumbers, provided that $\beta \ll 1$. For high values of k, this instability can be stabilized by dissipative mechanisms, either collisional or collisionless.

5.7 Nonlinear evolution of the Rayleigh–Taylor instability

In the linear phase of the RT instability development, when the initial perturbations are small compared to the wavelength, the small-amplitude perturbations grow exponentially with time. This early stage in the growth of the instability can be analyzed using the linearized form of the dynamical equations for the plasma. In the nonlinear phase of the instability, when the amplitude of the initial perturbation grows to a value of order of the inverse wavenumber, k^{-1}, substantial deviations from the linear theory are observed. This next stage is characterized by the development of structure on the spikes and interaction among the bubbles; further growth implies ejection of a significant mass from the bubble region to the spike. This can continue until the mass per unit area of the bubble becomes very small. The phenomena can originate from several sources: nonlinear interaction among initial perturbations of different frequency; bubble amalgamation, when larger bubbles absorb smaller ones, with the result that large bubbles grow larger and move faster; finally, from dramatic changes in the fluid properties, the development of nonlinear stage of the RT instability result in a regime of turbulence, mixing of the two fluids etc.

There is a considerable literature on the nonlinear evolution of the RT instability (for instance, see the book by Chandrasekhar, 1961; the reviews by Sharp, 1984 and Kull, 1991; and references therein). Understanding the RT instability, which develops at the interface where a low density fluid pushes and accelerates a higher density fluid, is important to the design, analysis, and ultimate performance of any application of Z pinches. The linear regime can be fully described with linear analysis based on 2-D eigenfunctions. Full 3-D simulations are very difficult; they are limited by computer time and size, if other physical processes are included (Manheimer et al., 1984; Orszag, 1984; Sakagami and Nishihara, 1990; Town and Bell, 1991; Dahlburg et al. 1993; Dunning and Haan, 1995). Two-dimensional numerical MHD simulations are expected to describe the behavior of such a system reasonably well (Deeney et al., 1993; Peterson et al., 1996; Sheehey et al., 1992; Hammer et al., 1996; Maxon et al., 1996). Unlike the situation usually encountered in fluid treatments of RT instabilities, the plasma represents a medium of finite thickness, and thus bubbles can eventually burst through into the inner vacuum region. When this occurs, magnetic flux may be carried into the central vacuum by very low density plasma ahead of the main plasma, decreasing the effective acceleration of the plasma. The magnetic field also diffuses into a resistive plasma, spreading the acceleration over a distance characterized by the magnetic scale length. Current flow results in Joule heating of the plasma, which changes its resistivity and, in turn, the magnetic scale length.

The observation of bright spots following the stagnation of Z pinches involving gas puffs, wire arrays, and foils has been reported in the literature (Nash et al., 1990a; Burkhalter et al., 1977; Stallings et al., 1976; Fraenkel and Schwob, 1972). The statement, which usually accompanies this observation is that "the $m = 0$ sausage instability is associated with the bright spot." The sausage instability is identified by the outer boundary "necking off" and the RT instability by its characteristic bubble-and-spikeshape. A reasonable assumption is that the bright spots are formed by the RT instability, since substantial acceleration takes place during the implosion. The pinch begins to radiate as the bubble mass first reaches the axis, and it continues to radiate while the mass that is entrained within the spikes and within unperturbed parts of the shell also arrives on axis. This model relates the time at which the bubble arrives on axis to an initial wavelength and amplitude of a single mode of the RT. On the contrary, the 1-D calculations typically show collapse of the entire mass of the pinch to a radius of tens of microns and x-ray bursts of extreme power and short duration (less than 1 ns), while experiments show pinch radii of order several hundreds of microns, much less power, and much longer pulse time (of order 10–30 ns). The 2-D simulations for most loads show strong RT growth, causing the formation of bubbles and spikes from initial density perturbations of order 10^{-2}. The spikes contain most of the mass and become extended over a significant fraction of the initial radius. The radially distributed mass stagnates and converts kinetic energy to x rays over a much longer period than observed in the 1-D simulations. Direct evidence for the growth of RT modes on SATURN can be found in experiments on gas-puff implosion (Spielman et al., 1985b; Wong et al., 1995). The development of the nonlinear phase of the hydromagnetic RT instability may be the dominant mechanism, leading to a lengthening of the radiation pulse over that which would be expected from 1-D simulations.

In the nonlinear phase of the instability, when the amplitude of the perturbed mode is of the order of its wavelength, further growth implies ejection of significant mass from the bubble region of the instability into the spike. This can continue until the mass per unit area of the bubble becomes very small (Birkhoff and Carter, 1957; Garabedian, 1957). In particular, thin-shelled bubbles acted upon by a constant force per unit area will run away from the spike, reaching the axis early when this mass per unit area becomes sufficiently small. Hussey et al. (1995) have developed a simple heuristic model for the early nonlinear stage of the RT instability in thin cylindrical-shell Z-pinch implosions, which, based on the model for the nonlinear RT, was first proposed by Baker and Freeman (1981).

In reality the process is complicated, since the implosion feeds back into the circuit through the term for time-varying inductance and by distortion of the field–plasma interface, and enhances field penetration into the plasma. A relatively small amount of azimuthal asymmetry may explain the reduced-stagnation density. Alternative possibilities include deviation from the classical resistivity, which can affect both the linear growth rates and nonlinear saturation. Wall instabilities may arise because of energy transfer from the imploding plasma to the electrode surface, boundary effects at the electrode surface, and because the drive current transitions from radial to axial flow at the electrode plasma boundary. In a water-line-driven gas-puff Z pinch (short pulse), the current tends to flow as a skin current outside of the gas shell. Consequently, the current flow follows the divergence of the gas shell, which results in the magnetic pressure being at the highest near the nozzle. For a constant mass per unit length, the shell will collapse first near the nozzle and then subsequently along the remaining pinch length. This is the characteristic zipper effect (Stallings et al., 1979; Hussey et al., 1986; Hsing and Porter, 1987). By tilting the nozzle inward, the zippering can be eliminated and the imploding shell rapidly thermalizes uniformly over entire length (Spielman et al., 1985a).

Stability of thin shells that are accelerated by the pressure of a low-density fluid or the thin shells that are accelerated radially inward is of special importance for the ICF problem and for magnetically imploded liners. For ablatively accelerated targets, the problem was studied in the most detail in the context of laser-fusion research (e.g., see Bodner et al., 1978; Verdon et al., 1982; Henshaw et al., 1987; Emery et al., 1982, 1988, 1989; Dahlburg et al., 1993). Two-dimensional numerical simulations indicate that the most damaging modes for thin-shelled systems are those with wavelengths that are a few times larger than the shell unperturbed thickness, $\lambda \cong (3 - 7)d$. This means that we can use the thin-layer approximation, where a fluid layer can be approximated as an infinitely thin sheet with a finite surface density. Two-dimensional numerical simulations for an incompressible fluid (Manheimer et al., 1984) have shown that the thin-layer approximation is reasonably good, not only in the limit $\lambda \gg d$, but even for wavelengths as small as $\lambda = 2\pi d$.

Consider a thin horizontal layer of fluid in the x-y plane, accelerated from below in the z direction by a constant pressure P of another massless fluid. The dynamics of an infinitely thin layer having a finite surface density $\sigma = \sigma(x, y)$ is governed by the following equations (we shall consider 2-D planar modes in the x-z plane):

$$\frac{\partial^2 x}{\partial t^2} = -\frac{P}{\sigma_0} \frac{\partial z}{\partial \xi}, \tag{5.62}$$

$$\frac{\partial^2 z}{\partial t^2} = \frac{P}{\sigma_0} \frac{\partial x}{\partial \xi}, \tag{5.63}$$

where $x(\xi, t)$ and $z(\xi, t)$ are coordinates of an element of the perturbed layer and ξ is the Lagrangian coordinate, which coincides with the coordinate x in the unperturbed state of the layer. The unperturbed layer density σ_0 is related to the perturbed density as follows:

$$\sigma(\xi, t) = \frac{\sigma_0}{((\partial x/\partial \xi)^2 + (\partial z/\partial \xi)^2)^{1/2}}. \tag{5.64}$$

Equations (5.62) and (5.63) are linear in x and z, and their unstable eigenmodes for a constant pressure are (Ott, 1972)

$$x = \xi + a_0 e^{\Gamma t} \sin(k\xi), \tag{5.65}$$

$$z = \frac{1}{2} gt^2 + a_0 e^{\Gamma t} \cos(k\xi), \tag{5.66}$$

where a_0 is the amplitude of the initial perturbation, and

$$\Gamma = \sqrt{gk} = \sqrt{Pk/\sigma_0} \tag{5.67}$$

is the growth rate of the RT instability. The solution, (5.65) and (5.66), describes the evolution of this initial profile, while $dx/d\xi > 0$, for any ξ. The remarkable fact is that expressions (5.65) and (5.66) describe the evolution of unstable modes as much for the early linear phase, when $a_0 k e^{\Gamma t} \ll 1$, as for the nonlinear regime, when $a_0 k e^{\Gamma t} \gg 1$.

The characteristic time for solution (5.65) and (5.66), when the flow becomes essentially nonlinear and cusps appear along the sheet surface at $k\xi_c = \pm\pi, \pm 3\pi \ldots$, is

$$t_c = \frac{1}{\sqrt{gk}} \ln(1/ka_0), \tag{5.68}$$

From $t = t_c$, the flow is essentially nonlinear, not even described fully by the exact solution (5.65) and (5.66), though for some parts of the solution, where $dx/d\xi > 0$, it is still valid. The dependence $z(x)$, given by (5.65) and (5.66), is an equation of a cycloid in a parametric form. At $t = t_c$, cusps are formed on the cycloid (Fig. 4.8), since then the solution (5.65) and (5.66) becomes invalid near the cusps because it predicts a multivalued shape of the profile. A self-consistent way to extend the solution is to assume that matter from the unphysical loops forms spikes that move by inertia in the laboratory frame of reference and falls freely in the accelerated frame associated with the unperturbed sheet motion. Equations (5.62) and (5.63) do not provide a possibility for collision between adjacent sheet sections; thus, we must assume that the spikes are represented by vertical walls inside the unphysical loops. The moment $t = t_c$ identifies the onset of the free-fall phase for spikes in the accelerated frame of reference. In the laboratory frame, this moment marks the termination of acceleration for the bulk of the initial sheet mass.

The nonlinear evolution changes the perturbation profile: the minima of $z(x)$ become broader, the maxima sharper (e.g., see Fig. 4.8). The characteristic feature of the nonlinear flow is broad bubbles of massless fluid rising up and penetrating the heavy fluid, which falls down in sharp spikes. The surface density at the top of the rising bubbles decays exponentially as

$$\sigma_b(t, 0) = \frac{\sigma_0}{1 + \exp|\Gamma(t - t_c)|}. \tag{5.69}$$

Note that the surface density and the asymptotic laws of bubbles rupturing the shell are essentially different for 2-D and 3-D bubbles: $\sigma/\sigma_0 \propto \exp(-t/t_c)$ for 2-D bubbles, but $\sigma/\sigma_0 \propto (1 - t/t_c)^2$ for 3-D bubbles. The characteristic vertical velocity of the bubble top at the moment, t_c, is

$$u_b = \zeta_2 \sqrt{g/k}, \quad \text{where } \zeta_2 = 1/\sqrt{2\pi} = 0.399. \tag{5.70}$$

The next, essentially nonlinear stage of the flow, is different for the cases of a thin layer and a half-space filled by a heavy fluid. The study of the bubble dynamics, including various forms of bubble rise and bubble interaction, provides a basic approach to understanding the mixing layer process. Of the most practical interest is the motion of bubbles, because these bubbles break through the shell, destroying its integrity. The spikes of heavy fluid represent sinks where the mass of the heavy fluid is partly ejected. The rate of this process is the most important: the less mass is left, the faster the bubbles rise. Ott's solution, (5.65) and (5.66), predicts exponential growth of the bubble velocity. A similar result was obtained numerically for a shell of small but finite thickness d, where mass redistribution is taken into account (Hussey, 1984). In a cylindrical geometry of a thin-shell compression, a bubble is represented by a neck of radius r_b, and the mass driven by it is $M_b = (\overline{\rho}A)/k$, where $A = 2\pi r_b d$ is the bubble area, $\overline{\rho}$ is the average density of the heavy fluid, d is the thickness of the shell, and k is the wavenumber. The mass outflow from the bubble can be estimated as

$$\frac{dM_b}{dt} = -\overline{\rho}AV_{\text{out}}, \qquad (5.71)$$

where V_{out} is the velocity of axial flow of mass from bubble to spike. The 2-D numerical calculations of Hussey (1984) show that V_{out} tends to a constant value. Thus, the bubble mass is exponentially decreasing, the rate being of order of kV_{out}, and as long as the external magnetic pressure remains constant, its velocity also grows exponentially. When the increased magnetic pressure in the neck becomes noticeable, its inward motion would be even faster (Sec. 4.6).

Another flow pattern arises when a bubble penetrates into the heavy fluid. Most of the results obtained for this case refer to a 2-D nonlinear evolution problem. Note that the quantitative characteristics of a 3-D flow pattern may significantly differ from those predicted by a 2-D theory. However, the 2-D results can be used for the rough estimates discussed in this section. Free surface bubbles may be classified according to their topology as open or closed, and according to the relevant flow pattern dimensions as 2-D or 3-D. Closed, underwater bubbles have been studied experimentally by Davis and Taylor (1950). Open-surface bubbles rise as an approximately periodic 2-D structure in the nonlinear RT instability. They approach the form of open-ended broad columns that are separated by narrow sheets of falling fluid (Lewis, 1950; Emmons et al., 1960).

If the perturbed 2-D flow is exactly periodic with constant wavelength λ, then the bubble velocity tends to a constant value (5.70) (Birkhoff and Carter, 1957; Garabedian, 1957; Baker et al., 1980). Calculations yield values of the coefficient ζ_2 between 0.2 and 0.3. But the periodic motion turns out to be unstable (Birkhoff and Carter, 1957; Garabedian, 1957): the amalgamation of bubbles is advantageous, because greater bubble radius corresponds to an effectively increased wavelength, λ, and hence, to higher velocity. If motion along the initial interface plane is possible (i.e., if the distance to the boundaries is much greater than the bubble radii), then the bubble velocity increases due to their amalgamation. If we approximate the shape of the bubble head by a sphere of radius $R(t)$, then the effective wavelength can be estimated as the distance between neighboring bubbles, $\lambda_{\text{eff}}(t) = 2R(t)$. With the aid of (5.70) we obtain (Henshaw et al., 1987):

$$\frac{dR}{dt} = \zeta_2(g\lambda_{\text{eff}}(t))^{1/2} = \sqrt{2}\zeta_2(gR(t))^{1/2}, \tag{5.72}$$

and hence,

$$R(t) = \zeta_3 g t^2, \tag{5.73}$$

where the value of the dimensionless coefficient $\zeta_3 = \zeta_2^2/2$ varies from 0.02 to 0.45. Both 2-D numerical calculations (Youngs, 1984) and experiments (Read, 1984) yield a slightly higher, different values of ζ_3: from 0.05 to 0.7, respectively. The difference may be due to the fact that $\lambda_{\text{eff}}(t)$ differs from $2R(t)$ by a factor of order unity.

Equation (5.73) can be regarded as one of the results of nonlinear RT evolution studies, which are most important for the problems related to plasma acceleration. Since the integrity of the layer is destroyed due to the bubbles penetrating it at the moment when $R(t) = d$, and the motion of the layer itself in the laboratory frame of reference is given by $z(t) = gt^2/2$, we find from (5.73) the limitation on the aspect ratio of the distance L, over which the layer can be accelerated before it breaks:

$$\frac{L}{d} < \frac{1}{2\zeta_3}. \tag{5.74}$$

Substituting the values of ζ_3 given above, we find the estimate $\max\{L/d\} \leq 20$, which agrees reasonably well with experiments. Of course, one must be cautious in applying these results directly to the RT instabilities developing in electromagnetically accelerated plasmas, as was mentioned above. In particular, a breakdown of a plasma layer where a current flows should be accompanied by major changes in the whole flow pattern, including the profile of the effective gravitational acceleration.

With the aid of (5.70) and (5.73) one can render the criterion more concretely (Henshaw et al., 1987). Let the perturbation growth during the linear stage be expressed by $x \cong (a_0/2) \exp[(gk)^{1/2}t]$. Determining the instant $t = t^*$ when transition to the nonlinear stage takes place from

$$(a_0/2)\exp((gk)^{1/2}t) = \begin{cases} \zeta_2(g/k)^{1/2}t \\ \zeta_3 g t^2/2 \end{cases} \tag{5.75}$$

where the upper and the lower lines in the right-hand side of (5.75) correspond to the cases of bounded and unbounded medium in the horizontal direction, respectively, one finds that solutions of both equations (5.75) for t exist only if a_0 is bounded from above. Sometimes even more restrictive limitations on the perturbation amplitude are given (Verdon et al., 1982). Therefore, the linear stage of development of short-wavelength perturbations appears not to be observable at all. The nonlinear effects become dominant very soon, and the energy of the short-wavelength perturbations is either dissipated by collisional mechanisms or pumped to large-scale perturbation modes.

The spikes of heavy fluid penetrate the massless fluid at a depth that grows proportionally to gt^2, i.e., faster than the bubbles rise, though slower than the free fall would be $gt^2/2$ (Fermi, 1951). The slowing down of spikes is mainly caused by their "mushrooming" due to development of the Kelvin–Helmholtz instability at the boundary between the heavy fluid in the spike and the light fluid. The typical shape of the spike at this stage, calculated by Sharp (1984) for a perturbation in a bounded medium (that is, with a fixed wavelength), is shown in Fig. 5.4. A similar mushroom structure is shown in Fig. 4.9 above, where the evolution of a local sausage perturbation of a Z pinch is presented. Further nonlinear development of the mixing zone near a spike leads to its atomization: it decays into separate drops of the heavy fluid.

We see that the flow pattern in the nonlinear stage is determined to a large extent by the nonlinear interaction of perturbations with different wavelengths λ, by energy transfer from small-scale to large-scale perturbations. Though the nonlinear evolution of a perturbation with a fixed wavelength was studied in detail, the picture of the development of initially chaotic perturbations is not yet clear. Though most qualitative features of a 2-D flow remain unchanged, the bubbles of the light fluid rise up and the spikes of heavy fluid fall down, many important issues require further study. In particular, it is not yet known whether the size and velocity distributions of the bubbles tend to universal functions or depend on the initial conditions. It should be taken into account that even when the flute, purely interchange-perturbation modes are studied, for which the magnetic field, supporting or accelerating the plasma, behaves in an ideal MHD model as a massless fluid, the nonlinear evolution may be affected by some factors specific for electromagnetic acceleration and beyond ideal MHD. For instance, saturation of the growth of short-wavelength perturbations in an accelerated Z-pinch plasma, shown in Fig. 5.5 from Hussey et al. (1980), is due to diffusion of the magnetic

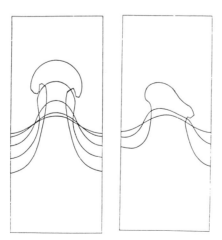

FIGURE 5.4. Nonlinear growth of a fixed-wavelength perturbation (Sharp, 1984).

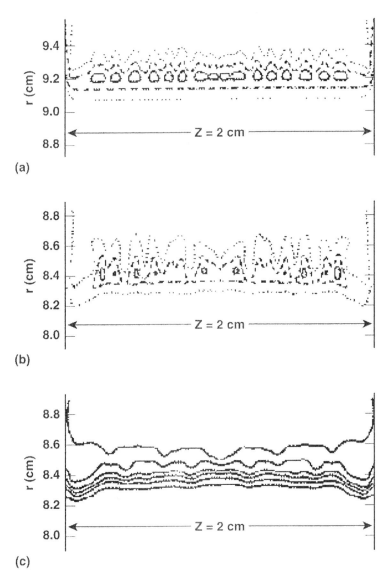

FIGURE 5.5. Development of initial short-wavelength perturbations in an imploded liner (Hussey et al., 1980). (a) Election number density contours. Solid, dashed, and dotted lines correspond to $3 \times 10^{20}, 3 \times 10^{19}$, and 3×10^{18} cm^{-3}, respectively. (b) The same after emergence of a longer dominating perturbation wavelength. (c) Contours of $r B_{\phi}$ at the same time as (b).

field between neighboring nonuniformities. When the short-wavelength mode $\lambda = \lambda_{min}$, which is the fastest at the linear stage, is saturated, another mode, whose wavelength is two to four times longer than λ_{min}, becomes dominant and eventually breaks the current shell.

We see that the conclusions related to the nonlinear evolution of the RT instability are even more sensitive to the physical mechanisms affecting its development than the results obtained for the linear stage. Another example of it is presented in Figs. 5.6(a) and (b) from Huba et al. (1987), where numerically calculated flow patterns typical for the nonlinear evolution of the flute RT modes are presented: one corresponds to the ideal MHD model (a), and the other to one-fluid Hall MHD model of a LLR plasma (b), where the destabilizing influence of the Hall effect is important (see Chap. 4). When an unperturbed compressible plasma is at rest in a gravitational field $\mathbf{g} = -g\widehat{\mathbf{e}}_z$, a random initial perturbation is introduced at the level of 1%. The unperturbed density profile has a maximum $n = 0.73$ at $z = z_m = 0.2$ in the adopted units, so that the instability initially develops in the plasma volume only for $z < z_m$. Figure. 5.6(a) is plotted for $t = 5.5(h/C_f)$, where h is the characteristic scale of density decrease for $z < z_m$ and C_f is the characteristic sound velocity, represents a typical picture of the nonlinear RT evolution. The bubbles rise up, the spikes fall down and decay. The perturbation vanishes in the stable region, where $(dn/dz) < 0$. The flow shown in Fig. 5.6(b) for an earlier time $t = h/C_f$ is quite different. In agreement with the linear theory (see Chap. 4.2), the perturbation develops much faster. It has a specific spatial structure, the "fishbone-like" flow pattern shown in Fig. 5.6(b), and destroys the unperturbed density profile in a much wider layer. Thus, the destabilizing Hall effect in the

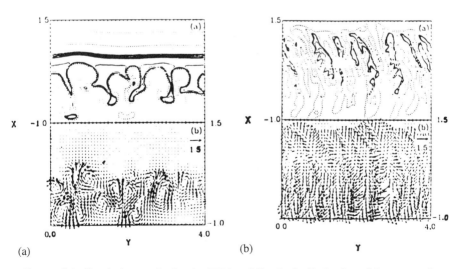

(a)   (b)

FIGURE 5.6. Simulation results for the RT instability in the limit of small Larmor ragius (Huba et al., 1987); (a) density contours; (b) velocity vector field.

LLR limit can drastically change the scenario of the nonlinear evolution presented above for the evolution of the RT instability, which is even more sensitive to the physical mechanism affecting its development than the results obtained for the linear stage.

6

Stability of Dynamic Z-Pinches and Liners

6.1 The thin-shell model

6.1.1 Instabilities of a collapsing cylindrical shell

The thin-shell model (see Sec. 3.2.1) is a simple, but very useful, model of a dynamic annular Z pinch, even more so because it permits a quite simple stability analysis with the results being expressed in elementary functions (Harris, 1961). This section is very relevant to modern experiments because liner and wire-array Z-pinch configurations are well represented by a thin-shell model. These experiments are dominated by instabilities described herein.

We consider a perfectly conducting cylindrical shell of infinitesimal thickness collapsing toward the axis, under the pressure of an azimuthal magnetic field produced by a current $I = I(t)$ carried by the shell parallel to the axis. Let the unperturbed motion of the thin shell be described by (3.2). Let $\boldsymbol{\xi}(\varphi, z, t) = (\xi_r, \xi_\varphi, \xi_z)$ be a small displacement of the particle, which was on the initial shell surface. Taking into account the azimuthal and axial symmetry of the problem under consideration, the dependence of the displacement vector on φ and z can be taken in the form $\exp(im\varphi + ikz)$. The equation of motion of an element of area of the shell is

$$\frac{\partial}{\partial t}\left[dM\left(\widehat{\mathbf{e}}_r \frac{dR}{dt} + \frac{\partial \boldsymbol{\xi}}{\partial t}\right)\right] = -P\mathbf{n} \cdot d\boldsymbol{a}, \qquad (6.1)$$

where $dM = \sigma_0 R d\varphi dz$ is the mass of the area element, σ_0 being the mass per unit area of the unperturbed shell; $\mathbf{n}(\varphi, z)$ is a unit vector normal to the shell surface; $d\boldsymbol{a}$ is an element of area of the perturbed surface. Then

$$\mathbf{n} \cdot d\boldsymbol{a} = [\widehat{\mathbf{e}}_\varphi(R + \xi_r)d\varphi + \boldsymbol{\xi}(z, \varphi + d\varphi) - \boldsymbol{\xi}(z, \varphi)]$$
$$\times [\widehat{\mathbf{e}}_z dz + \boldsymbol{\xi}(z + dz, \varphi) - \boldsymbol{\xi}(z, \varphi)]$$
$$= R d\varphi dz \left[\widehat{\mathbf{e}}_r\left(1 + \frac{\xi_r}{R} + \frac{im}{R}\xi_\varphi + ik\xi_z\right) - \widehat{\mathbf{e}}_\varphi \frac{im}{R}\xi_r - \widehat{\mathbf{e}}_z ik\xi_r\right], \quad (6.2)$$

where $R = R(t)$ is the radius of the unperturbed shell surface. The perturbation of the magnetic field $\mathbf{B}_1$ in the vacuum outside the shell can be easily derived, taking into account that $\nabla \cdot \mathbf{B}_1 = 0$ and $\nabla \times \mathbf{B}_1 = 0$, and $\mathbf{B}_1$ can be taken as $\mathbf{B}_1 = \nabla \Psi$, where a scalar potential Ψ is a solution of Laplacian equation $\nabla^2 \Psi = 0$ that is limited at infinity. Therefore, we find $\Psi = C K_m(kr) \exp(im\varphi + ikz)$, where $K_m(kr)$ is the modified Bessel function.

Let $P = P_0 + P_1$ be the magnetic pressure on the perturbed shell surface. The boundary condition on the perfectly conducting shell surface is that the perturbed magnetic field remains parallel to the perturbed surface. This condition results in $\mathbf{n} \cdot \mathbf{B} = \mathbf{n}_0 \cdot \mathbf{B}_1 + \mathbf{n}_1 \cdot \mathbf{B}_0 = 0$, where

$$\mathbf{n}_1 = \left(0, \, -\frac{1}{R}\frac{\partial \xi_\varphi}{\partial \varphi}, \, -\frac{\partial \xi_r}{\partial r} \right) \tag{6.3}$$

is the perturbation in the unit normal vector. Determining the constant C in $\Psi = C K_m(kr)$ from the boundary condition, we find

$$P_0 + P_1 = \frac{B_\varphi^2}{8\pi}\left[1 - \frac{2\xi_r}{R} - \frac{2m^2 K_m(kR)}{k K_m'(kR)}\frac{\xi_r}{R} \right], \tag{6.4}$$

where $B_\varphi = 2I(t)/cR(t)$ is the unperturbed azimuthal magnetic field at the shell surface.

The zero-order terms of (6.1) reproduce (3.2), whereas the first-order terms are the linearized equations for small perturbations:

$$\frac{\partial^2 \xi_r}{\partial t^2} = g\left[\frac{\xi_r}{R}\left(1 + \frac{2m^2 K_m(kR)}{kRK_m(kR)} \right) - \frac{im}{R}\xi_\varphi - ik\xi_z \right], \tag{6.5}$$

$$\frac{\partial^2 \xi_\varphi}{\partial t^2} = \frac{img}{R}\xi_r - \frac{g}{R}\xi_\varphi, \tag{6.6}$$

$$\frac{\partial^2 \xi_z}{\partial t^2} = ikg\xi_r, \tag{6.7}$$

where $g \equiv -(d^2 R/dt^2)$ is the effective gravitational acceleration.

Using the WKB approximation (5.1) and substituting $\Gamma^2 \xi$ instead of $d^2\xi/dt^2$, and assuming the conventional azimuthal and axial dependence of perturbations, $\exp(im\varphi + ikz)$, one easily obtains a dispersion relation to determine the instantaneous growth rate $\Gamma(t)$:

$$\Gamma^4\left(\Gamma^2 + \frac{g}{R} \right) - \frac{g}{R}\left(1 + 2\frac{m^2 K_m(kR)}{kRK_m'(kR)} \right)\Gamma^2\left(\Gamma^2 + \frac{g}{R} \right)$$

$$- g^2\frac{m^2}{R^2}\Gamma^2 - g^2 k^2\left(\Gamma^2 + \frac{g}{R} \right) = 0. \tag{6.8}$$

Equation (6.8) demonstrates that instability exists for any wavenumbers k and m not equal to zero simultaneously: one positive root of (6.8) always exists. The dependence of Γ on the wave numbers k and m is shown in Fig. 6.1. This spectrum does not change its shape in time, only the scaling factor varies, being proportional to $\sqrt{g/R} \propto I(t)/R(t)$. To obtain a better estimate of the perturbation growth, one

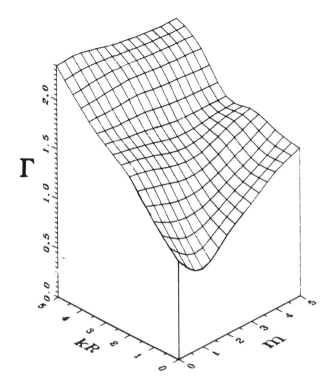

FIGURE 6.1. Instability growth rate versus wavenumbers for a thin shell imploded by the pressure of an azimuthal magnetic field.

can solve (6.5)–(6.7) without any further approximations for a given profile of the current pulse. The results of such calculations for the sausage perturbation modes (Ruden et al., 1987) are presented in Fig. 6.2. It was found that the perturbation growth is not very sensitive to the pulse shape, provided that the pulse duration was matched to the implosion dynamics.

We see that if kR is not too small, the sausage $m = 0$ mode dominates, for which

$$\Gamma^2 = \frac{g}{R} \left[\frac{1}{2} + \left(\frac{1}{4} + k^2 R^2 \right)^{1/2} \right]. \tag{6.9}$$

For $kR \gg 1$, when the effect of cylindrical geometry is not important, the classical expression for the RT growth rate is reproduced. The sausage instability mode in this case is quite similar to the classical RT mode. This is not surprising because the perturbations do not bend the force lines of the magnetic field, and hence the magnetic field acts just as a massless fluid. For filamentation instability modes ($k = 0$, $m \gg 1$), this is not the case, and it is natural to expect a decrease in

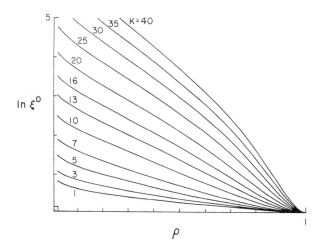

FIGURE 6.2. Estimate of the RT growth (in a logarithmic scale) of radial perturbations for a smooth current-pulse profile (Ruden et al., 1987).

the growth rate due to the additional energy required to bend the magnetic lines. Indeed, introducing an effective azimuthal wavenumber $k_{\text{eff}} = m/R$, we obtain in this limit

$$\Gamma = (\sqrt{2} - 1)^{1/2}\sqrt{|gk_{\text{eff}}|}. \qquad (6.10)$$

Thus, the decrease in the growth rate from (6.9) is given by the factor $(\sqrt{2} - 1)^{1/2}$, whereas its dependence on the wave number and g is not changed (see Fig. 6.1). Is this just a property of the thin-shell model? In a sense, the answer is yes. Let us consider the RT instability of a plane layer of conducting fluid of finite thickness d, supported in a gravitational field $\mathbf{g} = g\hat{\mathbf{e}}_z$, $g < 0$ by the pressure of a magnetic field $\mathbf{B} = B\hat{\mathbf{e}}_x$. Assuming ideal conductivity and incompressibility, we obtain the following equation for the growth rate of instability (Harris, 1961):

$$\Gamma^4 - 2gk_{\|}^2 d \coth(kd)\Gamma^2 - g^2 k^2 (1 - 2k_{\|}^2 d/k) = 0, \qquad (6.11)$$

where $k_{\|} = (\mathbf{k} \cdot \mathbf{B})/B \equiv k_x$. For $k_{\|} = 0$, when the perturbations do not bend the magnetic lines, (6.11) is reduced to $\Gamma = \sqrt{gk}$, as it should be. On the contrary, if a perturbation wave vector is parallel to the magnetic field $k = k_{\|}$, the instability develops only for wavelengths $\lambda > 4\pi d$. In the long-wavelength limit $kd \ll 1$, (6.10) stems from (6.11). Therefore, the short-wavelength limit of (6.8) for $k = 0$, $m \gg 1$ yields the same results as the long-wavelength limit of (6.11) and corresponds to the wavelength range $d \ll \lambda \ll R$. For short wavelengths, the finite thickness of the plasma layer, on whose surface the current is concentrated, is an effective stabilizing factor. Of course, to make quantitative estimates valid for the whole range of perturbation wavelengths, the problem of the RT instability should be studied in its general formulation, without simplifying assumptions like ideal conductivity, incompressibility, etc.

A similar analysis can be extended for the case of a thin shell compressing an axial magnetic field. In this case, the unperturbed equation of motion is (3.11). Equations (6.1) and (6.2) remain valid, but the pressure on the perturbed shell surface differs from that given by (6.4) by a term, taking into account the perturbation of the axial magnetic field inside the shell. The equation for the instantaneous growth rate in the WKB approximation can be easily obtained by taking into account that now instead of (6.4), the disturbed pressure has the following form:

$$
\begin{aligned}
P_0 + P_1 &= \frac{1}{8\pi} (B_\varphi^2 - B_z^2) + \frac{1}{4\pi} \left(B_\varphi B_{1\varphi} + B_\varphi^2 \frac{\xi_r}{R} - B_z B_{1z} \right) \\
&= \frac{1}{8\pi} (B_\varphi^2 - B_z^2) + \frac{1}{8\pi} \frac{\xi_r}{R} \left[\left(\frac{2m^2 K_m(kR)}{kK_m'(kR)} + 1 \right) B_\varphi^2 \right. \\
&\quad \left. - B_z^2 \left(\frac{kR I_m(kR)}{I_m'(kR)} - \frac{m^2 K_m(kR)}{kK_m'(kR)} \right) \right].
\end{aligned}
\tag{6.12}
$$

The spectra of instantaneous growth rates given by (6.12) are shown in Fig. 6.3. Here, the very shape of the spectrum depends on time. Initially, the shell is accelerated inwards by the pressure of the azimuthal magnetic field, and the axial magnetic field is relatively weak. Thus, the flute instability modes dominate with respect to the outer shell surface, i.e., the sausage modes [see Fig. 6.3(a)]. At the final stage, when the shell is decelerated by the pressure of the compressed axial magnetic field, the flute modes with respect to the inner shell surface, i.e., the filamentation modes dominate [Fig. 6.3(b)]. This illustrates the mechanism of implosion stabilization by an axial magnetic field: the sausage and kink instability modes, responsible for breaking the symmetry of the shell without an axial magnetic field develop more slowly when the axial field is present, so that a greater compression of the cylindrically symmetric shell is possible. It should be noted, however, that the calculations based on the thin-shell model underestimate the stabilizing effect of the magnetic shear. A stability analysis of the 1-D flow is required to make even rough quantitative estimates of the axial magnetic field required to stabilize an implosion.

6.1.2 WKB approximation

Consider perturbations of a thin conducting shell compressed by a current pulse. The perturbation growth estimated with the aid of the WKB approximation is given by (5.1). Here, we suppose that the shell is imploded by a constant current profile, $I = $ constant, and limit ourselves to the study of comparatively short-wavelength flute perturbation modes, whose growth rates are given by (5.5). Then, the integration in (5.1) is easily performed and $n_{\text{eff}}(t)$ is given by (Hussey et al.,

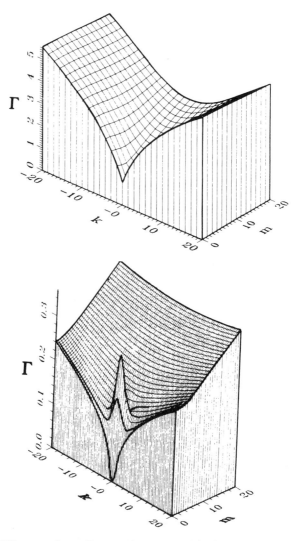

FIGURE 6.3. Instability growth rate Γ versus the wavenumbers for a thin shell imploded by an axial current and compressing an axial magnetic field trapped in the annulus; (a) initial stage (acceleration inwards); (b) final stage (deceleration).

1980):

$$dn_{\text{eff}}(t) = \Gamma(t)dt = |g(t)k|^{1/2}dt = |2\pi g(t)/\lambda|^{1/2} \left| \frac{dR}{u} \right|, \qquad (6.13)$$

where $g(t) = -\ddot{R}(t)$ and $u(t) = \dot{R}(t)$ are expressed via the inverse compression ratio $R_0/R(t)$ by (3.2) and (3.3), and this permits us to integrate (6.13), expressing the number, n_{eff}, of e-foldings via the inverse compression ratio $R_0/R(t)$:

$$n_{\text{eff}} = \pi \left(\frac{2R_0}{\lambda} \right)^{1/2} \text{erf} \left\{ \left[\frac{1}{2} \ln(R_0/R(t)) \right]^{1/2} \right\}. \tag{6.14}$$

If the value of n_{eff} is not very large, then the coefficient of the exponential, representing the corrections to the zero-order WKB approximation formula (5.1) may be important. A first-order estimate for the sausage instability modes (Hwang and Roderick, 1987) can be presented in the form

$$n_{\text{eff}} = \int_0^t \Gamma(t')dt' + \frac{3}{4} \ln(R(t)/R_0) - \frac{1}{2} \ln(I(t)/I_0), \tag{6.15}$$

where $I(t)$ is the current at time t.

Comparing (6.14) with the results from the estimates presented in Fig. 6.2, we see that the WKB approximation for a constant current predicts values of $n_{\text{eff}}(R_0/R(t))$ about 1.5 times greater than the results obtained for the case when the current rises in a smooth monotonic manner, even if the logarithmic corrections (6.15) are taken into account. Equations (6.14) and (6.15) and Fig. 6.2 may be thus regarded as estimates of n_{eff} from above and below, respectively.

An important feature of (6.14) is a very weak dependence of n_{eff} on the inverse compression ratio $R_0/R(t)$. When its value is varied between 4 and 20, the error function in (6.14) changes from 0.76 to 0.92. Thus, for all compression ratios of practical interest, the value of erf$(\ldots)$ $= 0.8$ is a good approximation, and an estimate of the perturbation growth in the course of compression can be taken as (Hussey et al., 1980)

$$n_{\text{eff}} \cong 3.6(R_0/\lambda)^{1/2}. \tag{6.16}$$

Suppose that the perturbation wavelengths responsible for destroying the shell are bounded from below, due to the stabilizing role of magnetic viscosity. Then, the characteristic value of $\lambda = \lambda_{\min}$ to be substituted in (6.16) is $(2\pi/k_0)f^{-2}$, where k_0 is determined by (5.50) with $D = \nu_M$ (the coefficient of magnetic viscosity), and f is a dimensionless factor of order unity. The fastest perturbation mode is selected during the acceleration phase when $R \cong R_0$, so that the value of ν_M corresponds to the conditions at the outer surface of the shell at this time, when the temperature there is of order 10 eV. The value of g is also estimated for $R \cong R_0$ with the aid of (3.2) as

$$g = \frac{\pi}{2} (R_0/\tau^2) = \frac{\pi}{2} (u_f/2.38\tau), \tag{6.17}$$

where τ is the effective implosion time and u_f is the final velocity of the shell at the end of implosion. With the aid of these expressions and (6.14), the following estimate for n_{eff} can be obtained (Hussey et al., 1980):

$$n_{\text{eff}} \cong 1.53 f \left(\frac{R_0^2}{\nu_m \tau} \right)^{1/3} = 1.53 f \left(\frac{0.42 R_0 u_f}{\nu_m} \right)^{1/3}$$

$$= 1.53 f \left(0.177 \frac{u_f^2 \tau}{v_m} \right)^{1/3}, \tag{6.18}$$

where according to the estimates of Hussey et al. (1980), f is between 0.8 and 1.1 (see also paper by Hammer et al., 1996).

According to (6.18), $n_{\text{eff}} \cong (R_0/\delta_0)^{2/3}$, where $\delta_0 = \sqrt{v_M \tau}$ is the skin depth. Thus, the stronger the skin effect at the initial stage of acceleration, the less stable is the implosion. Equation (6.18) also shows that for a fixed value of the characteristic implosion velocity u_f, which is determined mainly by Z-pinch applications, the acceleration is more stable for shorter characteristic times, τ, and smaller initial values of radius the R_0. This is a quite general feature of the RT instability (Rostoker and Tahsiri, 1977). Indeed, for the final velocity we have $u_f = \sqrt{2gL}$, where L is the distance over which the acceleration takes place, so that n_{eff} can be estimated as (Hammer et al., 1996)

$$n_{\text{eff}} \cong \sqrt{gk\tau} = \sqrt{2kL} = \sqrt{2k/g}\, u_f. \tag{6.19}$$

We see that the total number of e-foldings that the instability grows while the layer is accelerated to velocity u_f is reduced by increasing the acceleration. If $g \to \infty$ and $L \to 0$ with $u_f = $ const, then $n_{\text{eff}} \to 0$, and the acceleration to a final velocity u_f could be accomplished without an RT instability.

The 2-D numerical simulations of Hussey et al. (1980), performed for the conditions of SHIVA implosions, demonstrate that the instability of an accelerated shell is barely visible, if the estimate (6.18) with $f = 1$ predicts $n_{\text{eff}} \leq 9$, and the instability is manageable, if $9 \leq n_{\text{eff}} \leq 12$. Note that here implosions of foil liners are considered, so that the initial uniformity of the shell can be made high enough to be compatible with perturbation growth given by a factor $e^9 = 8.1 \times 10^3$, without destroying the cylindrical symmetry completely. In this case, the perturbation growth remains exponential during the nonlinear stage, too. The net effect of nonlinearity and diffusive stabilization makes the wavelength of the dominating instability mode 2–4 times longer than the wavelength corresponding to maximum growth rate during the linear stage. The plasma shells produced by the gas-puff Z-Pinch facilities have much greater nonuniformities, so that both linear and nonlinear stages of the RT instability development are essentially different.

6.2 Growth of the RT instabilities in a layer of finite thickness

We consider now some features of the growth of the RT instability related to the finite thickness of the accelerated layer. We consider a layer of incompressible liquid of width d and density ρ, being accelerated in the $+\hat{z}$ direction by the pressure of a massless fluid, P. In the frame of reference moving with the fluid layer, there will be a gravitational field, $\mathbf{g}$, in the $-\hat{z}$ direction. In the equilibrium state, the fluid is at rest and the force equilibrium equation is $P - P_0 = \rho g d$, where

P_0 is the pressure above the layer. The linearized equations for small perturbations are

$$ikv_x + \frac{\partial v_z}{\partial z} = 0, \tag{6.20}$$

$$\frac{\partial v_x}{\partial t} + \frac{1}{\rho} ik P_1 = 0, \tag{6.21}$$

$$\frac{\partial v_z}{\partial t} + \frac{1}{\rho} \frac{\partial P_1}{\partial z} = 0, \tag{6.22}$$

where $v_x(z, t)$, $v_z(z, t)$, and $P_1(z, t)$ are the perturbed velocities and pressure, respectively, and we look for solutions which are proportional to $\exp(ikx)$.

Let ξ be the displacement in the z direction and ξ_0 and ξ_d be the displacement of the upper and lower surfaces, respectively. At the top and lower surface we have

$$\frac{d\xi_0}{dt} = v_z(0, t), \quad \frac{d\xi_d}{dt} = v_z(d, t). \tag{6.23}$$

The condition that the pressure must be continuous gives the following two equations:

$$\xi_0 \frac{dP}{dz}(0, t) + P_1(0, t) = 0, \quad \xi_d \frac{dP}{dz}(d, t) + P_1(d, t) = 0. \tag{6.24}$$

We shall now examine two cases. First, let the initial displacement δ be at the lower (unstable) surface only, i.e.,

$$\xi_0(0) = \delta, \quad \xi_d(0) = 0. \tag{6.25}$$

Using the boundary conditions (6.23) and (6.24), we find the solution of (6.20)–(6.22) with the initial conditions (6.25), in the following form:

$$\xi_0(t) = \frac{\delta}{1 - \exp(-2kd)} [\cosh(\omega t) - \cos(\omega t) \exp(-2kd)], \tag{6.26}$$

$$\xi_d(t) = \frac{\delta \exp(-kd)}{1 - \exp(-2kd)} [\cosh(\omega t) - \cos(\omega t)], \tag{6.27}$$

$$v_z(z, t) = \frac{\delta \omega}{1 - \exp(-2kd)}$$
$$\times [\sinh(\omega t) \exp(-kz) + \sin(\omega t) \exp[k(z - 2d)]], \tag{6.28}$$

$$v_x(z, t) = \frac{i \delta \omega}{1 - \exp(-2kd)}$$
$$\times [- \sinh(\omega t) \exp(-kz) + \sin(\omega t) \exp[k(z - 2d)]], \tag{6.29}$$

where $\omega = \sqrt{kg}$. We next consider the case when the initial displacement δ is at the top (stable) surface only, i.e.,

$$\xi_0(0) = 0, \, \xi_d(0) = \delta. \tag{6.30}$$

The solution of (6.20)–(6.22) with the initial conditions (6.30) is

$$\xi_0(t) = -\frac{\delta \exp(-kd)}{1 - \exp(-2kd)} [\cosh(\omega t) - \cos(\omega t)], \tag{6.31}$$

$$\xi_d(t) = -\frac{\delta}{1 - \exp(-2kd)} [\cosh(\omega t) \exp(-kd) - \cos(\omega t)]. \tag{6.32}$$

Growth of the perturbations initiated at $t = 0$ as the "step" $\xi_0(0) = \delta sgn(x)$, $\xi_d(0) = 0$ and $\xi_d(0) = \delta sgn(x)$, $\xi_0(0) = 0$ are shown in Fig. 6.4. Note that perturbations at the lower surface are more dangerous for the distortion of the layer. The growth of the perturbations initiated at the upper surface is delayed by a time of about $\sqrt{d/g}$, while the more dangerous perturbations for distortion of

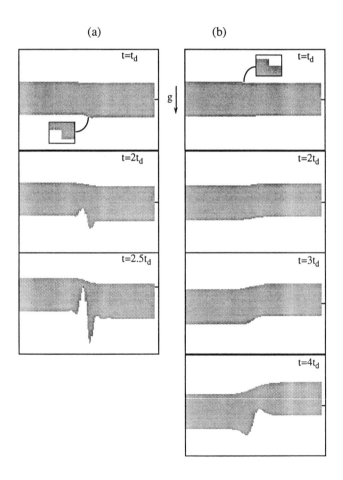

FIGURE 6.4. Growth of the perturbations initiated as a step function at the surface of a layer of finite thichness.

the layer start to grow exponentially after the time interval of about $1/\sqrt{kg}$, when the growing mode at the unstable surface starts to be predominant.

6.3 Rayleigh–Taylor instabilities in an imploding Z pinch: The snowplow model

Consider the implosion of a cylindrical plasma by the magnetic field generated by a current flowing in the axial direction. The current is switched on at time, $t = 0$, and penetrates into the plasma on the magnetic diffusion time scale. Thus, the current flows in a sheath on the outside of the plasma. The magnetic field accelerates and drives a shock wave into the plasma. We will use the snowplow model to analyze RT instabilities (Sec. 3.2.2). Two main processes are involved in the model: formation of a shock wave ahead of the moving piston and mass accretion. The first issue is when and how a shock wave is formed ahead of the piston. To clarify the physics, we first consider planar geometry for simplicity. When moving with acceleration, the piston irradiates sound waves that propagate ahead of the piston. The sound waves determine the resulting flow, which is a simple compression wave. The simple waves break and a shock wave is formed ahead of the piston. The time when this happens is determined by the equation (Landau and Lifshitz, 1987)

$$\omega t_s = \frac{C_0}{\pi(\gamma + 1)U} \exp\left(\nu \frac{\omega^2}{C_0^2} t_s\right),$$

where C_0 is the sound speed, U is the piston velocity, ν is the kinematic viscosity, and ω is the frequency of the sound. Estimating $\omega \cong g/U$, this condition can be presented in the form $U > \sqrt{g\lambda_{ii}}$, where λ_{ii} is the ion–ion mean-free-path. For typical Z pinches, $g \approx 10^{12} - 10^{14}$ cm/s^2, the mean-free-path in the plasma, $\lambda_{ii} \approx 10^{13}(T/\text{eV})^2/Z^4 N$, and we obtain (for $Z = 1$): U cm/s$\geq 10^5(T/\text{eV})$. The last formula says that the shock wave is formed ahead of the imploding current sheath in the Z pinch during the first stage of the implosion, when the plasma temperature inside of the imploding liner is low enough. At later stages, when the temperature of the inner plasma is raised, the shock wave is not necessarily formed. Indeed, as is well known, the implosion of a plasma shell in Z-pinch experiments has a shock during the very first stage of the implosion, whereas the later stages of the implosion are shock free.

   Thus, there are two possibilities: (1) a shock wave is formed ahead of the moving piston of any nature (a current sheath for the case of Z pinch) that sweeps up the mass of the resting plasma. In this case, two features must be taken into account when calculating the development of the RT instability: mass accretion by the imploding shell, and the fact that now the leading edge of the imploding shell is a shock wave instead of a free surface. (2) When the temperature inside the imploding shell is high enough, the formation of a shock wave becomes unlikely, but the resulting flow ahead of the shell means that the effective total mass of the shell increases. Since the density of the shell is typically much higher than the density of the resting plasma, we will take into account only mass accretion, neglecting the change of shell width.

6.3.1 Snowplow model: The effect of mass accretion

We will first consider the effect of mass accretion (without shock formation) on the development of the RT instability. Note that effect of mass accretion changes the dynamics of the liner acceleration. When the accelerated layer scoops unperturbed matter, entraining it into motion, the instantaneous growth rate of perturbations is smaller compared with the layer propagating into vacuum. However, the total perturbation growth, when the layer is accelerated up to a given energy increases the number of e-foldings (De Groot et al., 1997). Thus, the mass accretion can be a destabilizing effect in general, contrary to the conclusion in the paper (Golberg and Velikovich, 1993). Nevertheless, the changed dynamics of the liner acceleration in the presence of mass accretion may lead to a decrease in the number of e-foldings, and this effect is even more pronounced for the implosion of a cylindrical liner (De Groot et al, 1997). Consider a cylindrical plasma liner that is collapsing toward the axis, under the pressure of an azimuthal magnetic field, B_φ, produced by a pinch current, $I(t)$, and carried by the imploding Z-pinch shell along the z axis. The plasma liner is imploding toward the axis into the inner space filled by unperturbed resting plasma. In the spirit of the snowplow model, the cylindrical shell is assumed to be of infinitesimal thickness with an ideal electric conductivity (Sec. 3.2.2). The sharp increase in the pinch current drives the implosion of the current sheath with a shock wave propagating ahead of the current sheath at the beginning of the implosion. We consider the snowplow model after the shock has reached the axis, when mass accretion by the collapsing shell is the dominant effect. In the spirit of the snowplow model, the unperturbed equations of motion are given by (3.17, 3.20–3.22). As was seen in Chap. 3, the implosion dynamics of shells with the same final mass at late times are not sensitive to the precise shape of the current waveshape. However, it will be shown that the RT instabilities are sensitive to the amount of unperturbed plasma swept up by the collapsing shell.

Let us consider small perturbations in the 1-D implosion of the shell, including mass accretion of the unperturbed plasma (De Groot et al., 1997). We shall consider the case of constant current. For the limiting case of an infinitesimally thin shell collapsing into vacuum [$\mu = 0$, see definition (3.19)], the result must coincide with the analysis of the instability of a collapsing cylindrical shell of constant mass (Sec. 6.1). We introduce the displacement, ξ, of a particle which was originally on the surface of the shell at $\mathbf{r}_0(t) = \{R(t), \varphi, z\}$, and taking into account axisymmetry, we look for a solution in the following form:

$$\xi = \zeta(\tau) R_0 \exp\left[i\left(k\frac{z}{R_0} + m\varphi\right)\right], \qquad (6.33)$$

where $\zeta(t) = [\zeta_r(\tau), -i\zeta_\varphi(\tau), -i\zeta_z(\tau)]$ is a small displacement of the particle normalized to the initial radius of the shell. For convenience, we shall use the dimensionless coordinates (3.19). Let μ_1 be the perturbation in the mass normalized to the total mass of the system [see definition (3.19)]. Then, acting in the same way as in Sec. 6.1, we obtain the following linearized equations for the small perturbations μ_1 and ζ:

$$\frac{d\mu_1}{d\tau} = -2\mu\alpha \left[(\xi_r + m\xi_\varphi) \frac{1}{\alpha} \frac{d\alpha}{d\tau} + \frac{d\zeta_r}{d\tau} + k\xi_z \frac{d\alpha}{d\tau} \right], \qquad (6.34)$$

$$\frac{d}{d\tau} \left[\mu_1 \frac{d\alpha}{d\tau} + (1 - \mu\alpha^2) \frac{d\zeta_r}{d\tau} \right]$$

$$= \frac{1}{\alpha} \left[\frac{\zeta_r}{\alpha} + \frac{2m^2 K_m(k\alpha)}{k\alpha K_m'(k\alpha)} \xi_r - \frac{m}{\alpha} \xi_\varphi - k\xi_z \right], \qquad (6.35)$$

$$\frac{d}{d\tau} \left[(1 - \mu\alpha^2) \left(\frac{d\zeta_\varphi}{d\tau} - \frac{\zeta_\varphi}{\alpha} \frac{d\alpha}{d\tau} \right) \right]$$

$$= -\frac{m}{\alpha^2} \zeta_r - (1 - \mu\alpha^2) \frac{1}{\alpha} \frac{\delta\alpha}{d\tau} \left(\frac{d\zeta_\varphi}{d\tau} - \frac{\zeta_\varphi}{\alpha} \frac{d\alpha}{d\tau} \right), \qquad (6.36)$$

$$\frac{d}{d\tau} \left[(1 - \mu\alpha^2) \frac{d\zeta_z}{d\tau} \right] = -\frac{k}{\alpha} \zeta_r. \qquad (6.37)$$

If $\mu = \rho = M = 0$ and consequently $\mu_1 = 0$, then (6.34) is satisfied identically and the rest of (6.35)–(6.37) gives the solution of the stability problem for a thin shell of constant mass collapsing into vacuum (Sec. 6.1). In order to estimate the growth rate of the instabilities, we assume the WKB approximation, i.e., that

$$\zeta(\tau) = \zeta_{m,k} \exp(\Gamma_{m,k}\tau), \quad \mu_1(\tau) = \mu_{1m,k} \exp(\Gamma_{m,k}\tau), \qquad (6.38)$$

where $\zeta_{m,k}$ and $\mu_{m,k}$ are the time-independent amplitudes, and $\Gamma_{m,k}$ is the instantaneous growth rate of the perturbations.

For a simple analytical estimate, let us calculate the growth rate of instabilities at $t = 0$, when

$$\alpha = 1, \qquad \frac{d\alpha}{d\tau} = 0, \qquad \frac{d^2\alpha}{d\tau^2} = -\frac{1}{1 - \mu}. \qquad (6.39)$$

Taking into account (6.38) and (6.39), we find that a nontrivial solution of the system (6.34–6.37) exists, if

$$(1 - \mu)\Gamma_{m,k}^2 + \frac{2\mu}{1 - \mu} - 1 - \frac{2m^2 K_m(k)}{k K_m'(k)} = \frac{m^2}{(1 - \mu)\Gamma_{m,k}^2 + 1} + \frac{k^2}{(1 - \mu)\Gamma_{m,k}^2}. \qquad (6.40)$$

We shall consider solutions of (6.40) for special cases: $m = 0, k \neq 0$ and $k = 0$, $m \neq 0$. For $m = 0$, (6.40) becomes

$$(1 - \mu)^2 \Gamma_{0,k}^4 + (3\mu - 1)\Gamma_{0,k}^2 - k^2 = 0, \qquad (6.41)$$

which coincides with the equation for $\Gamma_{0,k}$, obtained above (Sec. 6.1) for $\mu = 0$ and $m = 0$. The solution of (6.41) is

$$\Gamma_{0,k}^2 = \frac{1 - 3\mu + \sqrt{(1 - 3\mu)^2 + 4k^2(1 - \mu)^2}}{2(1 - \mu)^2}. \tag{6.42}$$

In the limiting case of $m \gg 1$, and for $k = 0$, we find from (6.40)

$$\Gamma_{m,0} = \sqrt{(\sqrt{2} - 1)m/(1 - \mu)}, \tag{6.43}$$

which coincides with the limiting expression (6.10) for a thin shell (Sec. 6.1), if $\mu = 0$.

The instantaneous growth rates calculated according to (6.42) and (6.43) for a shell collapsing into vacuum ($\mu = 0$) and for a shell that sweeps up plasma in the course of the implosion ($\mu = 0.9$), indicates that the growth rate is greater in the case of the shell collapsing into a plasma, except for long wavelength instabilities with small wavenumbers m and k. This is because the acceleration of a shell is larger with mass accretion than for a constant mass shell collapsing into vacuum. As is seen, the instabilities are of the RT type, and the asymptotic dependence of the growth rate is $\sigma = \sqrt{gk}$, where $g \equiv -d^2\alpha/d\tau^2$ is the instantaneous acceleration of the shell.

Note that mass accretion can be stabilizing during the nonlinear stage of the RT instability. A simple stabilizing mechanism appears because the process of sweep-up conserves momentum (Book, 1996). In the nonlinear stage of the instability, the bubbles that protrude ahead of the mean shell radius have a lower-than-average mass per unit area. Consequently, they are decelerated more than the spikes, which lag behind and have a higher-than-average mass per unit area. Therefore, accretion tends to restore a uniform mass density and decrease the peak-to-peak variation.

6.3.2 Snowplow model: The effect of the shock wave

Another effect of the snowplow is the shock wave launched ahead of the liner. In this case, both the shock wave and mass accretion affect the instability growth. We shall consider the effect of a shock front for the simple case of planar geometry (DeGroot et al., 1997).

Let the imploding liner be a plane thin layer accelerated by magnetic pressure. In this case, the difference compared to a liner accelerated into vacuum is that the boundary of the layer is represented by a shock front propagating into a half-space filled by unperturbed plasma. In both cases, the condition for the development of the RT instability is satisfied. In the co-moving frame, the accelerated (contact) surface is resting, and we have the following unperturbed equilibrium equation:

$$\frac{dP_0}{dz} + \rho_0 g = 0. \tag{6.44}$$

Assuming incompressibility of the fluid behind the shock front (subsonic motion), we have the following linearized equations for small velocity perturbations $\mathbf{u} =$

$(v_1, 0, u_z + u_1)$ and pressure $P = P_0 + P_1$:

$$ikv_1 + \frac{\partial u_1}{\partial z} = 0, \tag{6.45}$$

$$\frac{\partial v_1}{\partial t} + \frac{ik}{\rho_0} P_1 = 0, \tag{6.46}$$

$$\frac{\partial u_1}{\partial t} + \frac{1}{\rho_0} \frac{\partial P_1}{\partial z} = 0. \tag{6.47}$$

Let ξ_0 and ξ_s be the displacements of the boundary at the piston and at the shock front, respectively, and D be the velocity of the shock wave. Then the boundary conditions for the perturbations at the piston surface $z = 0$ are

$$\xi_0 \frac{\partial P_0}{\partial z} + P_1(0, t) = 0, \tag{6.48}$$

$$\frac{\partial \xi_0}{\partial z} - u_1(0, t) = 0, \tag{6.49}$$

and at the front of a strong shock wave, at $z = d(t)$, are

$$\left[v_1 + \frac{2ikD}{\gamma + 1} \xi_s \right]_{z=d} = 0, \tag{6.50}$$

$$\left[u_1 - \frac{D}{\gamma + 1} \left(\frac{P_1}{P_0} + \frac{1}{P_0} \frac{\partial P_0}{\partial z} \xi_s \right) \right]_{z=d} = 0, \tag{6.51}$$

$$\left[\frac{\partial \xi_s}{\partial t} - \frac{D}{2} \left(\frac{P_1}{P_0} + \frac{1}{P_0} \frac{\partial P_0}{\partial z} \xi_s \right) \right]_{z=d} = 0. \tag{6.52}$$

We compare the time history of perturbations obtained from numerical solution of (6.45)–(6.52) with the perturbation growth in a plane accelerating layer with thickness d, that is initially at rest and accelerated up to the same velocity, u_z, with the same acceleration, g. The perturbation growth in the layer should be treated as an initial-value problem: the perturbations initiated by small disturbances of the boundary of the layer without initial velocity perturbations. If the perturbations in the layer are initiated by disturbances of this kind, all the eigenmodes (growing, damping, and oscillating) are initiated with almost the same amplitudes. Thus, the perturbations start to grow exponentially only after a time interval of about $1/\sqrt{gk}$, when the growing mode starts to predominate. The perturbation growth for both cases is compared with the perturbation growth in the case of the classical RT instability in Fig. 6.5. As is seen from Fig. 6.5, the perturbation growth is delayed in both cases by the time interval needed to establish a velocity distribution corresponding to the growing RT instability mode. In the case of a plane layer, this time interval may be estimated as $1/\sqrt{gk}$, and in the case of a shock wave, one should account for an additional delay of about $1/kD$, when a thick enough layer is accreted behind the shock front. Thus, the long-wavelength perturbations grow slowly compared with the classical RT instability, and the growth of the perturbations is defined by the dynamics of the shock. The stabilizing action of the shock is due to the stability of the shock front itself, and the presence of a shock

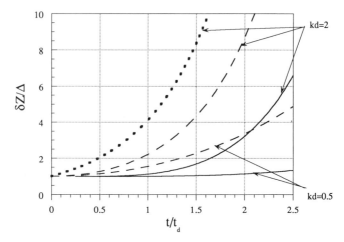

FIGURE 6.5. Growth of the perturbations because of RT instability. The dotted line corresponds to a growing eigenmode of "classical" RT instability. The dashed lines correspond to perturbation growth in a layer accelerated under vacuum, for an initial disturbance that does not coincide with the eigenmode. The solid lines correspond to growth of perturbations in the accelerating liner, with the shock wave at the leading edge. $t_d = \sqrt{d/g}$.

instead of a free surface is similar, in a sense, to a rigid wall at the inner boundary. As is seen, the presence of the shock wave at the leading edge of the accelerated liner has a stabilizing effect, despite the destabilizing action of the mass accretion. The stabilizing effect of a shock wave is especially effective in the case of a plasma focus. In this case, the radius of the implosion is large, and in fact, the "first stage" of the implosion involving a shock wave is the main stage of the plasma focus. The stabilizing effect of a shock wave may be responsible for the observed azimuthal symmetry of the plasma focus implosions with very high observed compression ratios—up to about 10^3. The filamentary instability modes dominate at the final stage of implosion (e.g., see Herold et al., 1989).

6.4 Imploding wire arrays

Imploding Z pinches employing wire arrays have applications for basic plasma physics and are of special interest as intense sources of x rays in a broad range. Developing an understanding of the detailed Z-pinch dynamics is critical for predicting scaling and optimization performance, and for designing loads that reach desired plasma and radiation conditions. Although many experiments have been performed on pulsed-power driven, wire-array Z-pinch implosions (Aivazov et al., 1988, 1990; Bekhtev et al., 1989; Baksht et al., 1989, 1993; Deeney et al., 1989, 1991, 1994, 1997; Spielman et al., 1989, 1994, 1998; Vikharev et al., 1991; Whitney et al., 1994; Sanford et al., 1996, 1997) and quite a comprehensive array of diagnostics have been used on these experiments, many facets of the problem still have not been studied and understood fully because the full 3-D MHD equations must be solved to understand wire-array implosions. A major motivation for carrying out the experiments was to determine how the experimental results from the present facilities would scale to higher currents and larger imploded masses.

Wire arrays are attractive because of the low fabrication cost and the high quality (axial and radial uniformity) that can be obtained. Typical parameters seen in modern experimental facilities are 2-cm-long arrays constructed of 8–300 wires, with array diameters of 1–5 cm; implosion times on the order of 50–100 ns; wire diameters 5–30 μm; and implosion velocities greater than about 30–50 cm/μs. Larger numbers of wires are used in an attempt to achieve maximum implosion symmetry and increased radiated power. The experiments have indicated (Spielman et al., 1994, 1998) that the pinch reaches a minimum diameter of as small as 1 mm (depending on the load parameters) and remains at approximately this diameter for many Alfvén transit times, indicating that the pinch is in force equilibrium. The pinch develops a complex, apparently helical, structure during an expansion phase after stagnation. The more uniform the final pinch (the smaller the final pinch diameter) the smaller the diameter of the helix that is formed. Spielman et al. (1994) emphasized that the integrated and apparent brightness of the images increases during the equilibrium phase, so that energy is dissipated in the pinch. Large-scale-length MHD instabilities are evident at 15 ns (Spielman et al., 1994), and x-ray emission greatly increases. One of the purposes of the experiments was to investigate optimum conditions for maximizing K-shell yields. For instance, the effect of varying the initial load diameter on the plasma and atomic physics of the Z pinch had only limited exploration. The experiments determined that the yield could be improved by reducing the initial array radius. The possible reason may be that an increasing density-squared-volume product more than compensated for the decreasing temperature.

Wire array Z-pinch experiments on the 20-MA Z accelerator (Matzen, 1997; Spielman et al., 1998) have demonstrated good stagnations ($\sim$ 2-mm diameter) but with clear evidence that instabilities limit implosion quality. (See Figs. 6.5a, b.) These experiments used tungsten wires in order to optimize total radiation output and maximized the number of wires in the arrays to keep pinch quality high. In these tungsten wire-array experiments > 200 wires were used for the first time because of the high currents generated by Z. In these experiments the array diameters were 2- to 5-cm and the array masses chosen to optimally couple to the 105-ns drive time of the accelerator. The highest x-ray powers were seen from 4-cm diameter arrays where a balance was achieved between lower implosion velocity (smaller array diameter) and instabilities (larger array diameters).

The excellent ($\sim$ ns) agreement among observed, experimental, and calculated implosion times indicates that the simple-circuit-model description of the wire-array implosion has some validity. This agreement suggests that the assumption that the vast majority of the mass of the wires implodes as a unit rather than being left behind or accelerated ahead is a good one. In this 0-D model, the pulsed-power generator is described as a time-dependent voltage source, $V(t)$, driving a

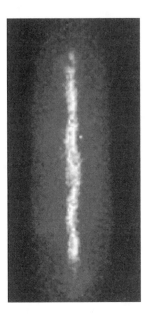

FIGURE 6.5A. A time-integrated x-ray pinhole camera picture of a 4-cm-diameter, 2-cm-long tungsten wire-array implosion. The camera has a spatial resolution of 75 μm and is viewing only photons > 1 keV.

line resistance, Z_0, a line inductance, L_p, and a dynamic imploding Z-pinch load with a time-dependent inductance, $L(r_\alpha(t))$, where $r_\alpha(t)$ is the radius of the array. The dynamic inductance of the load in this model is given by $L = -L_0 \ln(r_\alpha/R_0)$ (R_0 is the initial radius of the array). By using a slug-model description of the dynamic load to determine $r_\alpha(t)$, the load dynamics can be initially simplified to the problem of solving the following two equations (Katzenstein, 1981):

$$L_T \frac{dI}{dt} + \left(Z_0 - L_0 \frac{1}{r_\alpha} \frac{dr_\alpha}{dt} \right) I = V(t), \qquad (6.53)$$

$$\mu \frac{d^2 r_\alpha}{dt^2} = -\frac{L_0}{2I} I^2 \frac{1}{n r_\alpha}, \qquad (6.54)$$

where $L_T = L_p - L_0 \ln(r_\alpha/R_0)$, $L_0 = (2l/c^2)(1 - 1/n)$ in Gaussian units, $I(t)$ is the current flowing through the load, n is the number of wires in the array, l is the length of the array, and μ is the mass per unit length of each wire. Modern variations of Katzenstein's model are used at nearly all laboratories and provide the inputs for most 1- and 2-D radiation MHD computer codes.

Two important differences between theory and experiment are typically observed. First, the x-ray yields, as a function of mass or initial array radius, were not constant, but had a peak corresponding to a conversation of more than 100% of the 1-D radial implosion kinetic energy calculated for an infinitesimal thickness

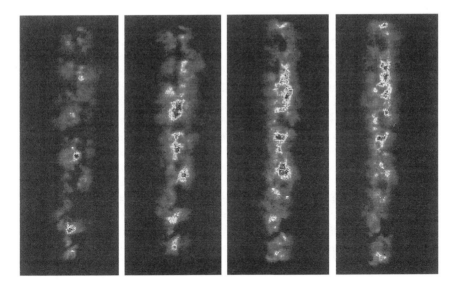

FIGURE 6.5B. Time-resolved x-ray pinhole photograph of a tungsten wire-array pinch. The gate time of the instrument is 100 ps and the interframe time is 2 ns. The resolution of the instrument is 75 μm. The peak radiation power occurs in the third frame. The initial wire array diameter was 30 mm and the length was 2 cm. The array had 180 10 μm diameter wires. The camera was filtered with 25.4 μm of Be to limit the x-ray energy to > 1 keV.

current sheath (the plasma diameter during the stagnation phase was assumed to be the measured plasma diameter, see Spielman et al., 1994, 1998). Second, very large initial perturbations (10–30%) were required in the 2-D MHD calculations to obtain agreement with the measured diameter of the stagnation plasma (about 1 mm). The first difference suggests that an additional mechanism (or mechanisms), besides 1-D radial implosion, converted magnetic energy to plasma energy. The development of 2-D structures such as results from the strongly driven RT instabilities or 3-D structures, such as results from strongly driven kink modes, would increase the magnetic energy converted to plasma energy. Two-dimensional calculations (Peterson et al, 1996) show that the additional energy comes from coupling of the $\mathbf{J} \times \mathbf{B}$ force into the finite-thickness sheath during the stagnation process. In addition, framing-camera images of the x-ray emission after stagnation (Spielman et al., 1994) indicate helical structures such as would result from kink modes. Another possible explanation is that the resistance is anomalously high. Also, the second difference could be due to anomalously high (presumably MHD-turbulence induced) viscosities and heat conductivities as well. Indeed, the phenomenological estimates utilizing enhanced electrical resistivities, viscosities, and heat conductivities seem to confirm these conjectures (Whitney et al., 1994). Phenomenological enhancements of the artificial electrical resistivity, viscosity, and heat conductivity by factors of about 20–40 were used in the calculations to "soften" the implosions

(Whitney et al., 1994). It is likely that both enhanced inductance due to the 2-D and/or 3-D structures, driven by RT and kink instabilities, and anomalous resistivity are both operating in these pinches.

The above concerns aside, 2-D MHD calculations of tungsten wire-array Z pinches have provided quantitative predictive capabilities (Peterson et al., 1996). These calculations show that the Raleigh–Taylor instability alone describes the vast majority of the pinch dynamics and that only a single variable (the initial perturbation level) is necessary to accurately model experimental data on Z. The calculations provide not only x-ray powers, energies, and pulse shapes but pinch sizes and shapes (see Fig. 6.5c). The detailed hydrodynamics in the calculations are able to duplicate the experimental energetic without resorting to modified or "anomalous" physics parameters.

It is important that solid wires be used in the experiments. Therefore, in order to unambiguously compare the calculated to the measured yield behavior, one must verify that the wire explosion and the array implosion dynamics are separable, and that the explosion approximately establishes the initial conditions that are used in

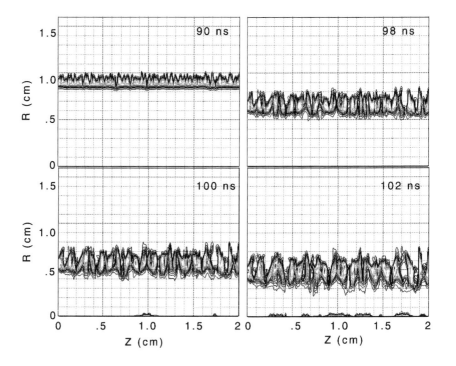

FIGURE 6.5C. A sequence of iso-density plots from a Eulerian MHD calculation is shown. The relative times for each plot is shown. Note the dramatic growth of the Rayeigh–Taylor instability in the implosion. This calculation was done for the parameters of Sandia's Z accelerator; $I_{max} = 18$ MA, $m_p = 5$ mg, $R_0 = 3$ cm, and $L = 2$ cm.

the 1-D MHD calculations. The initial conditions depend on the generator prepulse. An explosion model was developed (Bloomberg et al., 1980) that consisted of momentum and energy equations for a single wire that were coupled to a circuit equation for the pulsed-power generator. The current was assumed to uniformly penetrate the wire up to a skin depth; radiation losses were ignored and an *ad hoc* dependence of the ionization state of the wire on its average temperature was assumed. The worst assumption of this model was found to be in its use of the Spitzer formula for plasma resistivity, even though this does not apply at low temperatures (less than about 30 eV). However, the expansion dynamics were remarkably invariant to changes in the resistivity. The main effect of an artificially large resistivity was to slow the risetime of the current, but it did not otherwise basically alter the wire expansion dynamics.

All the implosions follow the same path, with energy released at stagnation and then later during an instability phase. Experiments (Baksht et al., 1989) have indicated that the x-ray yields near stagnation at the pinch radius, $r_f = 0.2$ cm, are several times larger than the kinetic energy of the imploded array. X-ray pinhole photos and laser shadow photos of aluminum arrays (Aivazov et al., 1988, 1990) also demonstrated a very well-pronounced spiral shape of the liner. The diameter of the pinch plasma, measured from x-ray pinhole photos, ranges from 1.5 to 3 mm. A typical spiral structure of the wire array liner is shown in Fig. 6.6 (courtesy of Baksht). In this photo, the pinch is recorded at 30 ns after the current through the pinch begins to flow ($I_{max} = 1$ MA).

It is possible that the whole picture is much more complicated, and the scenario of wire-array implosions depends strongly on the wire material being different for aluminum, copper, or gold wires. An alternative method for additional energy deposition is ohmic heating, as discussed above. The bounces can appear as the result of development of $m = 0$ instabilities (micropinches), with the final radius being less than 10 μm. Spitzer resistivity of a plasma with such a small radius would be extremely high, and the ohmic heating would meet the requirement for the energy balance, if the total number of micropinches is about 10 per centimeter (Baksht et al., 1993).

Another feature has been pointed out (Bekhtev et al., 1989) that may complicate the entire scenario of a wire-array implosion. It was found that part of the pinch current flows through the low-density plasma corona, which is accelerated towards the axis faster than the main mass of the wires and forms a plasma precursor on axis. Again, indications were found of the development of $m = 1$ instabilities, and sometimes $m = 0$ instabilities also. X-ray images show the formation of helical structures, with a characteristic scale of about 1 mm, that formed approximately 30 ns before the maximum yield.

Indeed, the stability analysis of the wire-array implosion indicates that the array develops a helical structure and generates an axial magnetic field, which will be frozen in the expanding plasma during the later stages of the implosion. The load inductance must increase as the effective plasma length increases. The appearance of an axial magnetic field may explain the observed plasma pulsation near the stagnation point. Evidently, the entire evolution of the collapsing, unstable plasma

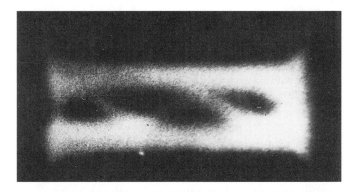

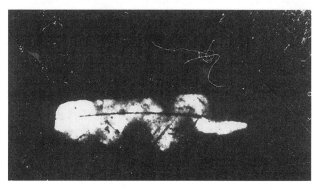

FIGURE 6.6. Top: Pinhole photograph of imploding array of six wires in visible light at 30 ns. Initial radius of the array is 1 cm, $I_{max} = 1$ MA, $t_{imp} \sim 100ns$. Part of the wires formed a spiral structure, the main mass of the array is still at rest. Bottom: Pulsation of a Z-pinch plasma during the later stage of implosion (courtesy of Baksht).

can be understood using 3-D MHD codes with realistic equations of state. However, we shall see that analysis of the dynamics and stability of the wire-array implosion, even in scope of the 0-D model, provides at least a qualitative understanding of the principal features of the experimentally observed implosion (Sanford et al., 1996).

6.4.1 Dynamics of an imploding wire array

Magnetic forces rapidly implode the wires in a high-current Z pinch. Both the explosion 1-D MHD model and the experimental data show that for a small number of wires (< 20), the wires continue to be separable and preserve themselves during the first stage of implosion while the spiral structure is formed. On the other hand, we see (Baksht et al., 1993) that the wavelength of the dominant mode is about 1 cm, of order of the height of the array and much greater than the radii of the wires, which are initially about 10 μm. After each wire is heated and expands, the wire diameter seems to remain fairly constant (about 1 mm) during most of the implosion of the array. Thus, the 0-D approach is indeed justified, and the wires can be regarded as infinitely thin current lines.

The dynamics of the unperturbed wire array is given by the same (3.2) [see also (6.53) and (6.54)]. The radius of the array varies in time, the equation of radial motion of the array being

$$\mu \frac{d^2 r}{dt^2} = -\frac{(n-1)}{n^2} \frac{I_{max}^2}{c^2 r} \Phi^2(t), \qquad (6.55)$$

where $I(t) = I_{max} \Phi(t)$ is the pinch current, $\Phi(t)$ is the current waveform, $r(t)$ is the radial coordinate of each wire, n is the number of wires, and c is the light speed. We assume that the return-current bars are mounted symmetrically and coaxially, and that the forces produced by the currents of the return bars can be neglected relative to the force produced by the other imploding wires. The wires are also assumed to be infinite in length (valid if the unstable wavelengths are long compared to the wire length).

It is convenient to use the dimensionless variables

$$R = r/R_0, \quad \tau = t/\tau_A, \qquad (6.56)$$

where R_0 is the initial radius of the array and $\tau_A = \mu^{1/2} c R_0 / I_{max}$ is the Alfvén transit time (see Sec. 3.2.1). In dimensionless variables, (6.55) takes the form

$$\frac{d^2 R}{d\tau^2} = -\frac{(n-1)}{n^2 R} \Phi^2(\tau), \qquad (6.57)$$

and for a constant current, $\Phi(\tau) = 1$, (6.57) can be solved exactly [see (3.3) and (3.4), Sec. 3.2.1]:

$$\frac{dR}{d\tau} = -\frac{n}{\sqrt{n-1}} [2\ln(1/R)]^{1/2}, \quad \tau = \sqrt{\frac{\pi n^2}{2(n-1)}} \, \mathrm{erf}[\sqrt{\ln(1/R)}]. \quad (6.58)$$

It follows from (6.58) that the characteristic run-in time of the array implosion is

$$\tau_{imp} = \sqrt{\frac{\pi n^2}{2(n-1)}}. \qquad (6.59)$$

Another feature of the pinch dynamics can be seen from the measured r-t diagram (Spielman et al., 1994) and from the 0-D model: the pinch radius does not change significantly during most of the implosion. In other words, most of the radial motion of the array occurs during the final stage ($\sim$ 20 ns) of the implosion. Felber and Rostoker (1981) have also found that the characteristic implosion time (6.59) is not sensitive to the precise shape of the current waveforms for the implosions. Therefore, in what follows, we shall assume $R(\tau) = R(0) = 1$ and $\Phi \equiv 1$ for analytical estimates.

6.4.2 Imploding wire array: Stability analysis

Consider the stability of the initially symmetric form wire array against small perturbations. The stability of a wire array was first considered by Felber and

Rostoker (1981). However, in their analysis, the authors neglected perturbations of the z component of the wire displacement. Therefore, the results obtained by Felber and Rostoker are correct only for the case of the displacement instability of pure radial perturbations. We shall consider two kinds of instabilities: (1) the displacement instability that causes each straight wire to move away from its canonical position in the imploding array; and (2) the instabilities with $k \neq 0$, in particular kink instabilities that cause the growth of sinusoidal perturbations. These instabilities cause braiding, twisting, and formation of a spiral structure, and are those most evident in x-ray pinhole photographs of imploding wire arrays.

Let us assume that each wire carries an equal current, I/n. Let S_j be a dimensionless parametric coordinate ($-\infty < S_j < \infty$) of the wire at a given instant of time, i.e., the dimensionless coordinate of a point at the wire is $\mathbf{R}_j(S_j, \tau) = \{x_j(S_j, \tau), y_j(S_j, \tau), z_j(S_j, \tau)\}$, where $j = 1, 2, \ldots$, is the number of a wire, and n is the number of wires in the array. In cylindrical coordinates (r, θ, z), we have the coordinates of each wire, $\mathbf{R}_j(S_j, \tau) = \{R_j(S_j, \tau), \theta_j(S_j, \tau), z_j(S_j, \tau)\}$. It is convenient to choose coordinate S_j, which coincides with the z coordinate of an unperturbed wire at the beginning of implosion, at $t = 0$. Then the unperturbed motion of j wire is given by

$$\mathbf{R}_j(S_j) = \{R_j(\tau), \theta_{0j}, S_j\}, \tag{6.60}$$

where $R(\tau)$ is given by the solution of (6.57), and $\theta_{0j} = 2\pi j/n$.

We introduce a small displacement of the wire from the equilibrium in the following form

$$\delta\mathbf{R}_j = \Delta \begin{pmatrix} A_r \\ A_\theta \\ A_z \end{pmatrix} \exp(\sigma_{mk}\tau + i(kS_j + m\theta_{0j})), \tag{6.61}$$

where $\Delta \ll 1$; A_R, A_θ, A_z are the amplitudes of the perturbation with the wavenumbers k (in the units $1/R_0$) and azimuthal wavenumber m ($m = 0, 1, 2, \ldots, n/2$).

The magnetic field (in units $B_0 = I_{\max}/ncR_0$) at the point $\mathbf{R}_i = \mathbf{R}_i + \delta\mathbf{R}_i$, generated by a wire segment $dl_j = l_j dS_j$, with the coordinate $\mathbf{R}_j = \mathbf{R}_j + \delta\mathbf{R}_j$, is given by the Biot and Savart law

$$d\mathbf{B}_{ij} = \frac{d\mathbf{I}_j \times \mathbf{R}_{ij}}{|R_{ij}|^3}, \tag{6.62}$$

where $\mathbf{R}_{ij} = \mathbf{R}_i - \mathbf{R}_j$. Integrating (6.62) to first-order terms in Δ, we obtain for the components of magnetic field B_{ij} in cylindrical coordinates

$$(\mathbf{B}_{ij})_R = \frac{2\sin(\theta_{ij})}{(\pi_{ij})^2} + \Delta_i \left\{ \frac{2A_\theta - 2A_R \sin(\theta_{ij})}{(\pi_{ij})^2} + \Delta_i e^{im\theta_{ij}} \right. \tag{6.63}$$

$$\times \left\{ A_R k^2 \sin(\theta_{ij}) K_0(k\pi_{ij}) - A_\theta \frac{2k}{\pi_{ij}} K_{-1}(k\pi_{ij}) + A_\theta \frac{k^2\pi_{ij}^2}{2} K_0(k\pi_{ij}) \right\},$$

$$(\mathbf{B}_{ij})_\theta = 1 + \Delta_i \left\{ \frac{2\cos(\theta_{ij})}{(\pi_{ij})^2} A_R \right\} + \Delta_i k^2 e^{im\theta_{ij}} \qquad (6.64)$$

$$\times \left\{ A_\theta \sin\theta_{ij} K_0(k\pi_{ij}) - A_R \frac{2}{k\pi_{ij}^2} K_{-1}(k\pi_{ij}) - A_R(1 + \cos\theta_{ij}) K_0(k\pi_{ij}) \right\},$$

where the following designations are used: $\Delta_i \equiv \Delta \exp[\sigma_{mk}\tau + i(kS_i + m\theta_i)]$, $\pi_{ij} \equiv 2\sin(\theta_{ij}/2)$, $\theta_{ij} \equiv \theta_{0i} - \theta_{0j}$, and $K_n(x)$ is the nth-order modified Bessel function. Note that the z component of the perturbed magnetic field $(\mathbf{B}_{ij})_z$ is of first order in Δ, so that it does not enter in the linearized equation of motion. The net magnetic field acting on the wire at the point $R_i(S_i)$ can be found by summing the field from over the all wires at $i \neq j$. After simple algebra, we obtain the components of the magnetic field in the following form:

$$(\mathbf{B}_i)_R = \Delta_i \{ A_R b + A_\theta f_{no} A_\theta b_\theta \}, \qquad (6.65)$$

$$(\mathbf{B}_i)_\theta = (n-1) + D_i \{ A_R f_{n1} + A_R b_R + A_\theta b \}, \qquad (6.66)$$

where

$$f_{nm} = \sum_{j=1}^{n-1} \frac{\cos(m\theta_{0j})}{1 - \cos(\theta_{0j})}, \qquad (6.67)$$

$$b = i \sum_{j=1}^{n-1} k^2 \sin(m\theta_j) \sin\theta_j K_0(k\pi_{jn}), \qquad (6.68)$$

$$b_\theta = -\sum_{j=1}^{n-1} \left[\frac{2k}{\pi_{jn}} K_{-1}(k\pi_{jn}) + (1 - \cos\theta_{0j})k^2 \pi_{jn} \right]$$
$$\times \cos(m\theta_{0j}), \qquad (6.69)$$

$$b_R = \sum_{j=1}^{n-1} \left[-\frac{2k}{\pi_{jn}} K_{-1}(k\pi_{jn}) + (1 + \cos\theta_{0j})k^2 K_0(k\pi_{jn}) \right]$$
$$\times \cos(m\theta_{0j}), \qquad (6.70)$$

Now, the equation of motion for the element $d\mathbf{l}_i = (\partial \mathbf{R}_i/\partial S_i)dS_i$ of the wire can be written as

$$\frac{d^2\mathbf{R}_i}{d\tau^2} = \frac{1}{n^2} \mathbf{l}_i \times \mathbf{B}_i. \qquad (6.71)$$

Substituting $\mathbf{R}_i$ and $\mathbf{B}_i$ from (6.60), (6.61), (6.65), and (6.66) into (6.71), one can find that (6.71) reproduces the equation for the unperturbed motion in zeroth order in Δ. To first-order terms in Δ, it gives

$$n^2\sigma_{mk}^2 A_R + (f_{n1} + b_R)A_R + bA_\theta + ik(n-1)A_z = 0, \qquad (6.72)$$

$$n^2\sigma_{mk}^2 A_\theta - bA_R - (b_\theta + f_{n0})A_\theta = 0, \qquad (6.73)$$

$$n^2\sigma_{mk}^2 A_z - ik(n-1)A_R = 0. \qquad (6.74)$$

The condition for the existence of a nontrivial solution of (6.72)–(6.74), i.e., setting the determinant of the coefficients (6.72)–(6.74) equal to zero, gives the dispersion equation determining the growth rate of the perturbations. In general, the dispersion equation is of third order in σ_{mk}^2. Below, we shall consider the development of the most unstable displacement and $k \neq 0$ instabilities, and the possibility of forming a spiral shape of the array due to the growth of instabilities. We shall consider also the nonlinear development of the fastest growing perturbations, by means of numerical modeling of the wire-array implosion.

6.4.3 Displacement instability

The displacement instability represents the limiting case of $k = 0$, when each wire may be displaced from its unperturbed position, remaining a straight line parallel to the z axis. For $k = 0$, we have from (6.67)–(6.70) $b_R = b_\theta = -f_{nm}$, and from (6.72)–(6.74), we obtain two solutions: either $A_z = A_\theta = 0$, $A_R \neq 0$

$$n^2 \sigma_{m0}^2 A_R = (f_{nm} - f_{n1}) A_R, \tag{6.75}$$

or, $A_z = A_R = 0$, $A_\theta \neq 0$

$$n^2 \sigma_{m0}^2 A_\theta = (f_{n0} - f_{nm}) A_\theta. \tag{6.76}$$

As is seen from (6.75) and (6.76), two different displacement modes grow independently. These are the radial mode, $A_R \neq 0$, $A_\theta \equiv 0$, and the tangential mode, $A_\theta \neq 0$, $A_R \equiv 0$. The growth rate of the radial mode can be obtained from (6.75):

$$\sigma_{m0}^R = \frac{1}{n} \sqrt{f_{nm} - f_{n1}} = \frac{1}{n} \sqrt{(1 - m)(n - m - 1)}. \tag{6.77}$$

For the tangential mode, $A_\theta \neq 0$, $A_R \equiv 0$, taking into account the identity $(f_{n0} - f_{nm}) = m(n - m)$, we find

$$\sigma_{m0}^\theta = \frac{1}{n} \sqrt{m(n - m)}. \tag{6.78}$$

In fact, the $m = 0$ radial mode is just the breathing mode, and it does not affect the symmetry of the implosion. This mode corresponds to equal displacements of all the wires in the radial direction, with the unperturbed acceleration. All other radial modes with $m \neq 0$ are either stable or marginally stable. The $m = 1$ radial mode is always marginally stable because it represents a translation of the entire array in the plane perpendicular to the axis.

Equation (6.78) demonstrates that an array is marginally stable against simple rotation ($m = 0$ mode), contrary to the result by Felber and Rostoker (1981), but that other azimuthal displacement modes are unstable. The maximum growth rate corresponds to $m = n/2$ or $(n \pm 1)/2$ for even and odd values of n, respectively. For large values of n, the maximum growth rate for the tangential modes can be estimated as

$$\sigma_{m0} \approx \frac{n}{2\sqrt{n - 1}}, \tag{6.79}$$

or in dimensional form this is

$$\overline{\sigma}_{m0} = \frac{n I_{\max}}{2c R_0 \sqrt{\mu(n-1)}}. \tag{6.79a}$$

It can be shown (Samokhin, 1988) that this estimate remains essentially unchanged, if a perturbation of the electrical current flowing through each wire due to a perturbed inductance is taken into account.

In effect, the instability of the azimuthal displacement modes is similar to the conventional filamentation instability. Its growth rate is found to be large enough to grow significantly during the implosion [approximately $\sqrt{n}$ times higher than the characteristic value $1/\tau_{\mathrm{imp}}$, cf. (6.59) and (6.79)].

As an assessment for the growth of perturbations, we take the value

$$\Gamma = \sigma \tau_{\mathrm{imp}}, \tag{6.80}$$

so that the growth of an initial amplitude of perturbation, A_0, is estimated as $A = A_0 \exp(\Gamma)$. As is seen below, this estimate is quite realistic and agrees qualitatively with the results of direct numerical modeling.

For the fastest tangential modes, (6.79), and for an array with $n \gg 1$, we obtain

$$\Gamma_{m0} \approx \frac{n^2}{(n-1)} \sqrt{\frac{\pi}{8}} \approx 0.6n. \tag{6.81}$$

This expression indicates that the amplitudes of the tangential perturbations that grow during the implosion become comparable with the average distance between the wires. For example, linear growth of a perturbation with an initial amplitude of about the initial wire diameter ($A_0 = 0.001$ cm) grows up to $A = 1$ cm at the end of the implosion, for an array with $n = 12$ wires. Nevertheless, linear instability of tangential modes does not affect the run-in times of the wires, nor does it change the value of the driving magnetic field, B_θ, because the magnitude of B_θ was shown to be independent of the angular position of wires at equal distances from the axis. Also, the displacements of the wires are still small in comparison with the pinch radius. Therefore, these perturbations essentially do not affect the motion of the wires. At a later stage of the implosion, when the wires coalesce into a dense plasma, these perturbations are responsible for the initial perturbations that bend the magnetic field lines. As is known, perturbations of this type are not very dangerous. An example of the wire array implosion through the stagnation stage is shown in Fig. 6.5a and 6.5b.

6.4.4 Instabilities with $k \neq 0$

The instabilities with $k \neq 0$ involve deformations of the wires, formation of a spiral structure, braiding, twisting, and turning of the wire plasmas in an imploded array, like those shown in Fig. 6.6.

The sausage perturbations, which are generally distinguished from kink perturbations, are included (Felber and Rostoker, 1981) in the same class of perturbations that are not simply displacements of vertical wires. The growth rate of the kink instability of a single wire of finite radius has been studied by Felber and Rostoker (1981). They found that the maximum growth rate for a single wire occurs at a wavelength about equal to two wire diameters. We shall consider the growth of the perturbations of a wire driven by its interaction with the magnetic field of all the other wires. This model breaks down in the short-wavelength limit, in which the radius of curvature of the kink is comparable to the wire radius.

Let us consider first the simplest case: $m = 0$, $k \neq 0$. In this case $b = 0$, and the system (6.72)–(6.74) reduces to separate equations for the tangential mode, $A_z = A_R \equiv 0$, $A_\theta \neq 0$:

$$(n\sigma_{0k}^\theta)^2 = b_\theta + f_{n0}, \tag{6.82}$$

and for the radial (sausage) mode, $A_\theta = 0$, $A_z \neq 0$, $A_R \neq 0$:

$$(n\sigma_{0k}^R)^4 + (b_R + f_{n1})(n\sigma_{0k}^R)^2 - (gk)^2 = 0, \tag{6.83}$$

where $g = (n - 1)/n^2$ is the dimensionless radial acceleration of the wires [see (6.57)]. Consider now the short-wavelength perturbations. This approach is applicable for perturbation wavelengths that are large compared to the curvature of the bent wires and to the distance between the wires. Taking into account that the distance between the wires is $2\pi/n$ in dimensionless units, we can write this condition as $k \gg n$. With the condition $k \gg n$, the system (6.72)–(6.74) reduces to an accuracy up to the terms of order $\exp(-2\pi k/n)$ to the equations

$$n^2\sigma_{mk}^2 A_R = -f_{n1}A_R - ik(n - 1)A_z, \tag{6.84}$$

$$n^2\sigma_{mk}^2 A_\theta = f_{n0}A_\theta, \tag{6.85}$$

$$n^2\sigma_{mk}^2 A_z = ik(n - 1)A_R. \tag{6.86}$$

The growth rate of the radial modes, $A_R \neq 0$, is obtained from (6.84) and (6.86)

$$\sigma_{mk}^R = \frac{1}{n}\left[\sqrt{\left(\frac{f_{n1}}{2}\right)^2 k^2 g^2} - \frac{f_{n1}}{2}\right]^{1/2}.$$

In what follows, we shall use the estimate $f_{n0} \approx 0.16n^2$, which is valid with good accuracy (other terms in the expression for f_{n0} are as small as $1/n$ in the power expansion for $n \gg 1$), and the functions f_{n1}, which have been tabulated for different n by Felber and Rostoker (1981).

For large enough wavenumbers, the expression is well approximated by

$$\sigma_{mk}^R \approx \sqrt{gk}.$$

Thus, this is an RT instability.

The growth of the tangential perturbation modes (Γ_{0k}^θ) for large wavenumbers can be obtained using (6.82) for σ_{0k}^θ. Using an asymptotic expansion for large x for the modified Bessel functions $K_0(x) \to 0$, $K_{-1}(x) \to 0$ in the expression (6.69) for b_θ, and using (6.83), we obtain

$$\Gamma_{0k}^\theta \approx 0.5\sqrt{n}. \tag{6.87}$$

Growth of the tangential, Γ^{θ}_{0k}, and sausage, Γ^{R}_{0k}, instabilities calculated using (6.82) and (6.83) is shown in Fig. 6.7 for a 24-wire array. As is seen from Fig. 6.7, the radial perturbations for the $m = 0$ mode grow faster than the tangential ones for large wavenumbers, and they have almost the same instability growth rates in the range of wavenumbers $2 \leq k \leq 5$. The net growth of tangential perturbations is asymptotically bounded (at large wavenumbers) by almost the value (6.87). The asymptotic dependence of the instability growth rate for radial perturbations is given by the expression $\sigma^{R}_{mk} \approx \sqrt{gk}$. Multiplying this expression by the implosion time, we obtain the instability growth during the implosion

$$\Gamma_k \approx \sqrt{k\frac{\pi}{2}}.$$

In fact, this equation indicates only that the highest relative growth of the amplitude corresponds to the perturbations with the shortest permissible wave length.

The stability analysis presented above is valid when the pinch radius is constant and the unperturbed wires are at rest, i.e., at the time instant $t = 0$. When the unperturbed wires start to move with finite velocity, the amplitudes of the tangential modes also grow because of the perturbations drifting with the unperturbed motion. This enhances the growth of the perturbations having angular displacements, so that the growth of the tangential perturbations for moderate wave numbers ($k < 10$) becomes larger than the growth of the radial modes very soon after the beginning the implosion. A consistent analysis taking into account the unperturbed motion of

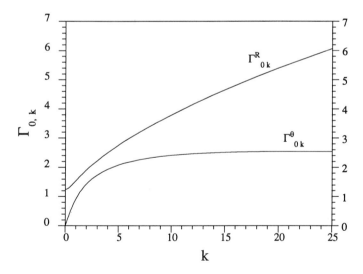

FIGURE 6.7. Growth of the tangential (Γ^{θ}_{0k}) and radial (Γ^{R}_{0k}) perturbations in a wire array of 24 wires during the linear stage of the development of instabilities calculated from (6.82) and (6.83).

the wires for all kinds of perturbations is a rather difficult task, but for the tangential mode, the growth rate can be obtained from the following equation [compare with (6.83)]

$$n^2(\sigma_{0k}^\theta)^2 - 2\sigma_{0k}^\theta n|\dot{R}| = b_\theta + f_{n0}. \tag{6.88}$$

An estimate for the perturbation growth, which follows from (6.88), gives values of Γ_{0k}^θ about 30–50% higher than the values obtained from (6.87). Indeed, numerical simulation confirms this conclusion: tangential perturbations with wavenumbers of order of unity start to grow faster than the radial perturbations very soon after the beginning of the implosion (see Fig. 6.8).

Numerical analysis of (6.72)–(6.74) demonstrates that for a given wavenumber k the fastest growing mode corresponds to $m = n/2$. Instability growth $\Gamma_{n/2,k}$ for this mode is presented in Fig. 6.9 for the arrays with different number of wires: $n = 12, 24, 48$. Figure 6.9 shows [see also (6.87)] that for perturbations with large wavenumbers the instability growth is almost independent of m (and the number of the wires) and the perturbations grow according to the asymptotic expression $\Gamma_k \approx \sqrt{k\frac{\pi}{2}}$ with rather good accuracy. Contrary to this, the long-wavelength perturbations with wave numbers of order unity grow faster for the array with a larger number of wires.

When considering the formation of a spiral structure of the wire array due to the growth of the instabilities, we should take into consideration that the maximum initial amplitude of the perturbations occurs to wavenumbers $k = 2\pi R_0/L$ (L is the length of the array, which is about the initial diameter of the array, $2R_0$) and that the initial perturbation amplitudes, $A_k(t = 0)$, decrease as $\propto k^{-2}$ for large wavenumbers according to the Fourier theorem. Thus, the typical amplitude of the perturbations is of order of the wire diameter, and the high relative growth of the perturbations does not mean growth to a higher final amplitude during the implosion. The diameter of a wire should be taken as the diameter of the exploded wire plasma, which seems to remain fairly constant at about 1 mm during most of the implosion of the array. Also, within the scope of a linear analysis, the perturbation amplitudes may be estimated using the exponential law (6.81), until the amplitude reaches the values of about $A_{mk} \approx k^{-1}$. For later times, only numerical simulations can provide a realistic picture to make the subsequent analysis valid for perturbations with moderate wave numbers, $k < 10$.

Flexible wires with constant radius are assumed to move only under the influence of the magnetic fields generated by their own currents and the currents of other wires in the array. At the same time, the growth of the instabilities results in twisting of the wires and the formation of an axial magnetic field inside the wire array. This occurs because the formation of a spiral structure during an earlier stage of the implosion causes an azimuthal component of the current still flowing through the wires. The axial magnetic field due to this kind of perturbation can be estimated as

$$B_z \approx \frac{A_\theta}{\pi} B_\theta. \tag{6.89}$$

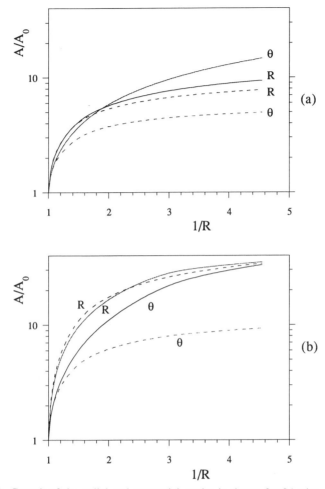

FIGURE 6.8. Growth of the radial and tangential modes is shown for 24-wire array as a function of the radial compression (a) for $k = 3$ and (b) $k = 10$. The solid lines are the result of numerical calculations; the dashed lines represent a linear stability analysis.

Therefore, the following scenario of the array implosion seems to be plausible. The wires are twisted during the first stage of the implosion because the fastest instabilities are the tangential perturbations with the wave numbers of order unity. Twisting modes, with initial amplitudes of about the diameter of the wire, grow exponentially. The initial stage of the twisted implosion of the array with small diameter wires establishes the initial conditions for forming a plasma at the later stage, with a frozen-in axial magnetic field, that reaches a magnitude of about $0.1 B_\theta$ at the later stage. During the subsequent compression when a dense plasma

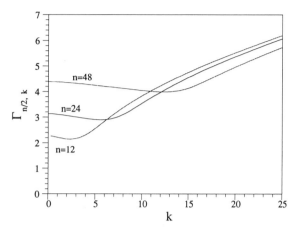

FIGURE 6.9. Growth of the fastest growing perturbations for an arrays of 12, 24, and 48 wires.

is formed, the magnetic field, B_z, is frozen-in and grows as R^{-2}, while the azimuthal magnetic field, grows as R^{-1}. Thus, at a compression $R^{-1} \approx 10$, both magnetic fields B_θ and B_z achieve almost the same magnitude. The magnetic pressure of the internal axial magnetic field may stabilize the pinch during the stagnation and post-stagnation phases (this is correct in the scope of an ideal MHD approximation). Figure 6.10 shows the results for different arrays of numerical calculation of the ratio of the azimuthal and axial magnetic fields, $\beta = B_z/B_\theta$, as a function of the radial compression, $1/R$ (here β is to be distinguished from the plasma β).

Concluding this section, it is interesting to note that, contrary to the usual gas-puff or plasma Z pinches, the presence of an external magnetic field, B_z^{ext}, has a destabilizing effect on the implosion of a wire array. This can be understood from a simple consideration of the Lorentz force, which increases the radial component of the perturbations if the velocity has an angular component, and increases the angular component of perturbation if the velocity has a radial component. Therefore, the perturbations with properly phased components grow. The analysis of the instabilities in an imploding array in an external magnetic field B_z^{ext} can be performed in a way similar to that in 6.4.2. The first-order terms in Δ give equations similar to equations (6.72)–(6.74), where the coefficient b must be replaced by $(b - iB_z^{\text{ext}})$. Again, the condition for the existence of a nontrivial solution of the equations, i.e., the determinant of coefficients is equal to zero, gives the dispersion equation determining the growth rate of the perturbations. In the limit of large wavenumbers, we find

$$\sigma \approx \sqrt{k[g^2 + (B_z^{\text{ext}}/B_\theta)^2]^{1/2}}.$$

As is seen, the external magnetic field enhances the instability of the RT type.

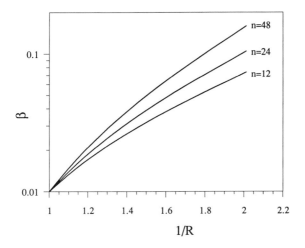

FIGURE 6.10. Results of the numerical calculation for the ratio of the azimuthal and axial magnetic fields, $\beta = B_z/B_\theta$, as a function of the radial compression, $1/R$.

6.4.5 Numerical modeling

A linear perturbation analysis suffices as long as the amplitude of the perturbation is much less than its wavelength. To confirm the qualitative conclusions, which have been reached on the basis of the linear analysis and which, in the scope of linear theory, can be treated rather as a tendency, the numerical simulation has been performed. In numerical modeling, the array is treated as being periodic with a given period along the parametric coordinate, which can be interpreted as a mass coordinate. The wires are assumed to be infinitely thin, and we consider the growth of the perturbations of the wires, driven by their interaction with the magnetic field of all other wires. Indeed, at array radii $kR \approx 1$, the effects of the other wires in the array dominate. Once the wires approach closely enough to interact, the instability modes have the largest growth rates and will establish themselves on the wires. The wavelength of these modes will be the wavelength established earlier in the implosion as the fastest growing self-induced mode, namely, about a few wire diameters.

The following numerical procedure was used. Each wire was considered as a set of finite-point masses, m_{ij}, having coordinates $\mathbf{r}_{ij}$. The coordinates of the points and their velocities, v_{ij}, are the unknowns of the problem. We assume that each wire carries equal current I/n. The direction of the current, at a given point of the wire, is the tangent to the curve represented by the points of the wire. A system of ordinary differential equations written for any point, i, of the wire, j

$$m_{ij}\frac{d\mathbf{v}_{ij}}{dt} = \mathbf{F}_{ij}, \quad \frac{d\mathbf{r}_{ij}}{dt} = \mathbf{v}_{ij} \tag{6.90}$$

was solved using a standard method (see, for example, Potter, 1978).

Our numerical modeling demonstrates that for wide range of initial conditions, the perturbations grow, being proportional to their initial amplitudes. Therefore, the time history of perturbations may be written in the form

$$\frac{A}{A_0} = G_{mk}(t), \qquad (6.91)$$

where the function $G_{mk}(t)$ is dependent upon the type of perturbations. In a sense, this expression is trivial, in the case of an infinitely small initial amplitude of the eigenmode. The perturbation amplitudes grow exponentially at the beginning of implosion, according to the expression (6.81). Figure 6.11 shows the growth of the perturbations [$G_{mk}(t)$ as a function of the radial compression $1/R$] for the radial ($A_R \neq 0$) and tangential ($A_\theta \neq 0$) modes ($m = 0, k = 10$) with different initial amplitudes of perturbations. The important feature is that the perturbations with larger initial amplitudes grow much faster than the perturbations with smaller initial amplitudes. This effect is even more pronounced for the tangential perturbations. Because of such self-induced acceleration of tangential modes, the twisting of the array becomes more important in the array dynamics.

Another nonlinear effect enhances redistribution of the perturbations energy in favor of twisting modes. Figure 6.12 presents the growth of the radial and tangential modes of finite initial amplitudes for two different initial conditions. The dashed lines are plotted for the case when radial and tangential perturbations grow independently, the solid lines show the result of nonlinear interaction between the modes. Owing to the nonlinear interaction, the tangential mode starts to grow much faster, while the growth of radial mode slows after the imploding array radius decreases about 1.5 times.

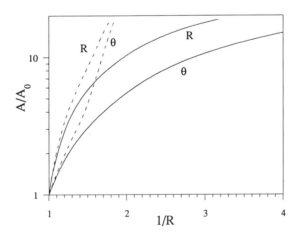

FIGURE 6.11. Nonlinear growth of the radial and tangential perturbations for the 24-wire array as a function of the radial compression ($m = 0, k = 10$). The solid lines are the result of numerical calculations for a small initial amplitude; the dashed lines are for finite initial amplitudes ($A_0 = 0.05$).

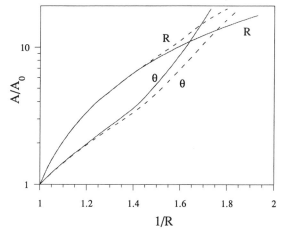

FIGURE 6.12. Nonlinear interaction between the radial and tangential modes ($m = 0$, $k = 10$; $A_0 = 0.05$). The dashed lines represent the growth of perturbations initiated separately; the solid lines demonstrate interaction of the simultaneously growing modes.

The nonlinear growth of perturbations results in the increase of the averaged kinetic energy of the wires. This may be another explanation for the large kinetic portion of the energy converted to thermal energy, and may explain the observed x-ray yields from imploding array. An excess of kinetic energy can be estimated as

$$\frac{\Delta E_{kin}}{E_{kin}} \approx \left(\frac{\sigma_k A}{\dot{R}}\right)^2 = \frac{1}{k}(kA)^2. \tag{6.92}$$

An estimate of (6.92) for the short-wavelength perturbations ($k \approx 30$), which reach the nonlinear stage ($kA \approx 1$) rather quickly, gives an energy excess of at least few percent. Figure 6.13 shows the excess of kinetic energy for 24-wire arrays because of the nonlinear growth of perturbations.

6.5 Ideal MHD model

6.5.1 WKB approximation

In this section we study the stability of 1-D plasma flows in dynamic Z pinches and liners. Consider the linear stage of perturbation growth in the course of a 1-D compression or expansion of a Z pinch. We will estimate the perturbation growth with the aid of a WKB approximation, as before, supposing that the inequality (5.2) holds, so that the rate of variation of the steady-state variables is much smaller than the instability growth rate. The problem of determining the slowly time-dependent, instantaneous growth rate is formally equivalent to a similar problem formulated

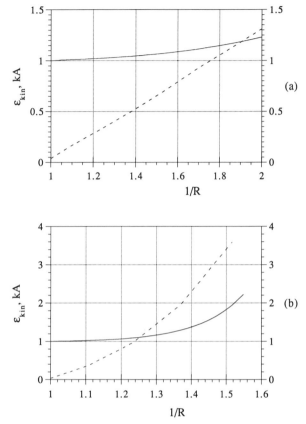

FIGURE 6.13. The solid line is the excess of kinetic energy of the perturbed array vs. the radial compression; the dashed lines are plotted for the function kA; (a) $k = 12$, (b) $k = 31$.

for a plasma equilibrium in an effective gravitational field. The equation-of-force equilibrium, taking into account the gravitational or inertial force at each instant of time, is

$$\frac{dP}{dr} + \frac{1}{4\pi}\left[B_z \frac{dB_z}{dr} + \frac{B_\varphi}{r} \frac{d}{dr}(rB_\varphi) \right] = \rho g, \qquad (6.93)$$

where

$$g(r, t) = -\left(\frac{\partial u_r}{\partial t} + u_r \frac{\partial u_r}{\partial r} \right)$$

is just the local plasma acceleration with the sign changed.

The growth rates are eigenvalues of the boundary-value problem for the same equation (4.30), where K is given by (4.31) and

$$L = \rho \Gamma^2 + F^2 + \frac{B_\varphi}{2\pi} \frac{d}{dr}\left(\frac{B_\varphi}{r} \right) + g \frac{d\rho}{dr}$$

$$-\frac{1}{D}\left[\rho^2 g^2\left(k^2 + \frac{m^2}{r^2}\right)(\rho\Gamma^2 + F^2) - \frac{\rho^2\Gamma^2 g}{\pi r}kB_\varphi\left(kB_\varphi - \frac{m}{r}B_z\right)\right]$$

$$+ r\frac{d}{dr}\left\{\frac{kB_\varphi}{2\pi r^2 D}\left(kB_\varphi - \frac{m}{r}B_z\right)\left[\rho\Gamma^2\left(\frac{\mathbf{B}^2}{4\pi} + \gamma P\right) + \gamma P F^2\right]\right.$$

$$\left. + \frac{\rho^2\Gamma^2 g F^2}{rD}\right\}, \tag{6.94}$$

F^2, D, and B^2 are determined above in Sec. 4.1.2. For an equilibrium plasma configuration ($g \equiv 0$), (6.94) is reduced to (4.31); and for the case of plane geometry, the equations of Sec. 5.2 are reproduced. The boundary conditions for (4.30) given in Sec. 4.1.3, remain valid; in particular, if the density vanishes at the plasma boundary and (ρg) is continuous there, the boundary conditions are (4.33) and (4.34). For a diffuse Z pinch without a distinct boundary, the perturbations of density, pressure, etc., should vanish at infinity. If (4.30) has a singular point at the boundary ($K = 0$), then the boundary condition is reduced to the requirement that $\xi_r(r)$ is regular at the boundary.

The spectrum determined by this boundary-value problem has the same general properties as those described in Sec. 4.1.3: the unstable eigenmodes correspond to discrete eigenvalues, so that for any wave numbers, m and k, a well-defined instantaneous growth rate Γ corresponds to the maximum eigenvalue. However, now the growth rates $\Gamma = \Gamma_{m,k}$ are time-dependent. Using (5.1) to calculate the value of n_{eff}, one should take into account that at different stages of motion the maximum growth rate may correspond to different combinations of m and k.

Note the difference between the time-dependent problem considered here and that discussed in Secs. 4.1.2 and 4.1.3. First, a steady-state plasma equilibrium may be stable, at least with respect to certain perturbation modes, as long as the respective criteria are satisfied [see (4.21) and (4.22)]. The RT modes are unstable during the implosion and develop at the surface and/or in the volume of the plasma. A stability criterion may be formulated only as a limitation imposed on n_{eff}. Second, only the instability modes that can develop during the characteristic time of the implosion, τ, i.e., the large-scale hydromagnetic modes, are really dangerous; the drift, kinetic, and other instabilities driven by slower mechanisms, which may be important for an MHD-stable steady-state plasma equilibrium (i.e., in the limit $\tau \to \infty$), are unimportant in most cases. Thus, one can use the ideal MHD approximation for the fast perturbation modes, provided that the conditions (3.45)–(3.51) are satisfied for them, whereas the dynamics of the unperturbed flow may be dissipation dominated.

To study the evolution of a perturbation, one must specify the unperturbed plasma flow in a cylindrically symmetric Z pinch. The best way to do this is to use the results of 1-D numerical simulations, which correctly describe the density, pressure, and magnetic-field profiles of the Z pinch, including the most important part of the flow, which represents the plasma–vacuum boundary where the main decrease in density takes place. We shall see below that the growth rates are very sensitive to the shape of the unperturbed density profile near the boundary.

Unfortunately, it is not quite clear yet which profiles are realized in the experiments. For this reason, the first papers on perturbation growth in the course of a Z-pinch compression used either the ideal MHD self-similar solutions with homogeneous deformation (Bud'ko et al., 1989a, b, 1990b) or time-dependent Z-pinch equilibria (Culverwell and Coppins, 1989). The interesting qualitative results are those that are not sensitive to the particular shapes of the self-similar profiles studied.

The results of Bud'ko et al. (1989a,b, 1990b) have shown that dynamic Z pinches exhibit much higher instability growth rates in comparison with steady-state Z-pinch equilibria, in agreement with the estimate (5.3). Figure 6.14 presents the eigenfunctions and the eigenvalues Γ corresponding to several of the lowest radial

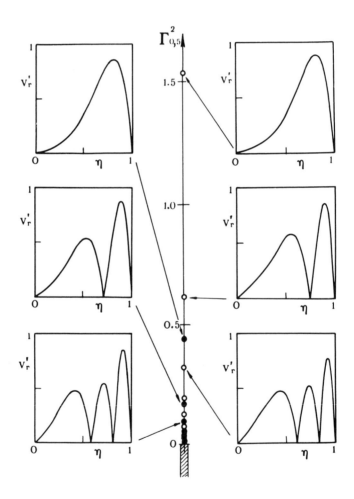

FIGURE 6.14. Radial velocity eigenfunctions, $v_r(r)$ (absolute values are shown), and the discrete eigenvalues Γ^2 for the cases of a steady-state equilibrium (left) and a Z-pinch implosion (right).

eigenmodes of the sausage instability ($m = 0, kR = 5$) of Z pinches with Gaussian density and pressure profiles. One is in a steady-state equilibrium (left), and the other is at rest at the initial moment of a self-similar compression, described by the equation of motion (3.73), with (the plasma β) $\beta = 0.1$ and $b = 0$, e.g., the total current is $\beta^{-1/2} = 3.16$ times greater, the current density profile being the same (right). We see that the profiles of the eigenfunctions are almost indistinguishable, but the growth rates are appreciably higher for a dynamic Z pinch.

Thus, the first conclusion is that if the right-hand side of (6.93) is of the same order as the maximum term on the left-hand side, then the RT instability develops and is the dominant hydromagnetic instability. The MHD stability criteria derived for steady-state equilibria are invalid here. In particular, sausage instability modes develop for any unperturbed density and pressure profiles, including those which satisfy the Kadomtsev criterion (4.21); the filamentation instability modes with $m \geq 2$, which do not develop in a steady-state diffuse Z pinch [see (4.21)], are present in a dynamic Z-pinch flow (see Fig. 6.15).

The second conclusion is that the shape of the spectrum of instability growth rates depends on the law of density decrease at infinity. In particular, for any power-law density profile at infinity $\rho \propto r^{-s}$ ($s > 2$, otherwise the line mass of the pinch diverges), the growth rate of the sausage instability modes is saturated for large values of k at the level

$$\Gamma_{0,\infty}(t) = (s - 2)^{1/2} V_A(t)/R(t), \tag{6.95}$$

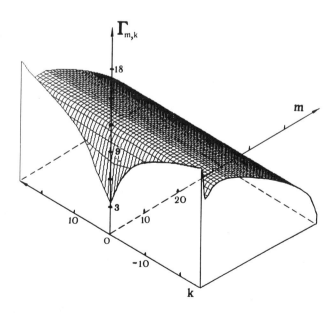

FIGURE 6.15. Spectrum of the instability growth rates for an imploding pure Z pinch with a truncated Gaussian density profile.

if the parameter β, characteristic of the thermal pressure in a Z pinch, is not small, then Γ saturates at a lower level. For a Gaussian density profile, the growth rates are not bounded in the limit $k \to \infty$. If we consider a truncated Gaussian profile, which is replaced by an arbitrary power-law profile at some density level $\rho = \rho_s$, then again, the growth rate saturates at the level

$$\Gamma_{0,\infty}(t) = \sqrt{2}[1 + \ln(\rho_m/\rho_s)]^{1/2} V_A(t)/R(t), \qquad (6.96)$$

which depends very weakly on the particular choice of ρ_s (here, ρ_m is the density on the axis). In this case, the dominant eigenmode of the sausage instability is localized in the volume of the plasma.

If the Z pinch has a sharp boundary, i.e., the density ρ vanishes at the surface $r = R$ [note that the function $\rho(r)$ can approach zero at $r = R$ in an arbitrarily smooth manner], then for large values of k, the local RT modes dominate, with the instantaneous growth rate given by (5.5). Thus, for the short-wavelength sausage instability modes (see Sec. 5.2) we have

$$\Gamma_{0,k\to\infty}(t) \cong |k\ddot{R}(t)|^{1/2}. \qquad (6.97)$$

The corresponding eigenmodes are localized at a distance of order k^{-1} from the pinch boundary [see (5.16) and Figs. 5.16 and 5.17]. Bulk convective modes are not localized near the surface of the Z pinch; for large wavenumbers they tend to be localized near some point inside the plasma cylinder (see Bud'ko et al., 1990a). This is illustrated in Fig. 6.16 where the eigenfunctions of the internal and RT instabilities are shown for an annular cylindrical liner. Finally, in the case in which the density goes to zero very rapidly, as $\rho \to \infty$, the discrete localized instability eigenmodes may be completely absent. Then, an arbitrary perturbation initially localized at $r \approx R$ would propagate to infinity, growing at the same time, and one has to deal with a case of a continuous spectrum of unstable eigenmodes (similar to an example considered in Sec. 4.2).

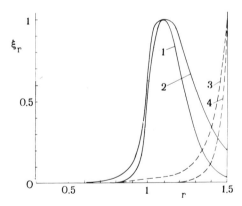

FIGURE 6.16. Eigenfunctions $\xi_r(r)$ of the internal and RT; solid lines: $m = 10$ and 5, $kR_0 = 10$ and 5, for curves 1 and 2, respectively; dashed lines: $m = 0$, $kR_0 = 5$ and 10 for curves 3 and 4, respectively.

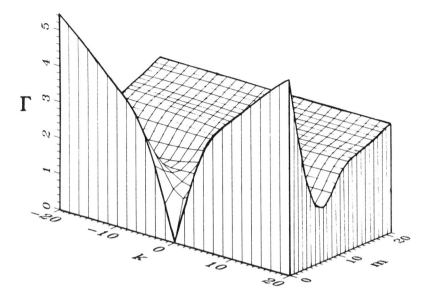

FIGURE 6.17. Spectrum of instability growth rates for a hollow, pure Z pinch of finite thickness at the initial moment of implosion.

In a sense, the third conclusion is complementary to the first one. We have seen that the RT instability is stronger than the hydromagnetic instabilities caused by curvature of the magnetic lines. But the inverse is also true: if the plasma acceleration and the density gradient have opposite directions [or to put it more exactly, if condition (5.25) is satisfied], and the right-hand side of (6.93) is of the same order as the maximum term on the left-hand side (this can take place, for instance, when the dynamic Z pinch or liner stagnates on axis, and the plasma is decelerated by its thermal pressure), then the plasma flow is stable. Not only the RT hydromagnetic modes are stabilized, but also the modes driven by the curvature of the magnetic lines. This effect is well known in plasma physics. For instance, an inertial-dissipative instability cannot develop if the curvature of the magnetic lines stabilizes the flute hydromagnetic instability modes (Mikhailovskii, 1977). The very hydromagnetic instability modes that would develop in a steady-state Z pinch are stabilized. Therefore, the expansion of a diffuse Z pinch, when the thermal pressure is greater than the magnetic pressure, is also stable. An example is given in Fig. 6.18, where the growth rates of the $m = 0$ and $m = 1$ instability modes are presented versus time for a self-similar Z pinch with a truncated Gaussian density profile [the parameters of (3.73) are $\beta = 0.1$, $b = 0$ and 0.01; in (6.96) $\rho_s/\rho_m = 8 \times 10^{-4}$]. The growth rates of all the instability modes, including those presented in Figs. 4.6 and 4.18, vanish when the equilibrium position is passed, the plasma pressure surpasses the magnetic pressure, and the plasma acceleration $R(t)$ changes sign. When an axial magnetic field is present (curve 3), the pressure

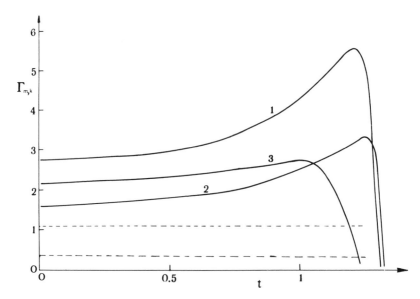

FIGURE 6.18. The instability growth rates versus time for a diffuse pure Z pinch, $m = 0$, $kR_0 = 5$ (curve 1); $m = 1, kR_0 = 5$ (curve 2); the same as (a) but with an axial magnetic field present (curve 3). The dashed line represents the growth rate of the $m = 0$ instability mode of a stationary Z pinch with the same shapes of current and pressure profiles.

decelerates the plasma growth faster with decreasing radius, $R(t)$, so that the equilibrium position corresponds to a greater radius and is achieved earlier.

Figures 6.15, 6.17, and 6.19 demonstrate the spectra of instantaneous growth rates obtained for the self-similar solutions of the same kind representing the unperturbed Z pinch with different density profiles. The cases of a Gaussian density profile truncated at the level $\rho_s/\rho_m = 8 \times 10^{-4}$ (Fig. 6.15) and of a power-law profile satisfying the Kadomtsev criterion (4.21) (Fig. 6.19) differ mainly in the role of the sausage instability modes. In the former case, these modes dominate. In the latter case, though the sausage modes are not stabilized, as they would be in a steady-state Z pinch according to the Kadomtsev criterion, they are not dominant either; rather, the filamentation modes with $m \geq 2$ dominate. Both spectra exhibit a local maximum of Γ as a function of m for a fixed k at $m > 1$; saturation of Γ for large values of k (for $m \geq 2$ the growth rates are almost independent of k); and stabilization of the instability modes with $m \geq \beta^{-1/2}$ due to the stabilizing role of finite thermal pressure. Owing to the saturation of the growth rates of the internal instability modes for large wavenumbers, a perturbation initially localized in the z or φ direction conserves its localized shape, growing exponentially (Goedbloed and Sakanaka, 1974), which may explain bands or strata observed in experiments with dynamic Z pinches.

In a Z pinch with a sharp boundary (Fig. 6.17), the short-wavelength sausage instability modes, i.e., the local RT modes with growth rates given by (6.97), dominate. The filamentation instability in this case grows slower, accompanying the dominating sausage instability development.

The number of e-foldings, $n_{\mathrm{eff}}(t)$, of the internal RT modes for the self-similar unperturbed solutions is estimated in the same way as in Sec. 6.1.2 [see (6.14)]. For the important particular case of a low decelerating pressure [$\beta + b \ll 1$ in (3.73)] we find

$$n_{\mathrm{eff}}(t) \cong C[\ln(R_0/R(t))]^{1/2}, \tag{6.98}$$

where the coefficient C is characteristic of the saturation of the growth rate at large wavenumbers and is taken from (6.95) or (6.96) for the cases of power-law and truncated Gaussian density profiles, respectively. Note that (6.98) predicts faster perturbation growth than (6.14). The right-hand side of (6.98) does not saturate in the limit $R_0/R(t) \to \infty$, but tends to infinity, though rather slowly, because the growth rates [(6.95) and (6.96)] scale as $R_0/R(t)$ (recall that for a constant current, V_A is constant in time) and not as $|g(t)|^{1/2} \propto \sqrt{R_0/R(t)}$, cf. (6.13). The higher is the compression, the shorter the wavelengths that correspond to saturation of the growth rate, if the stabilizing effect of finite resistivity is not taken into account. For typical values of $\rho_m/\rho_s = 10^3\text{--}10^5$, $R_0/R_f = 4-20$, the estimate (6.98) yields n_{eff} from 4.7 to 8.7. In particular, considering the stability of imploding gas-puff Z pinches, we must take into account both the effect of snowplow and smoothness of the initial density profiles produced by the gas-puff devices, which do not make the initial amplitudes of the small-scale nonuniformities too large. Thus, we find that for a six- to eight fold radial compression of gas-puff Z pinches, the instability growth is manageable, in agreement with the available experimental results (Shiloh et al., 1978; Wessel et al., 1986; Ruden et al., 1987; Spielman et al., 1995). Therefore the ideal MHD model explains the observed degree of stability of gas-puff Z-pinch implosions. It predicts also that the growth of the short-wavelength RT instability modes makes a stable 10- to 20-fold radial compression without additional stabilization hardly feasible. However, the best compression results to date ($\sim$ 30-fold compression, Spielman, et al., 1995) have been obtained with low density gas-puffs where kinetic effects (see Sec. 5.5 and 5.6) (De Groot et al., 1998) apparently reduce the growth rate (see Fig. 6.19a). The stability of implosions may be enhanced by an axial magnetic field (see below Sec. 6.6).

6.5.2 Ideal MHD model: Filamentation instabilities of dynamic Z pinches

The development of a small perturbation imposed on a time-dependent equilibrium state does not allow a rigorous definition of growth rate because time and space variables are generally not separated. Therefore, the particular and limiting cases when this can be done are of special interest.

Book and Bernstein (1979) considered the filamentation instability modes ($m \geq 2, k = 0$) of isothermal plasma shells ($\gamma = 1$). Han and Suydam (1982) considered this problem with an arbitrary value of γ. Both papers assumed that the plasma

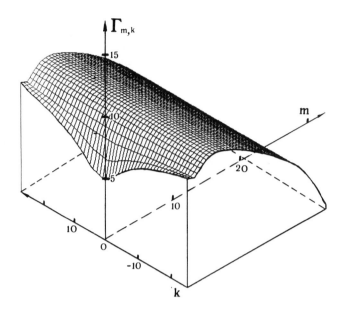

FIGURE 6.19. Spectrum of instability growth rates for a pure, diffuse Z pinch with a power-law density profile.

was perfectly conducting so the magnetic field accelerating the plasma shell does not penetrate it. The MHD problem thus reduced to a purely gas dynamic one, the magnetic pressure being represented by the pressure of a massless fluid. Hwang and Roderick (1987) studied the stability of an interface between heavy and light incompressible ($\gamma \to \infty$) fluids in cylindrical geometry with respect to the sausage and filamentation modes, the light fluid representing the magnetic field. Since the magnetic field behaves like a massless fluid only if the perturbations do not bend the magnetic lines, the study of filamentation instability modes in the papers cited above is adequate only for the plasmas accelerated or decelerated by an axial magnetic field, i.e., for a θ pinch or a magnetic flux compression device. Only the results of Hwang and Roderick (1987) on the sausage instability modes are thus directly applicable to a Z-pinch geometry. Exact solutions describing the linear behavior of small perturbations in the ideal MHD model can be obtained only for the filamentation instability modes in the limiting case of a cold plasma ($\beta = 0$), either for a Z pinch ($B_z \equiv 0$) or for a theta pinch ($B_\varphi \equiv 0$). (American notation for an azimuthal component is θ, however for a cylindrical co-ordinate system it is more convenient to use φ for the azimuthal component.)

An exact analytical solution of the stability problem can be obtained (Book and Bernstein, 1979, 1980; Han and Suydam, 1982; Kleev and Velikovich, 1990) assuming that the unperturbed flows are described by the self-similar solutions with homogeneous deformation (see Sec. 3.4.2). In this case, the variables are separable

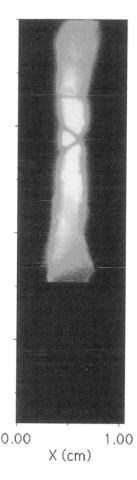

0.00   1.00

X (cm)

FIGURE 6.19A. Time-resolved x-ray pinhole camera picture of a uniform-fill krypton gas-puff implosion. The initial diameter was 6 cm and the final pinch diameter was 2 mm, giving a compression ratio of 30. The x rays are from the krypton L-shell near 1.6 keV.

when the displacement of a plasma particle from its unperturbed trajectory is taken in the form $\xi(\eta)T(t)\exp(im\varphi)$, where $\eta = r_0/R_0$ is the dimensionless Lagrangian coordinate (r_0 and R_0 are the initial values of the particle coordinate and the shell radius, respectively). A separation constant μ is found as an eigenvalue of the corresponding boundary-value problem for $\xi(\eta)$, and for each perturbation eigenmode, determines the time dependence $T(t)$ found from the equation

$$\frac{d^2T}{dt^2} = -\mu\alpha^{-2\gamma(t)}T, \tag{6.99}$$

where the time dependence of the unperturbed dimensionless radius $\alpha(t) = R(t)/R_0$ is given by the equation of motion (3.73), and the unit of time is R_0/V_A.

The cases of z and θ pinches in a cold plasma, when there is no skin effect, correspond to effective values of γ in (6.99), equal to 1 and 2, respectively. The solutions of (6.99) for these cases can be presented in a relatively simple form (Book and Bernstein, 1979). In particular, if the unperturbed flow is an implosion, then the solutions are

$$
T(t) = \begin{cases}
F\left(\frac{\mu}{2}, \frac{1}{2}, \ln \alpha(t)\right), \\
(-\ln \alpha(t))^{1/2} F\left(\frac{\mu+1}{2}, \frac{3}{2}, \ln \alpha(t)\right), & \text{for } \gamma = 1,
\end{cases}
\tag{6.100a}
$$

$$
T(t) = \alpha(t) \left[\frac{1+t}{1-t}\right]^{\pm \frac{i\sqrt{\mu-1}}{2}}, \quad \text{for } \gamma = 2,
\tag{6.100b}
$$

where $F(a, b, z)$ is a confluent hypergeometric function. For other values of the parameter γ, the solutions of (6.99) can be expressed using the Gauss hypergeometric function. Equation (6.99) demonstrates that the instability growth is larger for negative values of μ.

Consider first the filamentation instability modes of a θ-pinch. In this case, for any value of m, the global RT modes dominate, satisfying the conditions (5.17). The eigenfunctions for cylindrical geometry are

$$
\xi_r(\eta) = \eta^{\pm m^{-1}},
\tag{6.101}
$$

and the corresponding eigenvalues

$$
\mu = -m \pm 1, m = 2, 3, \ldots
\tag{6.102}
$$

Both the eigenfunctions and the eigenvalues are independent of the unperturbed profiles and the equation of state, as they should be for global RT modes. The signs $+$ and $-$ in (6.101) and (6.102) refer to the cases of a θ- pinch implosion and magnetic-flux compression during the accelerated and decelerated phase of imploding the cylindrical liner, respectively. The small-scale local RT modes with $m \gg 1$ are localized at a distance of order $R(t)/m$ from the outer or inner plasma surface in the former and latter cases, respectively. Estimating the instantaneous growth rate $\Gamma(t)$ of the filamentation instability with the aid of (6.99) as

$$
\Gamma(t) = (-\mu)^{1/2} \alpha^{-\gamma}(t),
\tag{6.103}
$$

and taking into account that the effective gravitational acceleration is $g = (V_A^2/R_0)\alpha^{-2\gamma-1}(t)$ [see (5.7) and (6.93)] and the effective wavelength of small-scale perturbations is $k_{\text{eff}} = m/R_0\alpha(t)$, we obtain from (6.102) and (6.103) a well-known expression (5.5) or (6.97).

The other internal or convective instability modes may be completely absent if criterion (5.25) is satisfied. Otherwise, they form a discrete set of instability eigenmodes, whose growth rates are smaller than those of the RT modes, i.e., for any given value of m, the corresponding eigenvalues μ are greater than (6.102). The growth rates of the internal modes depend both on the unperturbed profiles and on the thickness of the shell: the thinner is the shell, the greater are the values of $-\mu$ (however, they do not exceed $m \pm 1$). For $\gamma < 2$, standing sound waves are amplified in an imploding plasma shell. Their amplitudes, which would be constant in a stationary shell, decrease in an imploding shell, but the shell radius is decreasing even faster, so that a relative amplification factor can be large in the long-wavelength limit (Book and Bernstein, 1979; Han and Suydam, 1982). In the case of θ-pinch geometry ($\gamma = 2$), there is no such amplification. Equation (6.100) shows that the amplitudes of the oscillatory eigenmodes with $\mu > 1$ depend on time exactly as $\alpha(t)$.

Let us study now the filamentation instability modes of a Z pinch. Recall that the ideal MHD model predicts no filamentation instability of a steady-state Z-pinch equilibrium, and the observed filamentation is conventionally explained by some factors beyond this model (see Sec. 4.2). It was shown in Sec. 6.5.1 that the filamentation instability can develop in an accelerated plasma described by the ideal MHD model [first pointed out by Bud'ko et al. (1988) for a particular case of $m = 2$, $k = 0$ perturbations]. The filamentation instabilities due to plasma acceleration are hydromagnetic, and therefore, develop faster than filamentation instabilities driven by most other mechanisms. In the case of the implosion of a Z pinch or an annular plasma liner, one can study the linear development of filamentation perturbations with the aid of exact analytic solutions.

In the limit $\alpha(t) \rightarrow 0$, we find from (6.99)

$$\frac{T(t)}{\alpha(T)} = \frac{1}{\alpha(t)} (-\ln \alpha(t))^{\mu/2} \rightarrow \infty. \tag{6.104}$$

Therefore, all the eigenmodes of an imploding Z pinch are found to be unstable, even those with $\mu > 0$ that correspond to decreasing perturbations. Here, the geometric factor $1/\alpha(t)$ grows faster anyway, so that the relative amplitude of the perturbation also grows.

There are no global or local RT instability modes for Z-pinch geometry because the filamentation perturbations bend the magnetic lines. Thus, the spectrum of eigenvalues m is determined by the corresponding boundary-value problem and depends on the unperturbed plasma and current density profiles in a Z pinch. In some particular cases, exact analytic solutions of this problem can be found (Kleev and Velikovich, 1990). These solutions demonstrate that an asymptotic estimate of the eigenvalue corresponding to the fastest instability eigenmode in the short-wavelength limit is

$$\mu \cong -(\sqrt{2} - 1)m, \tag{6.105}$$

which agrees with the thin-shell model estimate of the instantaneous growth rate (6.10). Although the azimuthal magnetic field accelerating the shell is maximum at the outer boundary of the shell, the fastest growing instability mode is localized near its inner surface, just because the stabilizing effect of the azimuthal magnetic field on the filamentation perturbations is weaker there.

What happens in the case of a solid diffuse Z pinch with no inner plasma boundary? The closer to the axis, the smaller is the stabilizing magnetic field, and therefore, filaments tend to be near the axis. In this case, there is a continuum of instability eigenmodes. The linear evolution of an initial perturbation is represented by a Fourier integral over a continuous spectrum of eigenmodes (an example is given in Fig. 6.20). The perturbation growth is not local; the current filaments produced by the instability experience a relatively stronger $\mathbf{J} \times \mathbf{B}$ force directed to the Z-pinch axis than the neighboring plasma, and therefore, are attracted to the axis. This conclusion concerning the development of the filamentation instability in imploding solid Z pinches at this stage of implosion appears to be applicable to some experimental conditions, including a high-energy plasma focus after the shock passes through the plasma, so that the compression is uniform but the thermal pressure is still small compared to the magnetic pressure.

The filamentation instability of dynamic Z pinches and liners has been shown (Bud'ko et al., 1988, 1989 a,b, 1990a) to develop slower than the sausage instability modes in most cases. However, one can consider the limiting case of the final stage of the sausage instability evolution, when an almost uniform plasma column is rapidly imploding. The estimates (6.104) and (6.105) indicate that the relative linear growth of filamentation perturbations for moderate values of $m \leq 10$ and the degrees of compression $R_0/R(t)$ from 10^2 to 10^3 is of order $10^2 R_0/R(t)$. Evidently, the plasma compression in the neck may remain 1-D only if the filamentation instability is somehow stabilized. The conventional mechanisms of stabilization, dissipative and nonlinear, are insufficient to support this idea. In particular, an exact calculation of the nonlinear evolution of the $m = 2$ filamentation mode near the pinch axis (Bulanov, 1988) demonstrates that nonlinear effects do not cause a relative stabilization, making the perturbation growth slower than the exponential characteristic of the linear stage prediction. On the contrary, the $m = 2$ perturbation grows even faster, forming a quasi-1-D singularity (a plane current layer) in a finite time interval.

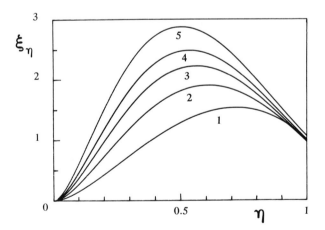

FIGURE 6.20. Evolution of an initial perturbation profile (curve 1) for an imploding Z pinch with uniform plasma and current densities, $m = 5, k = 0$. Curves 2–5 are plotted for the values of the compression ratio equal to 2, 4, 8, and 32, respectively.

6.5.3 Invariants and accuracy of the WKB approximation

The evolution of perturbations in dynamic plasma flows is often studied with the aid of the WKB approximation (4.59) (e.g., see Anisimov and Zel'dovich, 1977; Felber and Rostoker, 1981; Hattori et al., 1986; Hwang and Roderick, 1987), but its accuracy is not always easy to estimate. Here, this issue is discussed for dynamic Z pinches and similar plasma flows.

Are the estimates based on the WKB approximation close to those obtained with the aid of exact solutions of Sec. 6.5.2? Substituting (6.102) and (6.103) into (5.1) and calculating the integral with $\gamma = 1, 2$ [see (6.99) and (6.100)], we obtain for a Z-pinch implosion

$$F(t) = \exp[(2\mu \ln \alpha(t))^{1/2}], \qquad (6.106a)$$

and for a θ-pinch implosion

$$F(t) = \exp\left[\frac{1}{2}\sqrt{-\mu} \ln\left(\frac{1+t}{1-t}\right)\right]. \qquad (6.106b)$$

Comparing (6.106b) and (6.100b), we see that for $|\mu| \gg 1$, the WKB estimate (6.106b) is a very good approximation to the relative perturbation amplitude $T(t)/\alpha(t)$, especially as the eigenvalues m, corresponding to the dominating global RT modes in the WKB approximation, are calculated exactly. On the other hand, if $F(t)$ is regarded as an approximation to $T(t)$, then the dominant exponential factor is estimated correctly, whereas the coefficient of the exponential, which describes the effect of the geometric factor $\alpha(t)$, corresponds to the next order in the WKB approximation. In (6.106a), passing to the limit $(-\mu) \to \infty$ for a given value of $\alpha(t)$, we obtain the asymptotic estimate

$$T(t) \cong \sqrt{\alpha(T)} \exp[\sqrt{2\mu \ln \alpha(t)}], \qquad (6.107)$$

which demonstrates that the error in the WKB approximation in this case is estimated by the coefficient of the exponential.

For sausage perturbations of a liner compressed by an azimuthal magnetic field in the short-wavelength limit, $kR(t) \gg 1$, Hwang and Roderick (1987) obtained the following equation:

$$\frac{d^2\tilde{\xi}_r}{dt^2} + k\frac{d^2R}{dt^2}\tilde{\xi}_r = 0, \qquad (6.108)$$

where $\tilde{\xi}_r = \xi_r \sqrt{R_0/R(t)}$ and ξ_r is the displacement of the external boundary of the liner. It can be checked that the function $\phi(t) = p^{-1/2} \exp\left[\pm \int p(t)dt\right]$ is the first-order WKB approximation to the solution of a wave equation $\ddot{\phi} - p^2(t)\phi = 0$. For a thin shell $|\ddot{R}(t)| \propto I^2(t)/R(t)$, and hence (6.108) yields

$$\xi_r(t) \cong [R(t)/R_0]^{3/4}[I_0/I(t)]^{1/2} \exp\left\{\int_0^t \Gamma(t')dt'\right\}, \qquad (6.109)$$

where $\Gamma(t)$ is given by (6.97), I_0 is the characteristic value of current, and thus (6.15) is obtained. The calculations performed for typical conditions of experiments with a 1-MJ SHIVA system (Hwang and Roderick, 1987) show that the coefficient before the exponential may be essential for estimating the perturbation amplitude, particularly for sufficiently large degrees of compression.

We see that the WKB approximation, in most cases, yields reasonable estimates of the growth of any eigenmode. However, it was shown in Sec. 6.5.2 that it is not always possible to single out an isolated fastest-growing instability eigenmode. Consider now a more general problem: how can one characterize the growth of the fastest perturbations imposed on a given time-dependent unperturbed state, if an arbitrary initial perturbation cannot be decomposed into a discrete set of unstable eigenmodes, one of them dominant over all the others after a while? Let the initial perturbations of density, pressure and magnetic field be some known functions of r [their dependence on ϕ and z being given by $\exp(im\varphi + ikz)$, as before], and after a time interval, t, they are represented by some other functions of r. How can the perturbation at the moment t be compared with the initial perturbation? How can one be sure that another choice of the initial conditions would not provide a much faster (in some sense) perturbation growth? One possible approach to this problem, suggested by Bud'ko et al. (1989a), is described below.

Let us use the known Galerkin method (Fletcher, 1984) to approximate the linear system of perturbation equations in partial derivatives by a finite system of N first-order ordinary differential equations:

$$\frac{d\mathbf{x}}{dt} = \widehat{A}(t)\mathbf{x},$$ (6.110)

where $\mathbf{x} = \mathbf{x}(t)$ is an n-dimensional vector representing the perturbation at the instant t, $\widehat{A}(t)$ is an $n \times n$ matrix of the perturbation operator that is completely determined by the evolution of the unperturbed flow. All the possible sets of initial conditions in this approximation are represented by a n $\times n$ matrix, $\widehat{\mathbf{X}}(t)$, which satisfies a linear matrix equation similar to (6.110): $d\widehat{\mathbf{X}}/dt = \widehat{A}(t)\mathbf{X}$. Its solution can be presented in the form

$$\widehat{\mathbf{X}}(t) = \widehat{U}(t, 0)\widehat{\mathbf{X}}(0),$$ (6.111)

where the linear evolution operator of the system is

$$U(t, 0) = T \cdot \exp\left[\int_0^t A(t')dt'\right],$$ (6.112)

where the operator $T \cdot \exp(\ldots)$ has the same meaning of temporal ordering as one conventionally used in quantum mechanics (Berestetskii et al., 1989). Let $v_j(t)$, with $j = 1, \ldots, n$, be the eigenvalues of the operator $\widehat{U}(t, 0)$. Then, the value of

$$v_n(t) = \max_{1 \le j \le n} |v_j(t)|$$ (6.113)

is an estimate from above for the growth during the time interval $(0, t)$ of a perturbation, whose shape at the end of the interval is the same as at the beginning.

This estimate allows one to avoid explicit consideration of oscillatory perturbation modes, which are always present. If there is a limit

$$v(t) = \lim_{n \to \infty} [v_n(t)], \qquad (6.114)$$

then it represents an exact invariant of the perturbation equations, characteristic of the fastest perturbation growth.

To test the validity of this approach, periodic self-similar solutions of the ideal MHD equations were used to represent the unperturbed flow. The period of motion is denoted by 2τ, where τ is the time of compression or expansion. If $\widehat{A}(t)$ is a 2τ-periodic matrix, then the evolution operator can be presented in the form

$$\widehat{U}(t, 0) = \widehat{S}(t) \exp(\widehat{B}t), \qquad (6.115)$$

where $\widehat{S}(t)$ is a 2τ-periodic $n \times n$ matrix and $\widehat{S}(0) = \mathbf{I}$ ($\mathbf{I}$ is the $n \times n$ unit matrix). Here $\widehat{B}$ is a constant $n \times n$ matrix of the Lyapunoff operator of the system (6.110), $\widehat{M} = \exp(2\widehat{B}\tau)$ is the matrix of the monodromy operator of the system (Arnol'd, 1978). Equation (4.43) demonstrates that the perturbation growth averaged over a period of motion is exponential, whereas deviation from the exponential law can be important for time intervals smaller than τ. The eigenvalues of the Lyapunoff operator, Γ_j, for the considered periodic motion are exact analogs of the growth rates and eigenfrequencies. If

$$\Gamma_m(n) = \max_{1 \le j \le n} [\text{Re}(\Gamma_j)], \qquad (6.116)$$

then, for any instant of time, the following estimate of the perturbation growth is valid:

$$|\mathbf{x}(t)| \le \text{const} \cdot \exp[\Gamma_m(n)t] \cdot |\mathbf{x}(0)|, \qquad (6.117)$$

with the right-hand side of (6.117) being the exact upper bound of the perturbation growth during the time interval τ. If there is a limit

$$\Gamma_m = \lim_{n \to \infty} (\Gamma_m(n)), \qquad (6.118)$$

then its value is also an exact invariant of the perturbation equations characteristic of the fastest possible perturbation growth imposed on the periodic plasma motion.

Calculations of $v(t)$, $\Gamma_m(t)$ were done by Bud'ko et al. (1989a), for a periodic self-similar motion describing an unperturbed diffuse Z pinch with a Gaussian density profile. The results are shown in Fig. 6.21. Here, the solid line corresponds to the WKB estimate (5.1), the crosses are the values of $v(t)$, and the cross marked M is the value of $\exp(2\Gamma_m t)$, where Γ_m is given by (6.118). The WKB method is shown to provide a good approximation to the more rigorous estimates. The convergence of the Galerkin numerical method was found to be sufficiently fast, in particular to calculate $v(t)$ and Γ_m with numerical errors below 5%. Only five Chebyshev polynomials should be retained in the expansion for each of the perturbed values.

We conclude that the simplest estimate (5.1) of the perturbation growth may be refined by calculating the value of $v(t)$ at each instant of time; the result would be

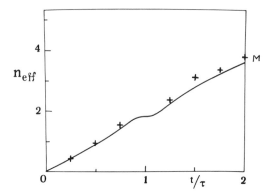

FIGURE 6.21. Comparison of the WKB (curve) and numerical (crosses) estimates for the growth of the fastest sausage perturbation mode ($m = 0$, $kR_0 = 5$) of an imploding diffuse pure Z pinch with a truncated Gaussian density profile.

invariant with respect to the arbitrary choice of initial conditions. This method can be used for any model of the unperturbed flow, be it 1-D, 2-D, ideal, or dissipative. However, a noticeable increase in accuracy may require considerable computer time. The results of the calculations indicate that the WKB approximation is particularly good for unperturbed motions that include acceleration and deceleration phases, when the plasma velocity is comparatively low and the perturbation growth is determined mainly by its acceleration.

6.6 Stability of gas-puff Z-pinch implosions

Since the early works by Latham et al. (1960) and Albares et al. (1961) many authors reported observations of the RT instability development in dynamic Z pinches. Modern diagnostic techniques made possible a detailed experimental study of this instability for gas-puff Z pinches. The imploding gas-puff experiments and 2-D modeling have been (Deeney et al., 1994) to study the zipper effect and to optimize the K-shell x-ray yields. The dynamics of the implosion and radiation phases of a gas puff was examined by Nash et al. (1990).

It was shown in Chap. 5 and Sec. 6.1 that the dominant short-wavelength instability modes of dynamic Z pinches are the sausage RT modes. This fact can be illustrated by the experimental results obtained on the University of California (Irvine) gas-puff Z-pinch device. The current rise time in these experiments was 1.25 μs, the maximum current was 0.45 MA, the initial radius of the puff was about 2 cm, and the test gas was helium (Ruden et al., 1987). It was possible to pre-ionize the gas up to a degree of ionization of approximately 10^{-5}, with the aid of an electron beam generated by a special circuit about 150 ns before the main current pulse. In the absence of the pre-ionization, the gas was ionized by the main current, and the stochastic nature of the breakdown process inevitably caused considerable initial nonuniformities of electron density in the gas column. A better initial uniformity is expected with pre-ionization.

The experimental studies by Spielman et al. (1985a, b) (see also Ruden et al., 1987) have shown definitely better stability of the gas-puff Z-pinch implosions with pre-ionization. They also demonstrated a gradual distortion of symmetric flow structure due to the development of RT instability. The dominant sausage mode was found to correspond to a wavenumber $k = 5 \text{ cm}^{-1}$. Various estimates [(6.14), (6.15), and (6.98)] yield values in the range 1.7–3, so that the perturbations observed in this experiment are 5–20 times greater than their initial values. Thus, the initial short-wavelength perturbations of a gas-puff Z pinch caused by the nonuniformity of the breakdown do not exceed a few percent. Elimination of these nonuniformities achieved with the aid of initial pre-ionization results in a greater degree of plasma compression. Then a relatively slow, long-wavelength instability mode turns out to be dominant, one caused by the nonuniformity of the gas-puff Z pinch in the axial direction. Its development is observed as a mass flow along the axis.

The RT instability modes behave in a similar way when gas-puff Z pinches are compressed by higher currents. Development of short-wavelength perturbations is clearly seen in the time-resolved x-ray pinhole photographs from Dukart et al. (1987) presented in Fig. 6.22: each frame is less than 2 ns duration, and the time interval between successive frames is 2 ns. The experiments of Dukart et al. (1987) were performed on the PROTO II generator, where a current pulse with FWHM of about 45 ns and amplitude above 6 MA was delivered to a gas-puff load with an initial radius $R_0 = 2$ cm. The plasma column was compressed to a final radius $R_f \cong 2$ mm. In general, the shape of the observed modes agrees with the conclusion that the sausage azimuthally-symmetrical modes should dominate, though signs of kink modes are also seen.

Assuming that at an early phase of the acceleration, when the dominant modes are formed during the linear stage, the temperature of the external plasma surface is low (about 5–10 eV), if we take in (6.16) $v_m = 10^6 \text{ cm}^2/\text{s}$, $\tau = 45$ ns, we find $n_{\text{eff}} \cong 5.4$. A close estimate for n_{eff} is obtained from (6.96) and (6.98). Taking the value of $\rho_m/\rho_s = 10^3$, we find $n_{\text{eff}} \leq 6.0$. Thus, the perturbation growth during the linear stage is estimated as $(2 - 3) \times 10^2$. The initial short-wavelength perturbations appear to have an amplitude of order 1% because the experiments on PROTO II include pre-ionization of the gas by a ultraviolet flashboard, so that breakdown does not produce considerable nonuniformities. Hence, the short-wavelength perturbations seen in Fig. 6.22 correspond to the beginning of the nonlinear stage of evolution, which agrees with their observed shape. Being short-wavelength, these perturbations are localized near the outer surface of the plasma column. This may explain why the perturbations observed in the photographs "did not severely damage the structural integrity of the target" (Dukart et al., 1987), a 2-μm-thick 2-mm-diameter parylene annulus (a "soda straw") placed inside the annular gas-puff Z pinch. Even when the initial perturbations are large enough (about 10%), plasma stagnation occurred across the straw in less than a nanosecond (Dukart et al., 1987).

FIGURE 6.22. Nanosecond, time-resolved x-ray pinhole photographs of a PROTO-II gas-puff Z-pinch implosion onto a 2-mm-thick parylene target (Dukart et al., 1987).

In Fig. 6.22, the radiating plasma cylinder is seen to grow in the axial direction from 7 mm to 20 mm in 14 ns. This is the result of the nonuniform implosion of a Z pinch, the so-called zippering effect due to divergence of the gas jet injected from the nozzle, which propagates from the cathode to the anode. The plasma column first stagnates near the cathode, where its initial radius is smaller, and then the stagnation phase propagates along the pinch at the implosion-phase velocity of about 2×10^8 cm/s in this experiment. This effect is illustrated by the results presented in Figs. 6.23–6.25 of a 2-D numerical simulation of Roderick and Hussey (1989a, b) that models the PROTO II experiments. Figure 6.23 shows the density contours of the injected gas at $t = 0$. Figure 6.24 shows the current lines at $t = 12$ ns, when the acceleration has just begun, and the current flows in the rarefied plasma at the periphery of the column. Figure 6.25 shows the density contours at $t = 62$ ns. We see that the implosion begins near the cathode and only then propagates to the anode; the plasma shell closes just like a zipper. The results presented in Figs. 6.23–6.25 correctly describe the large-scale behavior of a 2-D flow in a gas-puff Z pinch. Roderick and Hussey (1989a, b) also studied the possibility of reducing the axial zippering by shaping the injected gas-density profile and/or inclining the gas-puff nozzle to the axis. Both methods are shown to reduce the zippering time from 10 ns to 2–3 ns.

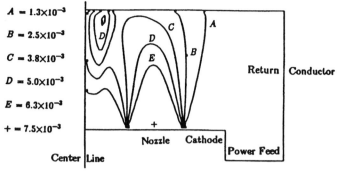

FIGURE 6.23. Density contours calculated for gas injection in PROTO-II geometry (Roderick and Hussey, 1989b). Curves A to E and + correspond to density levels from 1.3×10^{-6} to 6.3×10^{-6} and 7.5×10^{-6}. Exit Mach number of the nozzle is 2.5, exit density is 7.5×10^{-6}.

Note, Figs. 6.23–6.25 show no trace of the short-wavelength RT modes. Their stabilization can be related to the snowplow effect discussed in Sec. 6.3. Here, the relatively thick gas column acts as a source of mass accumulated by the imploding current sheath. No classical RT instability is described in Chap. 5, if the implosion of the current sheath is supersonic and the gas remains essentially unperturbed until the sheath reaches it. This hypothesis agrees with the results of Roderick and Hussey (1989a, b): the low-density plasma of the current sheath catches up to the main body of the puff in about 10 ns, which corresponds to a radial velocity of about 2×10^8 cm/s (the Alfvén velocity V_A for conditions of Fig. 6.24 is estimated as 1.5×10^8 cm/s). Of course, this is much greater than the sound velocity in the unperturbed gas, where no magnetic field has penetrated. The WKB approximation studied above becomes applicable only in the stage when the compression wave passes

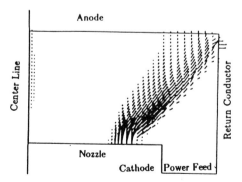

FIGURE 6.24. Current density at 10 ns showing current flow in the low-density gas region (Roderick and Hussey, 1989a).

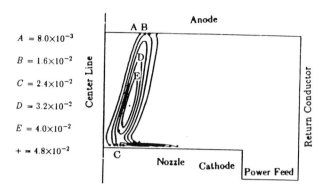

$A = 8.0 \times 10^{-3}$

$B = 1.6 \times 10^{-2}$

$C = 2.4 \times 10^{-2}$

$D = 3.2 \times 10^{-2}$

$E = 4.0 \times 10^{-2}$

$+ = 4.8 \times 10^{-2}$

FIGURE 6.25. Density contours calculated by Roderick and Hussey (1989b) at 62 ns showing annulus tilt and gas injection. Curves A to E and $+$ correspond to density levels from 8×19^{-6} to 4×19^{-5} and 4.8×10^{-5}.

through the whole gas puff and establishes an acceleration profile qualitatively similar to the linear one, i.e., to that given by the self-similar solution.

6.6.1 Stabilization of gas-puff Z-pinch implosions by an axial magnetic field

All the experiments on magnetic flux compression by imploding annular plasma liners demonstrated an unexpected degree of stability of the implosion process. In particular, the first experiments on the UCI gas-puff Z-pinch facility with a peak current 0.47 MA, have shown an enhanced stability (Wessel et al., 1986; Felber et al., 1988b). The experiments revealed that even a small amount of axial field, $B_{z0} \cong 1$ kG, produces a stabilizing effect on the pinch before and after pinching. Normally with $B_{z0} \cong 0$, the imploding plasma displayed an asymmetric radial profile approximately 75 ns before pinching, which is characteristic of small-amplitude RT instability growth. After pinching, the column became highly unstable and disassembled, usually in less than 25 ns. This indeed agrees with the estimates of instantaneous growth rates given in Sec. 6.5.3.

An axial magnetic field makes the column profiles smooth and uniform compared to the zero-field case. This is seen in the time-integrated pinhole photographs in Fig. 6.26, where the intensity and contrast of sequential photos were increased approximately 10 times from Fig. 6.26a to Fig. 6.26f. These pinhole photographs also reveal a lack of hot spots and unstable behavior as the initial axial field strength increase. Interferograms of the pinch (Fig. 6.27) also demonstrate its stabilization by the axial magnetic field. The enhanced stability of the pinch made it possible to observe multiple, periodic radial bounces of the plasma column due to strong compressional force, which remains after the first pinch. In the experiments on the SNOP-3 facility at Tomsk, Russia (Ratakhin et al., 1988; Sorokin, and Chaikovsky

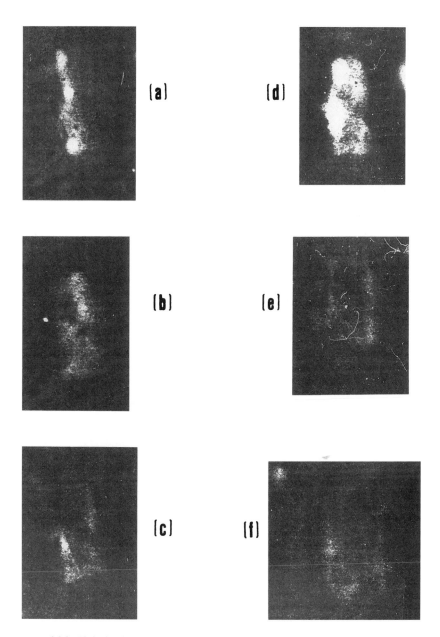

FIGURE 6.26. Pinhole photographs of an argon pinch as a function of increasing injected field strength (Felber et al., 1988).

1989) with $I_{max} = 1.4$ MA, $\tau = 100$ ns, where the peak axial magnetic field of 2.5 MG has been recorded, a considerable increase of the pinch homogeneity due to the axial magnetic field has been observed on the integral pinhole photographs.

Stabilization of gas-puff Z-pinch implosions has also been observed in the experiments on the PROTO-II pulsed power driver at Sandia National Laboratories, when both machine electrical diagnostics and x-ray pinhole photographs indicated that an initial axial magnetic field of 100 kG had been compressed to about 42 MG by an imploding Ne gas-puff Z pinch with $I_{max} = 7.5$ MA, $\tau = 60$ ns (Felber et al., 1988a). A stable 22-fold compression of the plasma column has been observed only in the experiments when a stabilizing axial magnetic field was present. The improved stability of the pinch was also demonstrated in long-time-scale XRD traces that showed radial bounces of the pinch on discharges with high B_{z0}.

The stabilizing effect of an axial magnetic field on gas-puff Z-pinch implosions can be explained with the aid of the ideal MHD model (Bud'ko et el., 1989a, b). In Sec. 5.5, the stabilizing role of magnetic shear on accelerated plasma layers of finite thickness has already been discussed. Stability of an implosion can be expected

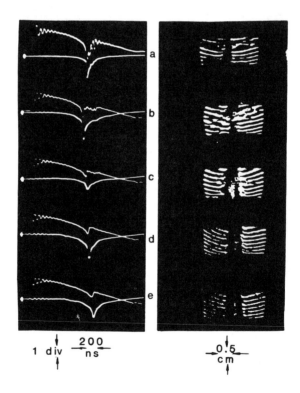

FIGURE 6.27. B-dot and XRD traces and laser interferograms as a function of increasing injected field strength (Felber et al., 1988).

if the final number of e-foldings of the dominant modes n_{eff} is shown to be small or, at least limited, in comparison with the case when no axial magnetic field is present. This stability analysis was performed by Bud'ko et al. (1989a) for diffuse imploding Z pinches with a frozen-in axial magnetic field and by Bud'ko et al. (1989b, 1990a) for imploding annular liners with sharp boundaries that compress magnetic flux in the annulus. Self-similar solutions with homogeneous deformation were chosen to represent the unperturbed 1-D plasma flow. The results for the two cases are quite similar. Figure 6.28 presents the spectrum of instantaneous growth rates calculated for an annular liner at time, $t = 0$, when the dimensionless radius $\alpha = 1$, and the dimensionless acceleration $\ddot{\alpha} = -0.9$ [see (3.68), where $b = 0.1$]. One can see that the growth rate of the local RT modes is not affected. At the same time, the growth rates of the internal modes are noticeably smaller than in the case when axial magnetic field is absent, although the dimensionless acceleration without an axial magnetic field is only 10% greater: $\ddot{\alpha} = -1$.

Figure 6.29 shows the spectrum calculated for $b = 0.01$ at an intermediate time when $\alpha = 0.103$, $\ddot{\alpha} = -0.64$. The dimensionless acceleration $\ddot{\alpha}$ is only 30% less than the acceleration at time $t = 0$ ($\ddot{\alpha} = -0.99$), but the axial magnetic field is large enough to stabilize the RT modes due to magnetic shear and to decrease the growth rates of the internal modes substantially, in agreement with the results of Sec. 5.5. In this case the value of the shear parameter χ is 4 [see (5.46)]. Therefore, Fig. 6.29 demonstrates the existence of a window of stability in the space of parameters of the accelerated motion. In the window of stability, the local RT modes, otherwise dominant during the acceleration phase, are fully suppressed, whereas the growth rates of the other modes are much smaller. It should be stressed that this is not only due to the decrease in acceleration, which remains of order unity.

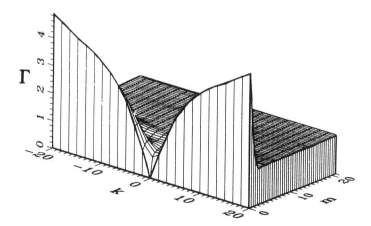

FIGURE 6.28. Spectrum of instability growth rates for an annular liner compressing an axial magnetic field at the initial moment of implosion.

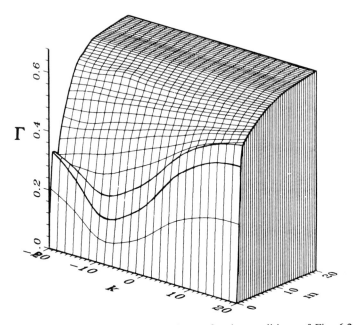

FIGURE 6.29. Spectrum of instability growth rate for the conditions of Fig. 6.28 at an intermediate stage of compression, when the magnetic shear comes into play. The most dangerous RT modes are suppressed.

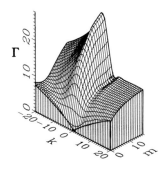

FIGURE 6.30. Spectrum of instability growth rate for the conditions of Fig. 6.28 at the deceleration stage.

Fig. 6.30 is plotted for the time when $\alpha = 0.164$, $\ddot{\alpha} = 16.5$, corresponding to the deceleration of the liner by the axial magnetic field; here $b = 0.1$. The local RT modes appear again, this time at the inner surface of the liner: $k = 0$, $m \gg 1$, the filamentation modes. In Fig. 6.30, the asymmetry of the $k > 0$ and $k < 0$ parts of the spectrum is very pronounced. The growth rates of the internal modes are saturated for large k and limited m.

Figure 6.31 shows the perturbation growth versus the radial compression, $1/\alpha$, for a diffuse Z pinch with truncated Gaussian density profile. It illustrates that even a relatively weak initial axial magnetic field stabilizes the compression, in some cases allowing several successive cycles of compression and expansion until growth of the instabilities destroys the pinch, as has been observed in the experiments.

Analysis of the perturbation growth in the WKB approximation, similar to that presented in Fig. 6.29, allowed Bud'ko et al. (1989a, b; 1990a) to make the conclusion that the instability growth is manageable for an initial axial field

$$B \geq (10\text{--}30 \text{ kG})\overline{I}(\text{MA})/R_0(\text{cm}), \tag{6.119}$$

where $\overline{I}$ is some average current during the implosion (for the analytical studies the current was chosen to be constant). Typically in pinch implosions, there is a maximum perturbation amplitude, beyond which the pinch symmetry is considered to be unacceptable. For a given acceptable perturbation amplitude, an optimal value of the initial axial field, B_{z0}, can be chosen to provide the highest radial compression (curve 4 in Fig. 6.31). This conclusion agrees with the results of experiments on high-current Z-pinch dynamics in an axial magnetic field. The experiments confirm the scaling (6.119), as is seen in Fig. 6.32. We see that an axial field of 15 kG is enough to stabilize 1.4-MA implosions (Sorokin and Chaikovsky, 1989), whereas

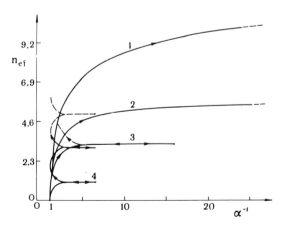

FIGURE 6.31. The number of e-foldings versus the dimensionless compression ratio for a Z pinch with a truncated Gaussian density profile: (curve 1) pure Z pinch, cold plasma; (curve 2) pure Z pinch, finite thermal pressure; (curve 3) optimal value of initial axial magnetic field; (curve 4) high axial magnetic field.

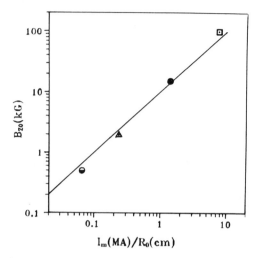

FIGURE 6.32. Initial magnetic field B_{z0} required for stabilizing the implosions: straight line: the lower boundary (3.169); data points are obtained in the experiments by Baksht et al. (1987) (◔); Felber et al. (1988b) (▲); Sorokin and Chaikovsky (1990) (●); Felber et al. (1988a) (▫).

for a 7.5-MA peak current, the field should be about 100 kG (Felber et al., 1988a). It is not surprising that stability is observed with axial magnetic fields near the lower bound given by the estimate (6.119), sometimes even below it. Indeed, the current pulse amplitude, I_{max}, overestimates the average current $\bar{I}$.

It should be noted that the theory indicates an interesting prospect of realizing an essentially stable compression regime. To do this, one must program the compression in such a way that the parameters of the plasma flow remain within the window of stability during the implosion, i.e., the magnetic shear parameter, χ, is kept sufficiently high. This is possible if the current flowing through the liner varies in time, so that the azimuthal magnetic field compressing the liner is close to the axial magnetic field frozen in it, i.e., $I(t) \sim R(t)^{-1}$. It is not yet clear whether a technical implementation of such a regime is feasible, though recent progress in developing the methods of the current pulse-shape programming (Ishii et al., 1986) seems to support the idea.

6.6.2 Stability of multilayer-shell implosions

It should be noted that the instantaneous growth rate for a dynamical system is not very informative (see Chap. 5). Consider for simplicity a planar liner of constant density ρ and thickness d that is accelerated to kinetic energy $E = \rho du^2/2$ by a magnetic field with constant pressure P over a time t. The important quantity is the number of e-foldings that the instability grows during the time t,

$$\sigma(E) = \int_0^t \Gamma(\tau)d\tau = \sqrt{2kE/P}, \tag{6.120}$$

where $\Gamma(t)$ is the instantaneous growth rate of the RT instability given by (5.5).
We see that the total number of e-foldings can be reduced by increasing the acceleration. Let us consider some conclusions that can be deduced on the basis of this
simple equation. Our purpose is to consider techniques to reduce the total growth
of RT instabilities so that the final target energy is maximized with the desired
uniformity. Consider the acceleration of multiple shells as an example of this simple idea to reduce the total instability growth. The enhanced stability of implosive
magnetic-field compression by multicascade magnetocumulative generators and of
double plasma-puff Z-pinch liners has been reported by many authors (Pavlovskii
et al., 1989; Sorokin and Chaikovsky, 1994; Baksht et al., 1994). To explain this
mechanism of enhanced stability let us consider, for simplicity, a double-cascade
configuration of two nested cylindrical shells with radii R_1 and R_2, and thicknesses
δ_1 and δ_2, respectively, that are imploded toward the axis by a constant current, I_0,
switched on at time, $t = 0$. Before the collision, the first shell implodes from R_1
to R_2, and its equation of motion is

$$2\pi R_1 \delta_1 \rho_1 \frac{d^2 R}{dt^2} = -\frac{I_0^2}{c^2 R}, \tag{6.121}$$

where $\rho_{1,2}$ are the densities of the shells. After the collision, the second shell
is accelerated by the pressure of the collision with the first shell, which can be
estimated as $P \approx \rho_2(dR/dt)^2/2$. Integrating the equation of motion (6.121), we
obtain

$$P = \frac{\rho_r I_0^2}{2\pi R_1 \delta_1 \rho_1 c^2} \ln(R_1/R_2). \tag{6.122}$$

Thus, according to (6.120), the stabilizing effect of the double cascade system is
effective, if the pressure of the collision exceeds the magnetic pressure at the time
of impact at $R = R_2$, i.e., if the following inequality holds

$$\frac{\rho_2 I_0^2}{2\pi R_1 \delta_1 \rho_1 c^2} \ln(R_1/R_2) > \frac{I_0^2}{2\pi R_2^2 c^2}. \tag{6.123}$$

Since optimal energy transfer takes place if the masses of the shells are
approximately equal, the optimal stabilizing effect corresponds to

$$\frac{\rho_2 R_2^2}{\rho_1 R_1 \delta_1} \ln(R_1/R_2) = \frac{R_1}{\delta_2} \frac{R_2}{R_1} \ln(R_1/R_2) > 1. \tag{6.124}$$

It is easy to see that the optimal condition in (6.124) for a fixed value of R_1/δ_2 is
achieved when $R_1/R_2 = e$, and for the maximum of the R_1/δ_2. Typically, the RT
instability limits the compression and has a significant effect at a radius ratio of
about $R_1/R_2 \approx 10$. Thus, for an optimal regime, the first liner should come safely
to the collision point, at $R_1/R_2 \approx e$, which is the optimal location for the first and
second shells.

It became possible to test the concept of multi-shell stabilization with the advent of the 20-MA Z accelerator (Spielman et al., 1998). Experiments were conducted in which wire arrays were used to approximate a multi-shell configuration. The outer shell consisted of 240 7.4-μm diameter tungsten wires located at a 4-cm diameter. The inner shells consisted of 120 tungsten wires on a 2-cm diameter. The diameter of the inner wires was varied to adjust the mass. The optimal results were obtained when the mass of the inner array was 0.25–0.5 × that of the outer array. These experiments conclusively demonstrated an increase in x-ray power (as high as 290 TW), a decrease in x-ray pulse width (as short as 4-ns FWHM), and a more uniform, apparently stable, stagnation. Figure 6.33 shows some of these data. A time-resolved x-ray pinhole photograph with 1-ns interframe times and 100-ps gate times shows the remarkable uniformity of the stagnation. The dark bands in the early frames are thought to be due to hole closure in the viewing slots. The imaged x rays are in the 200- to 300-eV energy band and constitute the bulk of the x-ray emission. These data can be compared with data taken on Z with a single wire array (Fig. 6.5b). The observed duration of the stagnated pinch (1.2-mm diameter) lasts for many Alfvén times. It is not clear that the quantitative details of the observed stabilization is fully explained by the preceding analysis.

6.7 Stabilization of long-wavelength sausage and kink modes of a Z pinch by radial oscillations

As we discussed above, a number of experiments performed in the last decade have demonstrated substantially improved stability of Z-pinch configurations. The improved stability with pinch lifetimes 10^1–$10^2\tau_A$ was reported for experiments with fiber-initiated dense Z pinches; with compressional, and gas-embedded Z pinches; with imploding gas-puff Z pinches; and the straight EXTRAP configuration. The striking stability with respect to sausage modes observed in the experiments with compressional and gas-embedded Z pinches can be explained, in principle, by ideal MHD theory (see Chap. 4), with the assumption of rapid plasma cooling after a peak current and slowly decreasing unperturbed pressure profile satisfying the Kadomtsev condition, as well as by finite plasma-conductivity effects for fiber-initiated pinches. The kink mode cannot be stabilized by choosing appropriate unperturbed profiles within the scope of the ideal MHD linear stability model. Investigations taking into account a finite-ion Larmor radius and carried out with the aid of the Hall MHD model and the aid of the Vlasov equation, which includes the drift-kinetic effects and the influence of resonant particles, demonstrated a significant decrease in the growth rates of the short-wavelength (compared to the pinch radius) kink modes (see Chap. 4). However, neither these models nor the analysis, including the effect of viscous damping, could explain the observed stability of long-wavelength kink modes. Thus, we conclude that the stabilization of the long-wavelength kink modes in Z pinches can be theoretically understood only by including an analysis of the nonlinear regime, which has been carried out only numerically under a very simplified skin-current assumption.

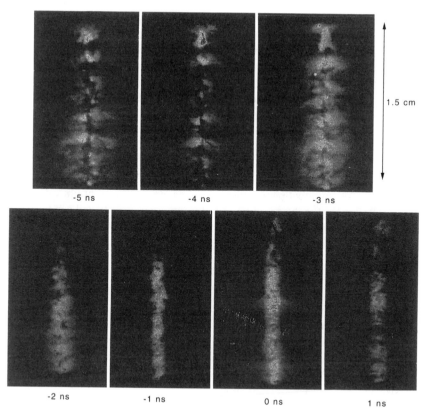

-5 ns   -4 ns   -3 ns

-2 ns   -1 ns   0 ns   1 ns

FIGURE 6.33. Time-resolved x-ray pinhole photograph of a multishell tungsten wire-array pinch. The gate time of the instrument is 100 ps and the interframe time is 1 ns. The resolution of the instrument is 75 μm. The peak radiation power occurs in the fifth frame. The initial wire-array diameter was 40 mm, and the length was 2 cm. The outer array had 240 7.1 $-$ μm diameter wires, the inner array has 120 8.9 $-$ μm diameter wires. The camera was filtered with 4μm of Kimfol to detect the x-ray energy band 200–300 eV.

Another possibility involves the idea of dynamic stabilization. It was demonstrated (Bud'ko and Liberman, 1992; Bud'ko et al., 1995) that the long-wavelength kink modes can be effectively stabilized by the radial oscillations of a Z pinch. A mechanical analog of this effect is a pendulum, whose point of support is subject to a high-frequency external force that makes the vertically upward position stable (Landau and Lifshitz, 1976). Usually Z pinches are considered to be a plasma column with a constant radius. In reality, the pinches are created by heating and ionization of an initially cold column of gas or solid, and this process is inevitably connected with transitional phenomena. In other words, the pinch should be treated as a substantially dynamical system. The duration of the current pulse that is used to produce the Z pinch is not large in comparison with the characteristic time scale of the pinch plasma dynamics, so oscillations of the plasma created by the discharge should persist during the experiment, although their effect on global thermodynamic characteristics may fall below the sensitivity of measurements.

In the analysis of the growth of a long-wavelength kink mode, we assume that the plasma flow becomes self-similar very early in the process. The equation of the radial motion of the pinch corresponding to the self-similar anzatz (3.62)–(3.67) is given by (3.73), with the corresponding energy integral (3.77). Let $\alpha(t) = R(t)/R_0$ be the instantaneous normalized pinch radius, where R_0 is the pinch radius in static equilibrium. The energy integral (3.77) can be rewritten in the form

$$\dot{\alpha}^2 = \frac{\beta}{\gamma - 1} (\alpha_{min}^{2-2\gamma} - \alpha^{2-2\gamma}) - 2 \ln \left(\frac{\alpha}{\alpha_{min}} \right), \tag{6.125}$$

where α_{min} is the minimal pinch radius normalized to the equilibrium value. The maximal radius α_{max} is the second root of the equation, $\dot{\alpha} = 0$, and can be easily obtained as well as the oscillation half-period T:

$$T = \int_0^T dt = \int_{\alpha_{min}}^{\alpha_{max}} \frac{d\alpha}{\dot{\alpha}(\alpha)}. \tag{6.126}$$

We consider small perturbations of the dynamic pinch equilibrium and introduce a small displacement vector, $\boldsymbol{\xi} = \partial \mathbf{u}/\partial t$.

Since the unperturbed variables in the linearized equations of the ideal MHD are time dependent, the perturbation evolution is not exponential and can hardly be traced analytically in the general case. We consider the limiting situation when the oscillation period is small compared with the perturbation growth time for the long-wavelength sausage and kink modes. Under this assumption, we can average the linearized equations over the oscillation period, taking into account that all terms in the linearized equations with unperturbed velocity vanish after averaging, with the following averaging procedure for any oscillating function, $f(r, t)$:

$$\langle f(r, t) \rangle = \frac{1}{T} \int_{\alpha_{min}}^{\alpha_{max}} f(\alpha, t(\alpha)) \frac{d\alpha}{\dot{\alpha}(\alpha)}. \tag{6.127}$$

Then, the linearized equations can be combined into the second-order equation of motion

$$\langle \rho \rangle \frac{\partial^2 \boldsymbol{\xi}}{\partial t^2} = \mathbf{F}(\boldsymbol{\xi}). \tag{6.128}$$

This equation is identical to the corresponding equation in the stationary-stability problem [see (4.15)], except for the averaged unperturbed profiles and the additional effective gravity term, $(\langle \rho \rangle \nabla \cdot \boldsymbol{\xi} + \boldsymbol{\xi} \cdot \langle \nabla \rho \rangle) \langle \mathbf{g} \rangle$, in the force operator, $\mathbf{F}(\boldsymbol{\xi})$, where $\mathbf{g} = d\mathbf{u}/dt$ is the particle acceleration. The crucial point in the stability analysis is that the averaged values of particle acceleration $\mathbf{g} = d\mathbf{u}/dt$, and force acting on a particle $\rho d\mathbf{u}/dt$, are not equal to zero and enter explicitly into the force operator, $\mathbf{F}(\boldsymbol{\xi})$. In a certain sense, this is equivalent to the presence of an effective-gravitation field, $\mathbf{g} = -d\mathbf{u}/dt$, in the local frame of reference associated with a moving particle. This nonuniform, effective-gravitation field produced by radial oscillations is characterized by the following quantities (Bud'ko and Liberman, 1992):

$$\Omega_1^2(r) = \frac{\langle \rho g \rangle}{r\langle \rho \rangle} = -\frac{\rho_0}{r\langle \rho \rangle T} \int_{\alpha_{min}}^{\alpha_{max}} \frac{\ddot{\alpha}}{\dot{\alpha}\alpha^3} N\left(\frac{r}{R_0\alpha}\right) d\alpha, \qquad (6.129)$$

$$\Omega_2^2 = \frac{\langle g \rangle}{r} = -\frac{1}{T} \int_{\alpha_{min}}^{\alpha_{max}} \frac{\ddot{\alpha}}{\dot{\alpha}\alpha} d\alpha. \qquad (6.130)$$

The sign of this function determines either the stabilizing or destabilizing effect of the corresponding terms in (6.128). Note that for any oscillation amplitude, $\Omega_2^2 < 0$. Indeed, in the liner regime, we have

$$\alpha(t) = 1 + \varepsilon \cos(\omega t), \qquad \omega = 2 + 2\beta(\gamma - 2), \qquad \varepsilon \ll 1, \qquad (6.131)$$

and

$$\Omega_2^2(\varepsilon) = \frac{\varepsilon \omega^2}{\pi} \int_0^\pi \frac{\cos x}{1 + \varepsilon \cos x} dx = -\frac{\varepsilon^2 \omega^2}{2} < 0. \qquad (6.132)$$

It can be shown that Ω_2^2 grows in absolute value and remains negative as the amplitude of the oscillations increases.

Numerical simulation of the problem (Bud'ko et al., 1995) demonstrated that even moderate radial oscillations cause reduction of the growth rate of long-wavelength sausage instabilities and complete stabilization of long kinks.

The growth rates of the kink modes in the long-wavelength limit can be obtained analytically with the aid of the perturbation theory for a weak gravitational field, assuming

$$(\Omega_1 \tau_A)^2 \ll 1, \qquad (\Omega_2 \tau_A)^2 \ll 1, \qquad k^2 \ll 1. \qquad (6.133)$$

Assuming that the perturbations have the usual exponential form $\xi(r) \exp(\Gamma t + im\varphi + ikz)$, (6.128) can be combined into one equation for the radial displacement, $\xi_r(r)$, with the boundary conditions (4.33) and (4.34) at $r = 0$ and at the outer pinch boundary ($\alpha = \alpha_{max}$), respectively. Expanding equations for $\xi(r)$ in small parameters (6.133), we obtain in zeroth order $\xi_{r0} = \text{const} = 1$, and the following equation for the first-order corrections, ξ_{r1}:

$$\frac{d}{dr}\left(r\langle B_\varphi \rangle^2 \frac{d\xi_{r1}}{dr}\right) = 4\pi(\Omega_2^2 - \sigma^2)r^2 \frac{d\langle \rho \rangle}{dr} - 3k^2 r\langle B_\varphi \rangle^2 + k^2 r^2 \frac{d\langle B_\varphi \rangle^2}{dr}, \qquad (6.134)$$

with the outer boundary condition

$$\frac{d\xi_{r1}}{dr}(\alpha_{max}) = (k\alpha_{max})^2 \ln \frac{1}{k}. \qquad (6.135)$$

The integral of (6.134) together with the boundary conditions (6.135) yields the growth rate

$$\Gamma^2 = \frac{1}{4M} k^2 \alpha_{max}^3 B_\varphi^2(\alpha_{max}) \ln\left(\frac{1}{k}\right) + \Omega_2^2(\alpha_{max}), \qquad (6.136)$$

where

$$M = 2\pi \int_0^{\alpha_{max}} \langle \rho \rangle r \, dr \qquad (6.137)$$

is the mass per unit length, and the terms of the order of unity are ignored compared with the logarithmic term.

The first term in (6.136) represents the growth rate of a stationary pinch (Nycander and Wahlberg, 1984), while the second one corresponds to the gravity effect and predicts a stabilizing correction for $\Omega_2^2 < 0$. It is important that this correction be independent of wavenumber k, which means that the kink mode becomes stable at a small wavenumber. The cut-off wavenumber, k_0, corresponding to the vanishing of the growth rate, is implicitly given by

$$k_0^2 \ln \left(\frac{1}{k_0} \right) = -\frac{4M\Omega_2^2(\alpha_{max})}{\alpha_{max}^3 B_\varphi^2(\alpha_{max})}. \qquad (6.138)$$

Figure 6.34 shows the cut-off wave number, k_0, as a function of the amplitude of oscillations, $A = (\alpha_{max} - \alpha_{min})/2$, calculated from (6.138). Figure 6.35 shows the cut-off wave number k_0 as a function of the amplitude of oscillations obtained by numerical simulation (Bud'ko et al., 1995).

6.8 Fiber-initiated Z pinches

Experiments at Los Alamos National Laboratory (Scudder, 1985; Hammel and Scudder, 1987; Hammel, 1989; Scudder et al., 1994), Naval Research Laboratory (Sethian et al., 1987, 1989), Imperial College (Haines, 1994; Niffikeer et al., 1994), and Dusseldorf University (Kies et al., 1991; Stein et al., 1994) have demonstrated

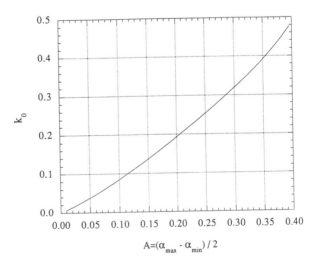

FIGURE 6.34. The cut-off wavenumber as a function of the amplitude of oscillations (6.138).

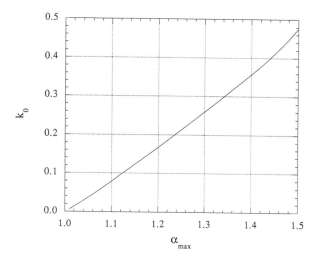

F<small>IGURE</small> 6.35. The cut-off wavenumber as a function of the amplitude of oscillations obtained numerically.

that very interesting plasmas can be created by discharging a modern high-voltage pulsed-power generator through frozen deuterium fibers or solid fibers. Each institution has developed different diagnostics and come to different conclusions about the behavior of fiber-initiated pinches. Enhanced stability of a dense Z-pinch plasmas formed from a frozen deuterium fiber (Scudder, 1985; Sethian et al., 1987) for a current linearly rising up to the peak value of 0.64 MA in 130 ns, is one of the most intriguing experimental facts in the field of modern Z-pinch research. If this property of fiber-initiated dense Z pinches permits scaling to smaller diameter fibers and greater currents, then it would be appropriate to reconsider the simple Z pinch as a possible approach to fusion.

Sethian et al., (1987, 1989) have interpreted the very rapid increase in visible light emission, as recorded on a streak photograph, and the rapid onset of neutron production as indicating the onset of the $m = 0$ instability. Experimental observations over a range of parameters have led to the hypothesis that fiber-formed plasmas are stable until the current reaches the maximum value, i.e., until $dI/dt = 0$. The first experiments on fiber-formed high-density Z pinches, with a peak current of 250 kA, demonstrated stability of the pinches for many Alfvén transit times τ_A. The conclusion (Sethian et al., 1987) was that the pinch remains stable as long as the current is increasing, and becomes unstable only after the current peaks. The pinch expansion is very slow until the current peaks, then a rapid expansion follows. The onset of neutron yield is at this very instant. It appears that before the current peaks, a time-dependent Z-pinch equilibrium, like those described in Secs. 3.5.1 and 3.5.2, is in effect, whereas after peak current, a violent development of sausage instabilities begins (see Fig. 6.36). It was quite natural to

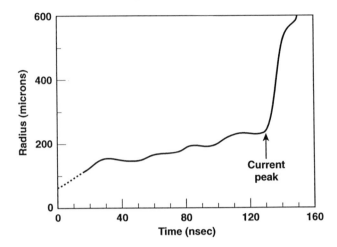

FIGURE 6.36. $R - t$ diagram of a fiber-initiated Z pinch (streak photograph from Sethian et al., 1987).

relate the observed neutron yield to accelerated ions, which are produced in the course of development of the $m = 0$ instability. On the other hand, no single 0-D or 1-D model of a Z pinch predicts such a rapid change in expansion velocity. The 0-D models can hardly provide anything better than (3.102) for the radius of a pinch in a time-dependent equilibrium. Results of a 1-D numerical simulation (Lindemuth et al., 1989) are in better agreement with the same (3.102) than with the experimental streak record. The simulation of the initial slow expansion of the plasma column is satisfactory, but the expansion following it is much slower than that observed. Thus, the flow after the current peak is most likely 2-D or 3-D, formed by instability development.

In contrast, LANL shadowgraph photography shows visible evidence of $m = 0$ behavior significantly earlier than the peak current. Similarly, and perhaps also in contrast with the LANL observations, $m = 0$ behavior has been observed at Imperial College at a time very early in the discharge, well before peak current, and even prior to the time when the fiber has been completely ionized.

Two-dimensional computations (Lindemuth, 1990; Sheehey et al., 1992) predict that the presence of a cold, solid central fiber throughout the duration of the experiments may play an important role in the experimental observables, and therefore, should be considered in any interpretation of the experimental diagnostics. The computations used cold-start initial conditions; a nonideal equation of state; a nonuniform, computed temperature distribution; and a self-consistent electrical source to compute the behavior of a fiber Z pinch from $t = 0$. The cold-start initial conditions are a solid, cryogenic deuterium fiber (radius $\sim 15 \ \mu$m, temperature 0.001 eV) surrounded, to about twice the fiber radius, by a low-density, warm halo plasma (density $= 0.001$ solid), temperature 1 eV), which provides an initial current-conduction path. The computational results were found to be insensitive to the details of the halo plasma after about 10 ns. Two-dimensional simulations of HDZP-I discharges show significant expansion and $m = 0$ instability development before the fiber has become fully ionized, which occurs in the simulations at 30–50 ns. The computational result was that the rapid expansion of visible light and onset of neutron production corresponds to complete fiber ablation and not to the onset of instability, which occurs significantly earlier. The $m = 0$ behavior began quite early in the discharge at the outer radial boundary of the hot, coronal plasma, which is formed by ablation from the cold central fiber. The $m = 0$ behavior enhances the fiber ablation rate, and by 45 ns, only approximately one-half of the material on axis remains at high density and the remainder has expanded and reached a relatively high temperature. Most of the emission prior to the minimum, at ~ 50 ns, occurs off axis in hot spots that occur at the current channel constrictions, which result from the $m = 0$ behavior. Neutron production occurs abruptly at 50 ns and corresponds to the time of complete fiber ablation on axis. The computations (Lindemuth, 1990) lead to a new interpretation of the behavior of fiber-formed pinches. According to the computations, unstable $m = 0$ behavior occurs very early in the discharge, prior to complete ionization of the fiber, in the coronal plasma that forms outside the cold central core. This behavior leads to enhanced fiber-ablation rates and enhanced expansion of the plasma ablated from the fiber. The early-time behavior is indicated by x-ray emission off axis in the coronal region. Ultimately, the material on axis is transformed into plasma, with densities and temperatures that are high enough to produce significant neutron yield. Neutron production, on-axis x-ray production, and rapidly expanding visible light signify complete fiber ablation, not the onset of unstable $m = 0$ behavior, which begins significantly earlier than neutron production.

Theoretical and computational investigation into the reported anomalous stability of the early, low-current experiments has concentrated on the connection between driving the current ramp and plasma profiles, and the stability of such profiles as modified by nonideal effects, such as resistivity. In particular, it has been found that relatively resistive, low-temperature plasma columns, with Lundquist number up to about 100, may be $m = 0$ stable. The plasma for which these results have been derived are constant-radius, uniform temperature, and near-solid density, i.e., the fiber is fully ionized, and expanded to no more than a few hundred μm. In contrast, low densities and high temperatures, which are seen in the simulations (Lindemuth, 1990; Sheehey et al., 1992), are factors that would raise the effective Lundquist number of such a plasma from "safe" values of order 1 to beyond the critical values around 100, even at low-current, early stages of the discharge. Once instabilities begin to grow in the corona, expansion of the column is enhanced, leading to still lower densities, higher radii, and still higher Lundquist numbers.

It is not easy to explain why the sausage instability starts to develop only after the current peak and is not observed before it. Most of the authors who have studied this problem believe that resistivity is the mechanism that stabilizes the pinch for low currents and low temperatures, but fails to do so after some critical value of the

Lundquist number, L_{ucr}, is exceeded; this value is estimated as 100 by Culverwell et al. (1989b) and 160 by Cochran and Robson (1989). Other stabilizing effects, noted by Lampe (1989), are based on the skin effect and resistive plasma heating, which are both stronger in the neck regions, especially when the current is rising. If the stabilizing mechanism in the experiment of Sethian et al. (1987) is indeed resistivity, with a critical Lundquist number of about 100, then the prospects of producing a fusion plasma with $L_u \approx 10^3$ (see Table 1) in a fiber-initiated Z pinch are not very optimistic. However, the above results do not yet constitute a consistent theory. Some of the experimental results, and particularly the very rapid switching on of the instability, have not been explained.

Comparison of simulation and experimental results by LANL leads to the conclusion that the anomalous stability reported for the thin-fiber HDZP-I experiments appears limited to the earlier, low-current stages of the discharge. Even then, instability may develop in the outer corona, which is only faintly reflected in the shadowgram images. Increased instability and expansion noted in the later, high-current stages are apparently related to the high temperatures, low densities, and higher effective Lundquist numbers. In any case, the coronal plasma created by preionization has a tendency to expand, leading to higher Lundquist number and instability.

There is some experimental evidence obtained in the Kalif experiments on CD_2 and frozen fibers (Karlsruhe) at a peak current of 0.8 MA that multiple micropinches (diameter < 50 μm) are embedded in a core of the plasma column, which otherwise looks quite stable (Decker and Kies, 1989). Hence, the fiber-initiated Z pinches may not be definitely stable after all, in which case the perturbation growth, up to a certain instant, somehow turns out to be compatible with their structural integrity. All this requires further study.

6.9 Two-dimensional simulation of magnetically driven Rayleigh–Taylor instabilities in cylindrical Z pinches

We have seen that simple imploding Z pinches are unstable to the magnetically driven RT instability. In the present experiments, the RT modes are predicted to grow to large amplitude ($n_{\text{eff}} \gg 1$). MHD simulation codes have been used to calculate the plasma dynamics of these unstable imploding plasmas. To date, extensive calculations have been performed with 2-D (mainly r-z) MHD codes. These codes are computationally intensive, since not only the MHD equations (3.36)–(3.41) are solved, but the codes use computational packages to calculate the equation of state, radiation transport, and the distribution of atomic states.

Hammer et al. (1996) have used a 2-D MHD Lagrangian code (Zimmerman, 1973; and Nielson, and Zimmerman et al., 1981) to simulate the dynamics of imploding aluminum wire-array and Ne-Ar gas-puff implosions on the Saturn pulsed-power accelerator, and compared these results with measurements. The code uses a multigroup diffusion model to calculate radiation transport with frequency-dependent emissivity and opacity determined by the average atom model (Lokke and Grassberger, 1977). Local thermodynamic equilibrium (LTE) is used to calculate the ionization level of the atoms and atomic level populations. The equation of state is based on the Thomas–Fermi–Cowan model (More et al., 1988) and the Lee and More (1984) extension of Spitzer resistivity and electron thermal conductivity is included.

Sanford et al., (1996, 1997) present measurements and 2-D MHD simulations of imploding aluminum wire-array implosions driven by the Saturn accelerator. The simple model of the growth of the classical RT instability (Chap. 5) gives the total number of e-foldings as $n_{\text{eff}} \sim 15$ for a 1-mm wavelength. Thus the RT instability is predicted to grow to large amplitude early in the implosion. The simulations show that the RT instability grows at short wavelengths (initially $\ll$ 1 mm, with even larger values of n_{eff}), seeds longer wavelengths, and produces the classic bubble-and-spike structure during the implosion. The main x-ray pulse begins when the bubble reaches the axis, the implosion energy is converted to thermal energy, and the electrons in the bubble region are heated to high temperature. The radial smearing of the mass from the breakup into bubble-spikes sets the timescale for the stagnation on axis to $t_{\text{stag}} = t_{\text{spike}} - t_{\text{bubble}}$, where t_{spike}, t_{bubble} are the arrival times for the spike tip and bubble, respectively. Since the radiation is produced when kinetic energy is converted to thermal energy, the output-radiation pulse also occurs in a time of order t_{stag}.

In the simulation, the radiating region contains $\sim$ 10–20% of the total mass in a region of very small size ($\sim$ 10s of μm), with the bulk of the material at scales of order 500–1000 μm. The FWHM of the radiating region near maximum emission is $\sim$ 50 μm in the simulations, compared to a measured value of about 600 μm, closer to the bulk-plasma radius found in the simulation. The emission tends to be localized near the position on axis where the spike plasma converges, as opposed to the experimental images, which show comparatively weak axial structure. Radiation cooling plays an important role in allowing the convergence of part of the mass to high density, in analogy with the more extreme radiative collapse observed in 1-D MHD simulations, and is sensitive to the atomic physics model used. In particular, the LTE approximation may overestimate the cooling rates and associated collapse behavior.

The small-diameter region of highly compressed plasma may be due to the assumption of azimuthal symmetry implicit in the 2-D calculations. A very small azimuthal nonuniformity would have little effect on the run-in dynamics and RT development, but could give imploding mass elements angular momentum that would prevent high convergence. The most important sources are the residual azimuthal nonuniformity after the individual wires have merged, the perturbations from the eight return current posts, and the RT instability in the azimuthal direction. A fluid element with angular momentum, $\delta M r_0 v_\theta$, conserved during the implosion would have its azimuthal velocity amplified as $v_\theta \sim r_0/r_f$. The fluid element ceases to converge when $v_\theta^2 = v_{\text{imp}}^2$. Resistive effects may also play a role. The simulations show behavior late in the implosion similar to the saturation model of Roderick and Hussey (1984), where material from the spikes diffuses into the "throat" of a bubble and partially shorts out the accelerating current. The bubble throat develops flow similar to a MHD nozzle. Enhanced resistivity would increase mass transport into the bubbles and may reduce the current density near $r = 0$ that helps drive radiative collapse. The MHD nozzle effect is also involved in the increased coupling to the circuit found in 2-D versus 1-D simulations, since the plasma acceleration and heating associated with the nozzle flow converts magnetic energy to kinetic and thermal energy. The magnitude of the increased coupling is not known accurately, due to the significant energy errors in the calculation.

In the course of the simulated stagnation, mass flows axially (in the $+\hat{z}$, $-\hat{z}$ directions) away from a spike impact point near $r = 0$. The axial flow from one impact point quickly encounters the opposing flow from a neighboring impact point, causing the flow in this interaction region to turn radially outward. The net result is a complex, circulating flow pattern. The stirring continues throughout the rest of the simulation, and acts to convert magnetic energy to thermal energy through PdV heating, producing about half of the long-lived "tail" on the radiation pulse. Of the remainder, 10–15% of the tail emission can be attributed to ohmic heating, with the rest due to decay of the stored electron and ion internal energy. The plasma relaxes to a state in near-pressure balance, with a dense core of order 500-μm diameter and a hot, low-density, weakly emitting halo (the reverse of the situation during the stagnation, where most of the mass was comparatively cool and at larger radius than the emitting region near the axis). Throughout the stagnation and subsequent near-equilibrium phase, most of the current flows at fairly large radius. This is due to the MHD-nozzle shorting effect, mentioned above, during the stagnation and to the current carried in the hot halo in the near equilibrium phase.

In both the primary pulse and the tail emission, thermalized energy is rapidly converted to radiation. This may explain why the simulation is capable of reproducing the observed radiation pulse while inaccurately modeling the spatial density distributions during stagnation. If the code correctly models the sheath acceleration and RT breakup, then the stagnation time and kinetic energy are determined. At stagnation, most of the kinetic energy will become thermalized through shock and PdV heating. As long as the stagnation density is sufficiently high, the radiative and electron-ion energy exchange rates are of order or faster than the stagnation time, and all of the thermalized energy is radiated with very little sensitivity to the details of the density distribution. In contrast, the energy distribution of the x-ray emission will depend on the details of the magnitude and distribution of density and temperature, which may explain some of the approximately factor-of-2 differences between simulated and experimental spectra.

Peterson et al. (1996) used an Eulerian 2-D MHD code to calculate the dynamics of an imploding-liner Z pinch, driven by the PEGASUS capacitor bank (capacity = 216 μF, resistance = 0.3 Ω, inductance = 36 nH, current rise time = 2 μs, and maximum current = 3 MA). Since the liner is initially solid, Peterson et al. (1996) used a 1-D MHD computer code to calculate the initial expansion and to obtain initial plasma parameters to use in the 2-D simulations. The RT instabilities are predicted to grow to large amplitude, $n_{\mathrm{eff}} \gg 1$. The fastest growing instabilities are sausage ($m = 0$, $k \neq 0$) modes so the nonlinear dynamics can be simulated with a 2-D MHD model. The RT instability is seen to grow to large amplitude early in the implosion. The usual bubble-and-spike nonlinear structure then develops. Measurements of liner breakup and healing of the liner are shown to agree with the calculations. The initial amplitude of the unstable modes are unknown, so the initial amplitude is chosen to obtain good agreement between measured and calculated x-ray emission (pulse shape, maximum power, and radiated energy). They find that mode coupling of the RT modes does not play a role in these implosions. One problem with these calculations is that shorter wavelength, faster growing modes are not resolved in these calculations (the shortest wavelength mode that is well-resolved [eight cells/wavelength is required] is ~ 0.4 cm, compared to a rough estimate of the fastest growing mode of ~ 0.001–0.01 cm).

Sheehey et al. (1992) used a 2-D MHD code to simulate deuterium-fiber-initiated Z pinches. They emphasize that the initial conditions are important in the development of the sausage instability. A low-density corona develops during the initiation process and the sausage instability develops in the corona. The calculations show the rapid development of the sausage instability when the fiber becomes fully ionized. The instabilities drive nonuniform heating and rapid expansion of the plasma column.

Maxon et al. (1996) used a 2-D Lagrangian MHD code (Zimmerman, 1973; and Nielson and Zimmerman, 1981) to study the dynamics of small-initial-diameter (0.4 cm) aluminum wire arrays driven by the Saturn accelerator. The calculations show that the RT instability dominates the pinch dynamics. The RT instability grows to the nonlinear bubble-and-spike phase that results the in the formation of high density and temperature islands near the bubbles. The calculations show that the x-ray emission is highly inhomogeneous giving rise to bright spots of dimension ~ 100 μm, in agreement with the measurements.

7

Applications of Z Pinches

7.1 Controlled nuclear fusion

The early attempts to realize controlled nuclear fusion, about forty years ago, were simple Z-pinch configurations driven on a μs timescale. It was found that Z pinches were very unstable, and the neutron yield observed in the Z-pinch experiments was produced by nonthermal rather than thermal ions. New interest in fusion applications of Z pinches is due to the development of fast pulsed-power drivers, capable of delivering multi-megampere currents to a Z-pinch load within times of tens of nanoseconds, which is important for producing uniform stable plasmas (Mehlman et al., 1986; Stephanakis et al., 1986); and to promising experimental results with pinches formed from frozen deuterium fibers. A number of other new ideas, which seem to be worth further study, use various modified Z-pinch schemes for controlled fusion (Freeman et al., 1972; Decker and Wienecke, 1976; Brownell and Freeman, 1980; Lindemuth and Kirkpatrick, 1984; Liberman and Velikovich, 1984; Yan'kov, 1985; Rahman et al., 1987a; Neudachin and Sasorov, 1988; Robson, 1988; Ikuta, 1988; Linhart, 1989; Golberg et al., 1989, 1990; Rahman et al., 1989). These new approaches to Z-pinch fusion differ from those of the early 1950s mostly in their parameter range. Characteristic lengths are of order of 10 mm, and characteristic times of order 10 ns.

There are two different, general concepts for a Z-pinch fusion systems today. One is based on magnetic confinement of time-dependent plasma equilibria. The other, representing inertial confinement or hybrid approaches, uses the dynamics of plasmas compressed by Z pinches (including the necks produced by the $m = 0$ mode instability development) or annular liners.

New perspectives on magnetic confinement were opened by the first successful experiments with frozen DT fibers. With an initially solid fiber it is possible to produce a dense plasma with $n_i \approx 2 - 5 \times 10^{21}$ cm^{-3}, at the same time avoiding the very unstable stage of radial compression. Indeed, the first experiments (Scudder, 1985; Scudder and Hammel, 1987; Scudder et al., 1994) with relatively small currents (current pulse amplitude, $I_{max} = 250$ kA) demonstrated striking stability of a fiber-initiated Z pinch. The experiments of Sethian et al. (1987, 1989) with $I_{max} = 640$ kA, current-pulse duration $t_{max} \approx 130$ ns, initial fiber radius $R_0 =$

62.4 μm, and length $L = 5$ cm, demonstrated that the pinch plasma was stable as long as the current was increasing ($dI/dt > 0$), and the fast development of instabilities begins near the peak of the current. In the experiments of Decker and Kies (1989) with the Kalif driver ($I_{\max} = 0.5$ to 1 MA, $t_{\max} \approx 60$ ns) and fibers ($R_0 = 25$ to 50 μm, $L = 3$ to 5 cm), no indications of pinch disruption by global $m = 0$ or $m = 1$ instabilities were found, although multiple micropinches were observed in the time-integrated soft x-ray pictures.

The neutron yield measured by Sethian et al. (1987) was 2×10^9 and 8.4×10^9 neutrons for $I = 500$ kA and 640 kA, respectively. A neutron yield $\approx 10^{10}$ was observed by Kies (1988) for $I = 500$ kA. The neutrons in the solid-fiber Z pinches are always generated near peak current, when the fast growth of sausage perturbations begins. The scaling of the neutron yield, $Y \propto I^8$, proposed by McCall (1989) for the mechanism of neutron production in the necks, is consistent with these experimental results. According to another explanation (Ikuta, 1988), the neutrons are produced in a narrow current channel near the axis, where the plasma is compressed to a degenerated state.

In the absence of fusion energy input to the plasma, the current is limited by $I_{PB} \approx 1$ MA. However, the energy delivered by fusion alpha particles, in addition to ohmic heating, is capable of delaying the onset of radiative collapse. A "striking analogy with stellar collapse" is stressed by Robson (1988) for about 100 ns, allowing one to pass a current rising linearly from 2 to 7 MA through a DT plasma. From the Bennett condition (2.8), we find that the ohmic heating is enough to heat a column of DT plasma produced from an initially solid fiber, whose radius is $R_0 \approx 25$ μm, to fusion temperatures $T \geq 10$ keV. Its expansion to $R \approx 100$ μm corresponds to an average ion number density $n_i \approx 3 \times 10^{21}$ cm^{-3}, and Lawson's criterion $n_i \tau \geq 10^{14}$ cm^3 yields a confinement time $t \geq 30$ ns. These values are consistent with both 0-D and 1-D calculations, if fusion heating and end losses are taken into account (Hartman, 1977; Hagenson et al., 1981; Haines, 1982; Miller, 1984; Robson, 1988). The energy gain Q expected in such a Z-pinch fusion system is small ($Q < 2$), if the magnetic energy associated with the current cannot be recovered. The limit on Q is explained by the low-conversion efficiency of magnetic energy to thermal energy in a thin equilibrium Z pinch. Indeed, since the thermal and magnetic energy densities near the pinch boundary are comparable, and outside the pinch, $\varepsilon_M = B_\varphi^2/8\pi \propto (r/R)^{-2}$ the total magnetic energy $\int \varepsilon_M(r)2\pi r\,dr$ is about $\ln(R_w/R) \geq 10$ times higher than the total thermal energy, R_w being the radius of return conductor. However, if 95% of the magnetic energy could be recovered, then Q can be as large as 33 [see Robson (1988)], which implies that a reactor with positive energy output is possible.

In any case, a breakeven experiment with a fiber-initiated HDZP appears to be feasible, provided that the plasma parameters can be kept close to their average values. This means that the effect of instabilities can be neglected during a confinement time τ. A fiber-initiated HDZP plasma has been observed to be stable as long as the current is rising (Sethian, 1987). This supports the hope that the plasma will be stable, if the energy source is powerful enough to keep the current rising linearly (Robson, 1988). However, this conclusion is not confirmed either by the experiments or by a stability analysis. In particular, the available 2-D numerical simulations (Lindemuth, 1989a, b; 1990), performed for similar experimental conditions show appreciable growth of perturbations. This problem certainly requires further study.

An alternative fusion scheme is based on the idea of producing ultrahigh temperatures and densities in a radiative collapse of a neck, resulting from the nonlinear development of a sausage ($m = 0$) instability mode (Vikhrev, 1978; Vikhrev et al., 1982; Yan'kov, 1985; Neudachin and Sasorov, 1987). Supposing that (a) the degree of radial compression is very high (from $R_0 = 0.5$ cm to $R \approx 1$ μm, about 10^4 times), and (b) the total current I continues to flow through the neck, without any redistribution, one finds ignition of a DT plasma by $I = 10$ MA. After ignition, the fusion reaction becomes self-sustaining, and the burning zone propagates axially outwards into regions of smaller temperatures and densities. This idea is based mostly on the results of studies of the bright spots in Z pinches, where large values of temperature, density, and $n_i \tau$ appear to be produced (see Sec. 2.3). However, one cannot be sure that such a strong compression would be possible in a deuterium or DT plasma, both of which radiate much less than those containing high-Z ions. Another major problem is that of passing a large current through a very thin channel of dense plasma, with some plasma always surrounding it. The current has a strong tendency to be redistributed to the outer plasma layers.

Inertial confinement Z-pinch fusion systems have at least one advantage: there is no limitation on I like I_{PB}. The current can be as high as the pulsed-power driver can deliver. In addition, the efficiency of magnetic energy conversion to implosion kinetic energy of a dynamic Z pinch or liner is estimated as $[1 + \ln(R_w/R_0)]^{-1}$, and with $R_0 \approx 1$ cm, is considerably higher than in the case studied above. Ikuta (1988) proposed using a Z pinch with current ≈ 10 MA as a driver accelerating an initially frozen annular DD target inwards with velocities up to $\approx 10^8$ cm/s, producing high compression and heating of the annulus when the target stagnates on axis. An annular liner accelerated by a Z pinch also can be used as an indirect driver, providing uniform irradiation and subsequent ablative compression of spherical fusion microtargets (Linhart, 1989). The method is similar to the hohlraum scheme of laser fusion (Yamanaka, 1983). Both a 1-D numerical simulation and simple analytical estimates indicate that for $I = 10$ MA, the collision of the accelerated liner with an inner annular target leads to the conversion of the liner kinetic energy into soft x-ray radiation with a power density of about 10^{15} W/cm near the microtarget for 2–3 ns, which is enough for igniting the target. There are also possibilities of using the radiation from a converging-current sheath in a gas-puff Z pinch or a plasma focus directly for ablative compression of microtargets placed at axis (Freeman et al., 1972; Decker and Wienecke, 1976; Brownell and Freeman, 1980). Calculations of Gratton et al. (1986) show that a plasma focus of the Poseidon class, with energy stored ≈ 250 kJ, is equivalent to a Nd glass laser facility with a pulse energy of 1 kJ and a pulse duration of 10 ns.

When a dynamic Z pinch is treated as an ICF system, its most important characteristic is the stability of compression. It was shown by Bud'ko et al. (1989a, b) that a relatively small axial magnetic field can significantly improve stability by suppressing the most dangerous sausage and kink modes of RT instability associated with the inward acceleration of an annular plasma by the azimuthal magnetic field. This allows a higher degrees of stable radial compression: 15–22 instead of the conventional 8–10, (see Felber et al., 1988a, b; Ratakhin et al., 1988; Sorokin and Chaikovsky, 1989).

The axial magnetic field can also provide confinement of the fusion α-particles, thus decreasing the ignition threshold (Lindemuth and Kirkpatrick, 1984; Liberman and Velikovich, 1984; Jones and Mead, 1986). Two schemes of a Z-pinch ICF system using an axial magnetic field were proposed recently. The first one, a Z-θ pinch, suggested by Rahman et al. (1987b), is based on compression of a frozen DT fiber in a θ-pinch geometry by a rapidly increasing axial magnetic field, which, in turn, is produced by magnetic flux compression with an imploding gas-puff Z pinch. An advantage of this scheme is that the dynamics of compression make the magnetic-pressure pulse acting on the fiber considerably shorter than the current pulse driving the puff. The other method is based on compression of a gaseous target by an annular liner in an axial magnetic field (Rahman et al., 1987a; Golberg et al., 1989, 1990). The following scaling was derived by Golberg et al. (1989) from the Lawson's criterion for breakeven conditions:

$$I_{max}(MA) = 550(m/M)^{1/4}\left[\frac{R_0(cm)}{(R_0/R_f)\ln(R_0/R_f)}\right]^{1/2}, \qquad (7.1)$$

where I_{max} is amplitude of a sinusoidal current pulse, m and M are masses of the compressed DT plasma and the liner, respectively, and R_f is the final radius of the liner. A 1-D numerical simulation of Golberg et al. (1989, 1990) confirms the estimate (7.1). The current pulse amplitude $I_{max} = 20$ MA and FWHM 40 ns are sufficient for breakeven, provided that $R_0 = 3$ mm, and the final compression ratio $R_0/R_f = 30$. This value is large, but not when, compared to the 22-fold compression already observed in gas-puff Z-pinch experiments with an axial magnetic field (Felber et al., 1988a) on the one hand, and to the much higher compression required by other Z-pinch fusion schemes on the other hand.

7.2 Z pinches as sources of x-ray and neutron radiation

The most important application of Z pinches at present is their use as sources of x rays and neutrons. The relative simplicity of design, high power density, and x-ray radiation energy output in a broad spectral range make Z-pinch sources unique for many applications, including x-ray lithography, spectroscopy and microscopy, etc. (Burkhalter et al., 1979; Bailey et al., 1982; Bruno et al. 1983; Spielman et al., 1985b; Pereira and Davis, 1988; Lebert et al., 1989). The efficiency of transformation of the kinetic energy of a liner and/or of the thermal energy of the pinch plasma into x-ray radiation can be as large as 100% (see Spielman et al., 1985a, b). Various Z-pinch systems and working regimes correspond to requirements in power, uniformity and spectral range of radiation, as required by their multiple applications. In particular, for fundamental research in atomic spectroscopy, intense radiation in resonant lines of high-Z ions is required, which is conveniently generated by exploding wires (see also the references on exploding wires in Chaps. 3 and 6).

Although the radiation power output and its spectral range are determined mainly by the energy delivered to the Z-pinch load and by the current rise time, the spectral distribution of the radiated energy is determined to a large extent by the dynamics of the Z pinch. In many cases, it cannot be even roughly estimated assuming Bennett equilibrium parameters for the pinch. The instability development which may result, in particular in the formation of bright spots, turbulent plasma leading to anomalous conductivity differing from the classical value by orders of magnitude, is a major factor in shaping the radiation spectra, and essentially depends on the dynamics of compression, implosion, or explosion. The dynamics, in turn, are influenced by the radial temperature and density profiles formed during the initial stages of the process, by the current rise time, and other dynamic characteristics of the pulsed-power driver. Even the electrode material can have a considerable effect on the initial perturbation growth, and consequently on the spectral distribution of emitted radiation (Gerusov and Imshennik, 1985; Bobrova and Razinkova 1985; Stewart et al., 1987). Although the formation of bright spots and micropinches that are responsible for most of the hard x rays and neutron emission in Z-pinch and plasma-focus experiments have been studied for a long time, many problems still remain.

Recent work at Sandia National Laboratories has emphasized that fast Z-pinch implosions can efficiently convert the stored electrical energy in a pulsed-power accelerator into x rays (see Matzen, 1997). Of course, hydrodynamic Rayleigh–Taylor instabilities and cylindrical load symmetry are critical limiting factors for the x rays that can be produced on these accelerators (see Chaps. 5 and 6). In recent experiments on the Z accelerator, these implosion nonuniformities have been minimized by using wire arrays with as many as 300 wires (see Secs. 6.4.1–5). Increasing the wire number produced significant improvements in the pinched plasma quality, reproducibility, and x ray output power. X-ray pulse widths of less than 5 ns and peak powers of about 80 TW have been achieved with arrays of 120 tungsten wires on Saturn. Similar loads have recently been fielded on the Particle Beam Fusion Accelerator (PBFA-Z, recently renamed Z), producing x ray energies of 2 MJ at powers in excess of 250 TW. The optimal, experimentally produced total x-ray energy output from several generations of pulsed-power accelerators at Sandia is plotted as a function of load current in Fig. 1.4 (Matzen, 1997).

One important Z-pinch radiation configuration is the vacuum "hohlraum"in which the return current conductor of the Z pinch confines the x rays generated by the pinch. Hohlraum temperatures approaching 140 eV have been demonstrated (Porter, 1997) using tungsten wire-array Z pinches on the Sandia Z accelerator. The hohlraum temperature is optimized by minimizing the wall area of the hohlraum and by using high-Z wall materials. This configuration is particularly valuable for radiating materials to study equations-of-state because of the very Planckian nature of the radiation field (compared with radiation directly from the pinch). Z pinches produce unique large volume ($\sim$ 5 cm^3) hohlraum sources because of the sizeable energies (MJ's) and long time scales (10 ns) that are naturally found in Z-pinch systems.

Z pinches, and particularly plasma-focus systems, are the most powerful laboratory neutron sources, capable of emitting up to $\sim$ 4 $\times$ 10^{12} neutrons per pulse (Schmidt, 1980; Hares et al., 1985, Spielman et al, 1991). Nevertheless, the mechanism of neutron production is still a subject of discussion. One opinion holds that the neutrons are produced in a neck formed as a result of the $m = 0$ instability development (Vikhrev and Braginskii, 1980; Vikhrev, 1986). This point of view is supported by results of a 2-D numerical simulation presented by Vikhrev (1986). The alternative hypothesis is based on the beam-target mechanism of neutron production due to the acceleration of nonthermal ions in the necks by strong induction electric fields (Trubnikov, 1958, 1986; Negus and Peacock, 1979).

The available experimental data are not sufficient to make the final choice between the two hypotheses; the proponents of both find their results consistent with the experiments. Scaling of the neutron yield in a plasma focus was studied by Decker and Wienecke (1976), Decker et al., (1979, 1980), and Brownell and Freeman (1980), and agrees with the experimental findings, though the validity of its extrapolation to energies above 300 kJ and currents above 3 MA is not yet established.

7.3 X-ray laser

The cylindrical geometry of a Z-pinch plasma with a small value of the radius-to-length ratio, R/L, makes it very attractive for using it both as a lasing medium and as a flash lamp for optically photopumping a lasant placed at axis (Hagelstein, 1983; Basov and Prokudin, 1988; Davis et al., 1988; Maxon et al., 1985; Spielman, 1985a, b; Felber et al. 1988b). The pinch plasma needed to pump the lasing medium should be sufficiently dense ($n_e \approx 10^{20}$ cm^{-3}), high-temperature ($T \approx 100$ eV), and symmetric to provide uniform irradiation of the lasant. At the same time, the lasant should be protected from destruction by the hot plasma surrounding it.

A soft x-ray laser with significant gain has been developed at several laboratories: Lawrence Livermore National Laboratory (LLNL) (Matthews et al., 1987), the Naval Research Laboratory (NRL) (Lee et al., 1987) and Princeton Plasma Physics Laboratory (PPPL) (Suckewer et al., 1985). At LLNL, the Nova Laser facility was employed to irradiate a metal foil. The foil exploded, and during expansion of the dense gas, population inversion was accomplished by collisional excitation and recombination. A neon-like scheme (molybdenum, germanium, selenium, and copper) produced a gain between 1.7 and 6 cm^{-1}, at wavelengths 106–285 Å. A magnetically confined plasma column has been employed at PPPL; population inversion was accomplished by fast recombination and photoresonant pumping. The laser involves hydrogen-like, helium-like, and lithium-like schemes and low z plasma, such as carbon, magnesium, aluminum, or calcium.

The neon-like lasers require an electron density of 10^{20}–10^{21} cm^{-3} and temperature 0.8–1 keV. In general, the plasma is produced by irradiation of a foil or fiber with a high-power laser pulse followed by rapid cooling by radiation or expansion. A confining magnetic field improves the control over density.

Besides achieving the proper ionization and population inversion, the plasma must be uniform enough to support proper beam propagation. The plasma created by a high-power laser pulse of long wavelength (1–10 μm) has satisfied these requirements. It is also possible to create a plasma by the pinch effect, with the appropriate density and temperature, and in a state of rapid cooling by radiation losses and/or expansion. Research programs during the past 10 years have attempted to produce a soft x-ray laser by means of a Z pinch used as a lasant material after compression. The conditions achieved, density, temperature, etc., were similar to those attained with the Nova laser at LLNL. However, the experiments at Physics International Co., Sandia National Laboratories, and NRL did not produce lasing, presumably because pinch instabilities result in a plasma that is too nonuniform for propagation of a beam of soft x-ray photons. It may be possible to use magnetic confinement, i.e., a trapped axial magnetic field, as in the PPPL experiments, to improve the uniformity sufficiently. An axial magnetic field trapped in an annular Z pinch can also be used to form a θ pinch (Rahman et al., 1989) along the axis, that has much better stability properties than a simple Z pinch, and much higher plasma densities are possible in the Z-θ pinch than in a Z pinch with a trapped magnetic field.

The most effort has been applied to the Sandia concept of a Z-pinch-pumped x-ray laser (Spielman et al., 1985a; Maxon et al., 1985), which has the goal of using the radiation of an imploded gas-puff Z-pinch plasma. The laser medium might be either a rod on axis or an annular foil layer. Something like a doped-foam cylinder (Matzen and Spielman, 1984) or a parylene soda straw (Dukart et al., 1987) is required to isolate the lasing medium from the imploding plasma and provide a time delay between the pumping radiation pulse and the arrival of the shocks and instabilities from the stagnating gas puff that may prematurely destroy the symmetry of the lasant. The experiments with gas-puff Z pinches in an axial magnetic field show that the compressed axial magnetic field can also be used to isolate the lasant from the plasma effectively, without absorbing the x rays emitted by it, at the same time improving the stability and uniformity of the imploding plasma and making it possible to tune the "hardness" of the implosion (Felber et al., 1988a, b). However, both isolation and enhanced stability can be obtained only at the expense of the energy required to compress the magnetic field, leading to a decrease in the total radiated power (e.g., see Baksht et al., 1987; Felber et al., 1988a, b).

Recently, considerable attention has been attracted to a sodium–neon x-ray laser using the transition between the $1s^2$ and $1s^1 4p^1$ levels of a helium-like ion, NeIX. Various Z-pinch systems are proposed to realize this scheme, including a conventional Z pinch, exploding wires, and more specific designs like a capillary discharge (Zigler et al., 1989). At present the 1-MA Z pinches in sodium plasmas are capable of generating more than 30 GW of radiation power in helium-like lines, with a pulse duration of about 20 ns.

7.4 Production of ultrahigh pulsed-magnetic fields

Radially imploding plasmas produced by gas-puff Z pinches can be used to entrain and compress an external axial magnetic field. This method for producing controlled ultrahigh magnetic fields up to the order of 100 MG has been proposed by Felber et al. (1985). The highest magnetic fields conventionally generated in a controlled manner were obtained in the fifties and sixties by Fowler et al. (1960), who reported production of fields ranging to 14 MG, and Sakharov et al. (1966), whose highest peak field was estimated as $\sim$ 25 MG, with the aid of explosive flux-compression generators. This is about the limit of capacities of the chemical explosive flux-compression technique (see Knoepfel, 1970; Herlach and Miura, 1985). The best results obtained today by this conventional method are very close to those of the sixties, and do not exceed $\sim$ 15 MG (Pavlovskii et al., 1989). Fields as high as $\sim$ 3 MG were produced by Alikhanov et al. (1968), with a Z pinch used as a driver compressing magnetic flux in an annular liner made of thin foil.

The gas-puff Z-pinch method has been tested in experiments with the University of California (Irvine) gas-puff Z-pinch device by Wessel et al. (1986), Felber et al. (1988b) (I_{max} = 0.47 MA, peak axial magnetic field produced B_{max} = 1.6 MG), at the Institute of High-Current Electronics (Tomsk, U.S.S.R.; I_{max} = 1.4 MA, B_{max} = 5 MG) by Ratakhin et al. (1988) and Sorokin and Chaikovsky (1989), and with PROTO-II accelerator at Sandia National Laboratories (I_{max} = 7.5 MA, B_{max} = 42± MG) by Felber et al. (1988a). Only the field measurement of 1.6 MG was directly measured, i.e., by Faraday rotation; all of other values of B_{max} were inferred from other measurements. Peak magnetic fields above 2 MG in these experiments were estimated indirectly by assuming flux conservation in the imploding liner. The enhanced stability of implosion and uniformity of the annular plasma liner allow one to obtain 1.5–2 times higher radial compression than the conventional value 6–10. Experimental results of 11 (Ratakhin et al., 1988; Sorokin and Chaikovsky, 1989), 13 (Felber et al., 1988b), 22 (Felber et al., 1988a) show that this radial compression factor is a well-reproducible, established feature. Stabilization of dynamic magnetic flux compression in gas-puff Z-pinch systems by magnetic shear, in agreement with theory (Bud'ko et al., 1988, 1989a, b), is a characteristic feature of this method, which is lacking in the conventional explosive systems. The stability analysis done by Bud'ko et al. (1989b) indicates that even higher degrees of radial compression than those observed are possible, if the implosion regime is specially designed to make proper use of the windows of stability which exist in parameter space if an axial magnetic field is present.

This confirms the conclusion made by Felber et al. (1985): the gas-puff Z-pinch technique of producing controlled ultrahigh magnetic fields is scaleable up to $\sim$ 100 MG, whereas the explosive technique is not. The 0-D estimate for B_m is

$$B_{max} = \left[\frac{\ln(R_0/R_F)^2}{1 - (R_0/R_f)^2} \right] \left(\frac{R_0}{R_f} \right) \frac{2I_{max}}{cR_0}$$

$$= (4.3 \text{ to } 9.8)(I_{max}/\text{MA})(R/\text{cm})^{-1}\text{MG} \qquad (7.2)$$

for radial compression R_0/R_f varying from 10 to 20, see Felber et al. (1988b). This estimate is confirmed by the results of analytic and numerical studies of a 1-D problem (Velikovich et al., 1985; Felber et al., 1988c, d).

Another important advantage of magnetic flux compression with gas-puff Z-pinch systems is the much higher repetition rate of the high magnetic-field pulses. Indeed, it is possible to produce many successive plasma pinches without breaking the vacuum (Felber et al., 1988b). The interval between the shots is determined mainly by the recovery time of the pulsed-power system, which varies in the range from $\sim$ 3 minutes for moderate Z-pinch devices ($I \approx 0.5$ MA, $B \leq 1 - 2$ MG) like those used in the experiments of Wessel et al. (1986) and Felber et al. (1988b), to several hours for large ones ($I = 5 - 10$ MA, $B \geq 20$ MG), like PROTO-II.

Possible applications of ultrahigh magnetic fields produced by gas-puff Z-pinch devices include high-energy particle acceleration (Terletskii, 1957), equation of state and material-property studies, and conversion of high-energy electrons to high-energy gamma radiation (Felber et al., 1985). Some other applications, like a hybrid scheme of ICF or an x-ray laser, were described above. The production of ultrahigh magnetic fields of order 100 MG could be important for studies of some fundamental effects of quantum electrodynamics, not observable otherwise, e.g., photon splitting, electrodynamic strong-coupling modifications, and vacuum-polarization Cerenkov radiation (Erber, 1966).

7.5 Focusing high-energy particles in an accelerator

The azimuthal magnetic field in a Z-pinch plasma column can be used as a magnetic lens to collect and focus high-energy charged particles in an accelerator. This idea was first suggested in the 1960s, when a Z-pinch plasma lens had been constructed and tested in the AGS synchrotron at Brookhaven National Laboratory (Forsyth et al., 1965). A plasma lens has some important advantages, including excellent beam-transport properties, negligible reabsorption of the high-energy secondary particles, high-poloidal magnetic-field gradients, simultaneous focusing of the particle in both transverse directions, and the ability of the plasma to remain unaltered in a highly radioactive area (Autin et al., 1987). Indeed, focusing of secondary particles by the BNL plasma lens had been successfully demonstrated, though the device lacked stability—it failed after a few hours of operation. No further development followed, perhaps due to the general decline of interest in Z-pinch systems in that time.

The renewed interest in Z-pinch plasma lenses is due both to the understanding of their advantages and to the demands of modern accelerator technologies. In 1983, a program was started in CERN aiming at developing a plasma lens to be used as a powerful collector lens for focusing the beam of 3.5-GeV antiprotons in the upgraded CERN antiproton source, ACOL (Christiansen et al., 1984; De Menna, 1984). The lens should operate more than 1000 h with a repetition rate of 0.4 Hz. The parameters of the Z-pinch device being developed to perform the focusing are the following: diameter and length of the plasma column; 4 cm and

30 cm, respectively total current 400 kA, and the plasma to remain stable for ~ 3500 ns (Autin et al., 1987).

The construction of a Z-pinch device with these parameters appears feasible. The single-shot experiments carried out with the tested prototypes have shown that the dynamics and stability of the pinch and the magnetic-field distribution can be made close to those satisfying all the requirements, when the Z-pinch implosion is sufficiently slow (current pulse duration ~ 45 μs). Long-term operation remains a major problem, just as 25 years ago, but a careful choice of the insulator material to minimize wall evaporation, and of a gas like hydrogen or deuterium that emits little radiation affecting the walls seems to be the right method for solution (Autin et al., 1987).

Conclusions

The objective of dense Z pinches is to obtain a plasma with a temperature of the order of kilovolts and as high a density as possible. The particle-collision frequency is usually of the order of or less than 0.1 ns, so that the particles have a thermal distribution. The radiation field is generally not in thermal equilibrium with the particles; if it were, it would require $\sim 10^{17}$ W of power to maintain a temperature of 1 keV.

A great deal of progress has been made by accelerating a hollow plasma shell of modest density to create a cylindrical implosion. Kinetic energy can be added to the particles over a relatively long period (tens of nanoseconds), and this energy is transformed into thermal energy in a time of order 1 ns. This makes it possible to reach high densities, provided that the plasma remains relatively cold during acceleration. Any heating during the acceleration phase will reduce the final density that can be achieved. This problem is similar to the problem of preheating in ICF that must be avoided if a high density is to be reached in the implosion of a capsule. As a result, high-density plasmas are only achieved for high-Z plasmas, where the radiation keeps the plasma cold during the acceleration phase. Such plasmas are mainly of interest for radiation sources. Hydrogenic plasmas of interest for controlled fusion do not compress well, because radiation cooling is weaker unless the plasma is seeded with high-Z impurities. Therefore, high-density pinches produced by the implosion of wire arrays or gas puffs are mainly of interest as radiation sources. Soft x-ray sources of very high intensity and total yield have been created with Z pinches during the past 20 years. There are numerous applications to simulation of nuclear weapons effects, weapons-related physics, ICF, lithography and x-ray microscopy, x-ray lasers, production of ultrahigh magnetic fields, etc.

The implosion of a hollow cylindrical plasma is a dynamic plasma problem. It is always characterized by instabilities: RT surface and internal instabilities, filamentation, etc. The progress made with plasma implosions is mainly determined by the extent to which these instabilities can be controlled. For example, the present achievements are mainly due to the fact that the instabilities can be suppressed for tens of nanoseconds, so that the plasma can be accelerated over several centimeters with present pulsed-power capabilities (i.e., power of 10^{13}–10^{14} W can be supplied over this time). A much higher temperature and density could be achieved, if stability could be maintained for a longer time, as the plasma could be accelerated over a greater distance. Pulsed-power systems are presently limited to a few megajoules, delivered in tens of nanoseconds. For example, capacitor banks could be employed directly instead of complicated Marx generators and pulse-forming lines and much more kinetic energy could be given to the plasma prior to implosion. For times like hundreds of nanoseconds, plasma-opening switches and capacitor banks could be employed for the same purpose. Improvements in power and total radiated energy from pinch sources depend on progress in controlling instabilities.

For hydrogenic plasmas, a different approach has been developed mainly at LANL and NRL. This approach consists of starting with a high-density cryogenic fiber of solid hydrogenic material and applying a fast-rising high voltage along the length of the fiber. The fiber is initially solid, and the objective is to delay expansion and possibly increase the density further by means of radiative collapse, which can take place if the radiative cooling reduces the plasma temperature so that the magnetic pressure dominates the plasma pressure. Theoretically, this takes place in a hydrogenic plasma at the Pease–Braginskii current of about 1.6 MA, and at a lower current in a plasma seeded with higher-Z impurities. Experimentally, it is clear that instabilities are of major importance. This is perhaps the only fact that is completely clear at this time, and the instabilities may not be the same as those discussed here. This is a relatively new field of research that does not have sufficient experimental documentation to distinguish among competing ideas. For example, it is not yet clear when complete ionization takes place, or the precise distribution of current within the pinch.

Additional magnetic fields can be quite effective in stabilizing a dynamic pinch, both during the implosion phase and after the impact of the accelerated plasma. Very large magnetic fields, up to tens of megagauss, can be produced by trapping frozen-in axial magnetic flux. Such high fields are of fundamental interest and they can be used to create more stable configurations, such as a high-density θ-pinch. Possible applications for improved Z-pinch implosions include high-density hydrogenic Z-θ pinches for fusion and configurations of improved stability for x-ray lasers.

References

Abramowitz M. and Stegun I. (1964) Handbook of Mathematical Functions, edited by M.Abramowitz M. I. Stegun (Government Printing Office, Washington, DC), Ch. 7.

Ågren O. (1988) Plasma Phys. Contr. Fusion **30**, 249.

Ågren O. (1989) Plasma Phys. Contr. Fusion **31**, 35.

Aivazov I.K., Vikharev V.D., Volkov G.S., Nikandrov L.B., Smirnov V.P., and Tsarfin V.Ya. (1988) Fizika Plasmy 14, 197 [Sov. J. Plasma Phys. (1988) **14**, 110].

Aivazov I.K., Bekhter M.B., Bulan V.V. et al. (1990) Fizika Plasmy **16**, 645 [Sov. J. Plasma Phys. (1990) 16, 373].

Åkerstedt H.O. (1988) Phys. Scr. **37**, 117.

Åkerstedt H.O. (1989) J. Plasma Phys. **41**, 45.

Åkerstedt H.O. (1990) J. Plasma Phys. **44**, 507.

Albares D.J., Krall N.A., and Oxley C.L. (1961) Phys. Fluids **4**, 1031.

Aleksandrov V.A., Koval'skii N.G., Lukyanov S.Yu., Rantsev-Kartinov V.A., and Stepanenko M.M. (1973) Zh. Eksp. Teor. Fiz. 64, 4 (Sov. Phys. - JETP (1973) **37**, 622].

Alikhanov S.G., Belan V.G., Ivanchenko A.I., Karasjuk V.N., and Kichigin G.N. (1968) J. Sci. Instr., Series 2 (J. Phys. E) **1**, 543.

Allen J.E. (1957) Proc. Phys. Soc. **70**, 24.

Alper B. (1990) Phys. Fluids B **2**, 1338

Anderson O.A., Baker W. R., Colgate S.A., Furth H. P., Ise J., Pyle R.V., and Wright R.E. (1958) Phys. Rev. **110**, 1375.

Anisimov S.I. and Zel'dovich Ya.B. (1977) Pis'ma v Zh. Tekh. Fiz. **3**, 1081 [Sov. Tech. Physics Letters (1977) in Russian].

Aranchuk L.E., Bogolyubskii S.L., Volkov G.S., Korolev V.D., Koda Yu.V., Liksonov V.I., Lukin A.A., Nikandrov L.V., Tel'kovskaya O.V., Tulunov M.V., Chernenko A.S., Tsarfin V.Ya., Yan'kov V.V. (1986) Fizika Plazmy **12**, 1324 [Sov. Journal of Plasma Physics (1986) **12**, 765].

Arber T.D. (1990) Phys. Fluids B **3**, 1152.

Arber T.D. and Coppins M. (1989) Phys. Fluids B **1**, 2289.

Arber T.D. and Coppins M. (1989) Dense Z-Pinches, Second International Conference., ed. by N.R. Pereira, J. Davis, and N. Rostoker (AIP Conference Proceedings, AIP, New York), p. 220.

Arnol'd V.I. (1978) Additional Chapters of the Theory of Ordinary Differential Equations (Nauka, Moscow), in Russian.

Artsimovich L.A., Andrianov A.M., Dobrohotov E.I., Lukyanov S.Yu., Podgornyi I.M., Sinitsyn V.N. and Filippov N.V. (1956) Sov. J. Atomic Energy 3, 76.

Artsimovich L.A., Mirnov S.V., and Strelkov V.S. (1964) Sov. J. Atomic Energy 17, 886.

Aspden H. (1985) Phys. Lett. A 107, 238.

Autin B., Riege H., Boggasch E., Frank K., De Menna L., and Miano G. (1987) IEEE Trans. Plasma Sci. 15, 226.

Bailey J., Ettinger Y., Fisher A., and Feder R. (1982) Appl. Phys. Lett. 40, 33.

Bailey J., Fisher A., and Rostoker N. (1986) J. Appl. Phys. 60, 1939.

Baker L.and Freeman J.R. (1981) J.Appl. Phys. 52, 655.

Baker G.R., Meiron D.I., and Orszag S.A. (1980) Phys. Fluids 23, 1485.

Baker W.L., Clark M.C., Degnan J.H., Kiuttu G.F., McClenahan C.R., and Reinovsky R.E.(1978) J. Appl. Phys. 49, 4694.

Baksht R.B., Kovalchuk B.M., Loskutov V.V., Luchinsky A.V., Mesyats, G.A. 10th European Conference on Controlled Fusion and Plasma Physics (10th European Conference on Controlled Fusion and Plasma Physics, Moscow, USSR, 14-19 Sept. 1981.) Petit-Lancy, Switzerland: European Phys. Soc, 1981. p. FP-5/615.

Baksht R.B., Datsko I.M., Korostelev A.F., Loskutov V.V., Luchinskii A.V., Chertov A.A. (1983) Fizika Plazmy 9, 1224 [Sov. Journal of Plasma Physics (1983) 9, 706].

Baksht R.B., Datsko I.M., Korostelev A.F., Loskutov V.V., Petin V.K., Luchinskii A.V., Chertov A.A. (1984) Ultrahigh Magnetic Fields. Physics. Techniques. Apllications, edited by V.M.Titov, G.A.Shvetsov (Moscow: Nauka), p. 360.

Baksht R.B., Velikovich A.L., Kablambaev B.A., Liberman M.A., Luchinskii A.V., and Ratakhin N.A. (1987) Zh. Tekh. Fiz. 57, 242 [Sov. Phys. - Technical Physics (1987) 32, 145].

Baksht R.B., Datsko I.M., Luchinskii A.V., Sukhov M.Yu., Ratakhin N.A., Faenov A.Ya., Fedyunin A.V., Feduschak V.F. (1989) Zhurn. Tekh. Fiz. 59, 57 [Sov. Phys. - Technical Physics (1989) 34, 163].

Baksht R.B., Bugaev S.P., Datsko I.M., Kim A.A., Koval'chuk B.M., Kokshenev V.A., Mesyats G.A., and Russkich A.G. (1993) Laser and Particle Beams 11, 587.

Baksht R.B., Datsko I.M., Luchinskii A.V., Oreshkin V.I., Fedyunin A.V., Korolev Yu.D., Shemyakin I.A., and Rabotkin V.G. (1994) Dense Z-Pinches, Third International Conference, ed. by M. Haines and A. Knight (AIP Conference Proceedings, vol. 299, p. 365, AIP, New York).

Baksht R.B., Labetskii A.Yu., Loginov S.V., Oreshkin V.I., Fedunin A.V., and Shishlov A.V. (1997a) Fizika Plazmy 23, 135 [Plasma Physics Reports 23, 119].

Baksht R.B., Fedunin A.V., Labetskii A.Yu., Rousskich A.G., and Shishlov A.V. (1997b) Phys. Plasmas **6**, 1.

Barenblatt G.I. (1979) Similarity, Self-Similarities and Intermediate Asymptotics (Consultants Bureau, New York).

Basko M.M. (1994) Phys. Plasmas **1**, 1270.

Basov Yu.G., Prokudin V.S. (1988) Izv. Sib. Otd. AN USSSR, ser. Tekh. Nauki, no.11, pt.3, p.61 [Sov. Journal of Applied Physics (1988) **2**, 75].

Bateman G. (1978) MHD Instabilities (MIT, Cambridge, Mass.).

Batjunin A.V., Bulatov A.N., Branitskii A.V. et al (1991) in Proc. of 8th Int. Conf. on High Power Beams, Vol.2, World Scientific, Singapore.

Bazdenkov S.V., Gureev K.G., Filippov N.V., and Filippova T.I. (1978) Pis'ma v Zh. Eksp. Teor. Fiz. 18, 199 [Sov. Phys. - JETP Letters (1978) **18**, 96].

Bekhtev M.V., Vikharev V.D., Zakharov S,V., Smirnov. V.P., Tulipov M.V., and Tsarpin V.Ya. (1989) Zh. Exp. Teor. Fiz. **95**, 1653 [Sov. Physics - JETP (1989) **68**, 955].

Belova I.V. and Brushlinskii K.V. (1988) Zh. Vych. Mat. Matem. Fiz. **28**, 72 [English Translation - USSR Comput. math. and Math. Phys. T07].

Bennett W.H. (1934) Phys. Rev. **45**, 890.

Berestetskii V.B., Lifshitz E.M., and Pitaevskii L.P. (1989), Quantum Electrodynamics (Pergamon Press, Oxford).

Bernard A. et al. (1979) Nucl. Fusion Suppl., p. 159.

Bernhardt P.A., Roussel-Duprw R.A., Pongratz M.B., Haerendel G., Valenzuela A., Gurnett D., and Anderson R.R. (1987) J. Geophys. Res. **92**, 5777.

Bernstein I.B., Frieman I.A., Kruskal M.D., and Kulsrud R.M. (1958) Proc. Roy. Soc. London A **244**, 17.

Bernstein I.B. and Book D.L. (1983) Phys. Fluids **26**, 453.

Bernstein M. (1970) Phys. Fluids **13**, 1858.

Bilbao L., Bruzzone H.A., and Kelly H.G. (1984) Plasma Phys. Contr. Fusion **26**, 1535.

Bilbao L., Bruzzone H.A., Kelly H.G., and Esper M. (1985) IEEE Trans. Plasma Sci. PS-13, 202.

Birkhoff G. and Carter D. (1957) J. Math. Mech. **6**, 769.

Blackman M. (1951) Proc. Phys. Soc. B **64**, 1039.

Bloomberg H.W., Lampe M., and Colombant D.J. (1980) J. Appl. Phys. **51**, 5277.

Bloomquist D.D., Stinnett R.W., McDaniel D.H., Lee J.R., Sharpe A.W., Halbleib J.A., Schlitt L.G., Spence, P.W., and Corcoran P. (1987) in Proceedings of the Sixth IEEE Pulsed Power Conference, Arlington, VA edited by P. J. Turchi and B. H. Bernstein (IEEE, New York,), p. 310.

Blumlein A. D. (1948) U. S. Patent No. 2,465,840 (March 29,1948).

Bobrova N.A. and Razinkova T.A. (1984) Institute of Theoretical and Experimental Physics, Preprint no. 11 (in Russian).

Bobrova N.A. and Razinkova T.A. (1985) Institute of Theoretical and Experimental Physics, Preprint no. 4 (in Russian).

Bobrova N.A. and Razinkova T.A. (1987) Fizika Plazmy 13, 92 [Sov. J. Plasma Phys. (1987) **13**, 53].

Bobrova N.A., Razinkova T.A., and Sasorov P.V. (1988) Fizika Plazmy **14**, 1053 [Sov. J. Plasma Phys. (1988) **14**, 617].

Bodner S.E., Boris J.P., Cooperstein G., Goldstein S.A. et al. (1978), in Proceedings of the Seventh International Conference on Plasma Physics and Controlled Nuclear Fusion, Innsbruck, IAEA, Vol. 3.

Boiko V.V., Mankelevich Yu.A., Rakhimov A.T., Suetin N.V., and Filippov S.S. (1988) Fizika Plazmy **14**, 1348 [Sov. J. Plasma Phys. (1988) **14**, 791].

Book D.L., Ott E., and Lampe M. (1976) Phys. Fluids **19**, 1982.

Book D.L. and Bernstein I.B. (1979) Phys. Fluids **22**, 79.

Book D.L. and Bernstein I.B. (1980) J. Plasma Phys. **23**, 521.

Book D.L. (1986) "Encyclopedia of Fluid Mechanics," Chap. 8 (Gulf Publishing Company, Houston, Texas).

Book D.L. (1996) Phys. Plasmas **3**, 354.

Bowers R.L., Nakafuji G; Greene A., Mclenithan K.D., Peterson D.L., Roderick N.F. (1996) Phys. Plasmas **3**, 3448.

Braginskii S.I. (1957a) Zh. Eksp. Teor. Fiz. **33**, 459.

Braginskii S.I. (1957b) Zh. Eksp. Teor. Fiz. **33**, 645 [Sov. Phys. JETP 6, 494 (1958)].

Braginskii S.I. (1958) Zh. Eksp. Teor. Fiz. **34**, 1548.

Braginskii S.I. and Shafranov V.D. (1958) in Plasma Physics and Problems of Controlled Thermonuclear Reactions (Moscow: Izdatel'stvo AN SSSR; in Russian), Vol. 2, p. 3; English translation (Pergamon, London, 1959), Vol. 2, p. 39.

Braginskii S.I. (1963) Voprosy teorii plazmy (Reviews of Plasma Physics, in Russian), edited by M.A.Leontovich, Vol. 1, p. 183; English translation (Consultants Bureau, New York) Vol. 1, p. 205.

Brownell J.H. and Freeman B.L. (1980) Appl. Phys. Lett. **36**, 193.

Brunsell P., Hellblom G., Karlsson P., Mazur S., Nordlund P. and Scheffel J. (1990) in Controlled Fusion and Plasma Heating (Proc. 17th European Conf., Amsterdam, 1990) Vol.14B, Part II, p.610, EPS, Geneva.

Brunsell P., Drake J.R., Mazur S. and Nordlund P. (1991) Phys. Scripta, **44**, 358.

Bruno C., Chevallier J., Delvaux J., Barbaro J., Bernard A., Wolff G., David J. (1983) in Proceedings of the Fifth International Conference on High-Power Particle Beams, San Francisco, CA, USA, p. 290.

Brushlinskii K.V. and Kazhdan Ya.M. (1963) Usp. Mat. Nauk **18**, 3 [in Russian].

Brushlinskii K.V. and Shatanov A. P. (1980) Preprint no. 150 Inst. Appl. Math. Acad. Sci. U.S.S.R (in Russian).

Bruzzone H. and Fischfeld G. (1989) Phys. Lett. **134** A, 484.

Bruzzone H.A. and Vieytes R.E. (1990) IEEE Trans. Plasma Sci. **18**, 689.

Bud'ko A.B., Felber F.S., Kleev A.I., Liberman M.A., and Velikovich A.L. (1988) Pis'ma v Zh., Tekh. Fiz. **14**, 1883 [Sov. Tech. Physics Letters (1988) **14**, 817].

Bud'ko A.B. (1989) Zh. Tekh. Fiz. **59**, 113 [Sov. Phys. - Technical Physics (1989) **34**, 914].

Bud'ko A.B. and Liberman M.A. (1989a) J. Plasma Phys. **42**, 205.

Bud'ko A.B. and Liberman M.A. (1989b) Zh. Prikl. Mekh. Tekh. Fiz., **30**, 3 [Journal of Applied Mech. and Technical Physics (1989) **30**, 831].

Bud'ko A.B., Felber F.S., Kleev A.I., Liberman M.A., and Velikovich A.L. (1989a) Phys. Fluids B **1**, 598.

Bud'ko A.B., Velikovich A.L., Liberman M.A., and Felber F.S. (1989b) Zh. Eksp. Teor. Fiz. **96**, 140 [Sov. Phys. - JETP (1989) **69**, 76].

Bud'ko A.B., Velikovich A.L., Liberman M.A., and Felber F.S. (1989c) Zh. Eksp. Teor. Fiz. **95**, 496 [Sov. Phys. - JETP (1989) **68**, 279].

Bud'ko A.B., Liberman M.A., and Velikovich A.L. (1989c) Dense Z-Pinches, ed. by N.R.Pereira, J.Davis and N.Rostoker (AIP Conference Proceedings, vol. 195; AIP, New York), p. 167.

Bud'ko A.B. (1990) private communication.

Bud'ko A.B., Liberman M.A., Velikovich A.L., and Felber F.S. (1990a) Phys. Fluids B **2**, 1159.

Bud'ko A.B., Liberman M.A., and Kamenets F.F. (1990b) Plasma Phys. Contr. Fusion **32**, 309.

Bud'ko A.B. and Liberman M.A. (1992) Stabilization of long wavelength sausage and kink modes of a Z-pinch by nonlinear radial oscillations, preprint UPTEC 92 058R.

Bud'ko A.B., Karlson E.T., and Liberman M.A. (1993) Phys. Fluids B**5**, 457.

Bud'ko A. B., Kravchenko Yu. P., and Uby L. (1994) Plasma Phys. Contr. Fusion, **36**, 833.

Bud'ko A.B., Kravchenko Yu. P., and Liberman M.A. (1995) Phys. Plasmas **2**, 792.

Bulanov S.V. (1988) Sov. Phys.-Lebedev Institute, Report No. 2, p.12.

Burkhalter P.G., Dozier C.M., and Nagel D.J. (1977) Phys. Rev. A **15**, 700.

Burkhalter P.G., Davis J., Rauch J., Clark W. Dahlbacka G., and Schneider R. (1979) J. Appl. Phys. **50**, 705.

Burns E.J.T., Degnan J. H., Reinovsky R.H., Baker W.L., Clark M.C.,and McClenahan C.W. (1977) Appl. Phys. Lett. **31**, 477.

Burtsev V.A., Kalinin N.V., and Luchinskii A.V. (1990) Electric Explosion of Conductors and its Applications in Electrophysical Devices, Energoatomizdat, Moscow (in Russian).

Butt E.D., Carruthers R., Mitchell J.J.D., Pease R.S., Thonemann P.C., Bird M.A., Blears J., Htrtill E.R. (1958) in Proceedings of the Second United National Conference on Peaceful Uses of Atomic Energy (United Nations, Geneva), Vol. 32, p. 42.

Bychkov V.V., Liberman M.A., and Velikovich A.L. (1990) Phys. Rev. A **42**, 5031.

Carruthers R. and Davenport P.A. (1957) Proc. Phys. Soc. **B70**, 49.

Chandrasekhar S. (1961) Hydrodynamic and Hydromagnetic Stability (Clarendon Press, Oxford).

Chen C.J. and Lykoudis P.S. (1972) Solar Phys. **26**, 453.

Chen F.F. (1984) Introduction to Plasma Physics and Controlled Fusion (Plenum Press, New York).

Chew G.F., Goldberg M.L., and Low F.E. (1956) Proc. Roy. Soc. London Ser. A **236**, 112.

Chittenden J.P. and Haines M.G. (1990) Phys. Fluids B **2**, 1889.

Choe W.H. and Venkatesan R.C. (1990) Laser and particle Beams **8**, 485.

Choe W.H. and Venkatesan R.C. (1992) Phys. Fluids B **4**, 1524.

Chittenden J.P., Power A. J., and Haines M.G. (1989) Plasma Phyis. and Contr. Fusion **31**, 1822.

Christiansen J., Frank K., Riege H., and Seebock R. (1984) Report CERN/PS/84-10 (AA).

Chuideri C. and Van Hoven G. (1979) Astrophys. J. Lett. **232**, L69.

Cilliers W.A., Datla R.V., and Griem H.A. (1975) Phys. Rev. A 12, 1408.

Clark R.W., Davis J., and Cochran F.L. (1986) Phys. Fluids **29**, 1971.

Cochran F.L. and Robson A.E. (1989) Dense Z-Pinches, ed. by N.R.Pereira, J.Davis and N.Rostoker (AIP Conference Proceedings, vol. 195; AIP, New York), p. 236.

Cochran F.L. and Robson A.E. (1990) Phys. Fluids B **2**, 123.

Cochran F.L., Davis J., and Velikovich A. L. (1995) Phys. Plasmas **2**, 2765.

Colgate (1957) University of California Lawrence Radiation Laboratory Rept. UCRL-4829.

Collins C.B., Anderson G.A., Davanloo F., Eberhard C.D., Carroll J.J., Coogan J.J., and Byrd N.G. (1989) Laser and Particle Beams 7, 357.

Coppi G. (1964a) Phys. Lett. **11**, 226.

Coppi B. (1964b) Phys. Lett. **12**, 213.

Coppi B. (1965) Phys. Lett. **14**, 172.

Coppi B., Greene J.M., and Johnson J.L. (1966) Nuc. Fusion **6**, 101.

Coppins M., Bond D.J., and Haines M.G. (1984) Phys. Fluids **27**, 2886.

Coppins M., Culverwell I.D, and Haines M.G. (1988) Phys. Fluids **31**, 2688.

Coppins M. (1988) Plasma Phys. Contr. Fusion **30**, 201.

Coppins M. (1989) Phys. Fluids B **1**, 591.

Coppins M.and Scheffel J. (1989) Dense Z-Pinches, ed. by N.R.Pereira, J.Davis and N.Rostoker (AIP Conference Proceedings, vol. 195; AIP, New York), p. 211.

Coppins M. and Scheffel J. (1992) Phys. Fluids B **4**, 3251.

Cowling T.G. (1957) Magnetohydrodynamics (Interscience Publishers, New York).

Culverwell I.D. and Coppins M. (1989) Plasma Phys. Contr. Fusion **31**, 1443.

Culverwell I.D., Coppins M., Haines M.G., Bell A.R., and Rickard G.J. (1989a) Plasma Phys. Controlled Fusion **31**, 387.

Culverwell I.D., Coppins M., Haines M.G. (1989b) Dense Z-Pinches, ed. by N.R.Pereira, J.Davis and N.Rostoker (AIP Conference Proceedings, vol. 195; AIP, New York), p. 246.

Culverwell I.D. and Coppins M. (1990) Phys. Fluids B **2**, 129.

Cox P.M. (1990) Plasma Phys. Contr. Fusion **32**, 553.

Dahlburg J.P., Gardner J.H., Doolen G., and Haan S.W. (1993) Phys. Fluids B **5**, 571.

Dangor A.E., Favre-Dominguez M.B., Lee S., and Kahan E. (1983) Phys. Rev. A **27**. 2751.

Dangor A.E. (1986) Plasma Phys. Contrl. Fusion **28**, 1931.

Davis J., Clark R., Apruzese J.P., and Kepple P.C. (1988) IEEE Trans. Plasma Sci. **16**, 482.

Davis R.M. and Taylor J. (1950) Proc. Roy. Soc. A **200**, 375.

Davies J. and Cochran F.L. (1990) Phys. Fluids B **2**, 1238.

Dagazian R.Y. and Paris R.B., (1986) Phys. Fluids **29**, 762.

Decker G. and Wienecke R. (1976) Physica 82C, 155.

Decker G., Herold H., Kaeppeler H.J., Kies W., Maysenholder W., Nahrath B., Oppenlander T., Pross G., Ruckle B., Sauerbrunn A., Schilling P., Schmidt H., Shakhatre M., Trunk M., Steinmetz K., Bruhns H., Ehrhardt J., Hubner K., Kirchesch P., Mechler G. (1979) Nucl. Fusion, suppl. 2, 135.

Decker G., Flemming L., Kaeppeler H.J., Oppenlander T., Pross G., Schilling P., Schmidt H., Shakhatre M., Trunk M. (1980) Plasma Phys. **22**, 245.

Decker G. and Kies W. (1989) Dense Z-Pinches, Second International Conference, ed. by N.R. Pereira, J. Davis, and N. Rostoker (AIP Conference Proceedings, vol. 195, p. 315, AIP, New York).

Deeney C., Nash T., LePell P.D., Krishnan M., and Childers K. (1989) Dense Z-Pinches, Second International Conference, ed. by N.R. Pereira, J. Davis, and N. Rostoker (AIP Conference Proceedings, vol. 195, p. 55, AIP, New York).

Deeney C., Nash T., LePell P.D., Childers F.K., and Krishnan M.(1990) J. Quantum Spectrosc. Radiat. Transfer **44**, 457.

Deeney C., Prasad R.R., Nash T., and Knobel N. (1990a) Rev. Sci. Instrum **61**, 1551.

Deeney C., Nash T., Prasad R.R., Warren L., Whitney K.G., Thornhill J.W., and Coulter M.C. (1991) Phys. Rev. A **44**, 6762.

Deeney C., LePell P.D., Roth I., Nash T., Warren L., Prasad R.R., McDonald C., Childers F.K., Sincerny P., Coulter M.C., and Whitney K.G. (1992) J. Appl. Phys. **72**, 1297.

Deeney C., LePell P.D., Cochran F.L., Coulter M.C., Whitney K.G., and Davis J. (1993) Phys. Fluids B **5**, 992.

Deeney C., LePell P.D., Failor B.H., Meachum J.S., Womg S., Thornhill J.W., Whitney K.G., and Coulter M.C. (1994) J. Appl. Phys. **75**, 2781.

Deeney, C., Nash, T.J., Spielman R.B., Seaman J.F., Chandler G.C., Struve K.W., Porter J.L., Stygar W.A., McGurn J.S., Jobe D.O., Gilliland T.L., Torres J.A., Vargas M.F., Ruggles L.E., Breeze S., Mock R.C., Douglas M.R., Fehl D.L., McDaniel D.H., Matzen M.K., Peterson D.L., Matuska W., Roderick N.F., MacFarlane J.J. (1997) Phys. Rev. E **56**, 5945.

De Groot J.S., Toor A., Golberg S.M., and Liberman M.A. (1997) Phys. Plasmas **4**, 737.

De Groot J.S., Deeney C., Sanford T.W.L., Spielman R.B., Estabrook K.G., Hammer J.H., Ryutov D., and Toor, A. (1998) in Proceedings of the Fourth International Conference on Dense Z-Pinches, edited by N.R. Pereira, J.

Davis, and P. E. Pulsifer (AIP Conference Proceedings, vol. 409, p. 157, AIP, New York).

De Menna L. et al. (1984) Report CERN/PS/84-13 (AA).

Deutsch R., Kies W., and Decker G. (1986) Plasma Phys. and Contr. Fusion **28**, 1823.

Deutsch R. and Kies W. (1988a) Plasma Phys. and Contr. Fusion **30**, 263.

Deutsch R. and Kies W. (1988b) Plasma Phys. and Contr. Fusion **30**, 921.

Dothan F. Riege H, Boggasch E. and Frank K. (1987) J. Appl. Phys. **62**, 3585.

Drake J.R. (1984) Plasma Physics **26**, 387.

Dukart R.J., Hussey T.W., and Roderick N.F. (1987) Megagauss Technology and Pulsed Power Applications, edited by C.M. Fowler, R.S. Caird, and D.J. Erickson (New York: Plenum Press), p. 183.

Dunning M.J. and Haan S.W. (1995) Phys. Plasmas **2**, 1669.

D'yachenko V.F. and Imshennik V.S. (1974) Voprosy Teorii Plazmy (Reviews of Plasma Physics, in Russian), edited by M.A.Leontovich, Vol. 8, p. 164; English translation (Consaltants Bureau, New York-London, 1980), Vol. 8, p. 199. .

Eckart C. (1960) Hydrodynamics of Oceans and Atmospheres (Pergamon, London).

Emery M.H., Gardner J.H., and Boris J.P. (1982) Phys. Rev. Lett. **41**, 808.

Emery M.H., Dahlburg J.P., and Gardner J.H. (1988) Phys. Fluids **31**, 1007.

Emery, M.H., Dahlburg J.P., and Gardner, J.H. (1989) Phys. Fluids **32**, 1256.

Emmons H.W., Chang C.T., and Watson B.C. (1960) J. Fluid Mech. **7**, 177.

Erber T. (1966) Rev. Mod. Phys. **38**, 626.

Ericksson G. (1987) Phys. Scr. **35**, 851.

Exploding Wires. Conference on the Exploding Wire Phenomenon, Proceedings, 1959–1968 (New York, Plenum Press) 4 V.

Faenov A.Ya., Khakhalin S. Ya., Kolomensky A.A., Pikus S.A., Samokhin A.I., and Skobelev I.Yu. (1985) J. Phys. D **18**, 1347.

Faghihi M. and Scheffel J. (1987) J. Plasma Phys. **38**, 495.

Felber F.S., and Rostoker N. (1981) Phys. Fluids **24**, 1049.

Felber F.S., (1982) Phys. Fluids **25**, 643.

Felber F.S., Liberman M.A., and Velikovich A.L. (1985) Appl. Phys. Lett. **46**, 1042.

Felber F.S., M.M.Malley, Wessel F.J.,Matzen M.K., Palmer M.A., Spielman R.B., Liberman M.A. and Velikovich A.L. (1988a) Phys. Fluids **31**, 2053.

Felber F.S., Wessel F.J., Wild N.C., Rahman H.U., Fisher A., Fowler C.M., Liberman M.A. and Velikovich A.L. (1988b) J. Appl. Phys. **69**, 3831.

Felber F.S., Liberman M.A., and Velikovich A.L. (1988c) Phys. Fluids **31**, 3675.

Felber F.S., Liberman M.A., and Velikovich A.L. (1988d) Phys. Fluids **31**, 3683.

Fermi E. (1951) "Taylor Instability of an Incompressible Fluid," Document AECU-2979, Part 1. Also published in The Collected Papers of Enrico Fermi, Vol. 2, E.Segre, ed. (University of Chicago Press, Chicago, 1965), p. 816.

Field G.B. (1965) Astrophys. J. **142**, 531.

Filippov N.V., Filippova T.I., and Vinogradov V.P. (1962) Nucl. Fusion Suppl. **2**, 577.

Filippov N.V. (1980) Pis'ma v Zh. Eksp. Teor. Fiz. **31**, 131.

Filippov N.V. (1983) Fiz. Plazmy **9**, 25.

Filippov N.V. and Yan'kov V.V. (1988) I.V.Kurchatov Institute of Atomic Energy, Preprint IAE-4740/6 (in Russian).

Finn B.S. (1971) British Journal for the History of Science **5**, 289

Fletcher C.A.J. (1984) Computational Galerkin Methods (Springer-Verlag, New York).

Foord M.E., Maron Y., and Sarid E. (1990) J. Appl. Phys. **68**, 5016.

Foord M.E., Maron Y., Davara G., Gregorian L., and Fisher V. (1994) Phys. Rev. Lett. **72**, 3827.

Forsyth E.B., Lederman L.M., and Sunderland J. (1965) IEEE Trans. Nucl. Sci. **12**, 872.

Fowler C.M., Garn W.B., and Caird R.S. (1960) J. Appl. Phys. **31**, 588.

Fraenkel B.S. and Schwob J.L. (1972) Phys. Lett. A **40**, 83.

Freeman B., Lace G., and Shalin H. (1972) in Proceedings of the Fifth European Conference on Controlled Nuclear Fusion and Plasma Physics, Grenoble, France, 1972.

Freidberg J.P., Morse R.L., and Ribe R.L. (1972) in Texas Symposium on the Technology of Controlled Thermonuclear Fusion Experiments and the Engineering Aspects of Fusion Reactors, November 20-22, Session 5, Paper 1.

Freidberg J.P. (1982) Rev. Mod. Phys. **54**, 801.

Freidberg Jeffrey.P. (1987) Ideal Magnetohydrodynamics, Plenum Press. New York & London.

Fruchtman A. and Maron Y. (1991) Phys. Fluids B **3**, 1546.

Fruchtman A. and Gomberoff K. (1993) Phys. Fluids B **5**, 2371.

Furth H.P. (1963) Phys. Fluids **6**, 48.

Garabedian P.R. (1957) Proc. Roy. Soc. A **241**, 423.

Garanin S.F. and Chernyshev Yu.D. (1987) Fiz. Plazmy **13**, 974 [Sov. J. Plasma Phys. (1987) **13**, 562].

Gary S. and Hohl F. (1973) Phys. Fluids **16**, 997.

Gary S. (1974) Phys. Fluids **17**, 2135.

Gasilov V.A., Grigor'ev S.F., Zakharov S.V., and Krukovskii A.Yu. (1989) Fizika Plazmy **15**, 966 [Sov. J. Plasma Phys. (1989) **15**, 559].

Gerusov A.V. and Imshennik V.S. (1985) Fizika Plazmy **11**, 568 [Sov. J. Plasma Phys. (1985) 11, 332].

Gerusov A.V. (1988) Fizika Plazmy **14**, 1487 [Sov. J. Plasma Phys. (1988) **14**, 871].

Gimblett C.G. (1988) Plasma Phys. Contr. Fusion **30**, 1853.

Glasser A.H. (1989) J. Comput. Phys. **85**, 159.

Glasser A.H. and Nebel R.A. (1989) Dense Z-Pinches, ed. by N.R. Pereira, J. Davis and N. Rostoker (AIP Conference Proceedings, vol. 195; AIP, New York), p. 226.

Goedbloed J.P. and Hagebeuk H.J.L. (1972) Phys. Fluids **15**, 1090.

Goedbloed J.P. and Sakanaka P.H. (1974) Phys. Fluids **17**, 908.

Goedbloed J.P. (1975) Phys. Fluids **18**, 1258.

Goedbloed J.P. (1984) Physica **12D**, 107.

Golant V.E., Zhilinskii A.P., and Sakharov S.A. (1977) Foundations of Plasma Physics (Atomizdat, Moscow; in Russian).

Golberg S.M., Liberman M.A., and Velikovich A.L. (1989) Dense Z-Pinches, ed. by N.R.Pereira, J.Davis and N.Rostoker (AIP Conference Proceedings, vol. 195; AIP, New York), p. 345.

Golberg S.M., Liberman M.A. and Velikovich A.L. (1990) Plasma Phys. Contr. Fusion **32**, 319.

Golberg S.M. and Velikovich A.L. (1993) Phys. Fluids B **5**, 1664.

Gol'din V.Ya. and Kalitkin N.N. (1970) Preprint Inst. Appl. Math. Acad. Sci. U.S.S.R. (in Russian).

Gol'ts E.Ya., Koloshnikov G.V., Koshelev K.N., Kramida A.E., and Sidel'nikov Yu.V. (1986) Phys. Lett. A **115**, 114.

Gomberoff K. and Fruchtman A. (1993) Phys. Fluids B **5**, 2841.

Gonzalez A.G. (1982) "Estudia de la Influencia de la Cizalladura del Campo Magnetico y de los Efectos Disipativos en la estabilidad de Rayleigh-Taylor." Thesis. Universidad de Buenos Aires, Argentina.

Gonzalez A.G., Gratton J., and Gratton F.T. (1989) Dense Z-Pinches, ed. by N.R. Pereira, J. Davis and N. Rostoker (AIP Conference Proceedings, vol. 195; AIP, New York), p. 280.

Gonzalez A.G. and Gratton J. (1990) Plasma Phys. Contr. Fusion **32**, 3.

Gorbunov E.P. and Rosumova K.A. (1963) Sov. J. Atomic Energy **15**, 1105 [in Russian].

Grad H. (1966) Phys. Fluids **9**, 225.

Grad H. (1973) Proceedings of the National Academy of Sciences, USA, **70**, 3277.

Graneau P. (1987) Phys. Lett. A **120**, 77.

Gratton J., Gratton F.T., and Gonzalez A.G. (1988) Plasma Phys. Contr. Fusion **30**, 435.

Gratton R., Piriz A.R., and Pouzo J.O. (1986) Nucl. Fusion **26**, 483.

Greifinger C. and Cole J.D. (1961) Phys. Fluids **4**, 527.

Guderley G. (1942) Luftfahrtforschung **19**, 302.

Gureev K.G., Filippov N.V., and Filippova T.I. (1975) Fizika Plazmy **1**, 120 [Sov. J. Plasma Phys. in Russian].

Gureev K.G. (1978) Fizika Plazmy **4**, 304 [Sov. J. Plasma Phys. (1978) **4**, 217].

Hagelstein P.L. (1983) Plasma Phys. **25**, 1345.

Hagenson R.L., Tai A.S., Krakowski R.A., and Moses F.W. (1981) Nucl. Fusion **21**, 1351.

Hain Von K. and Lust R. and Schluter (1957) Z. Naturforsch. **A12**, 833.

Haines M.G. (1960) Proc. Roy. Soc. Lond. **76**, 250.

Haines M.G. (1978a) J. Phys. D. Appl. Phys. **11**, 1709.

Haines M.G. (1978b) J. Phys. D. Appl. Phys. **11**, 1978.

Haines M.G. (1981) Phil. Trans. Roy. Soc. Lond. **A300**, 649.

Haines M.G. (1982) Physica Scripta **T2/2**, 380.

Haines M.G. et al. (1984) in Proceedings of the Tenth International Conference on Plasma Physics and Controlled Nuclear Fusion Research, London, September, Vol. 2, p. 647.

Haines M.G., Dangor A.E., Folkierski A., Baldock P., Challis C.D., Choi P., Coppins M, Deeney C., Favre Dominguez M.B., Figura E. and J.D.Sethian (1986) in Proceedings of the Eleventh International Conference on Plasma Physics and Controlled Nuclear Fusion Research, Paper IAEA-CN-47/D-iV-4.

Haines M.G., Bayley P., Baldock A., Bell A.R., Chittenden J.P., Choi P. Coppins M., Culverwell .I.D., Dangor A.E., Figura F.S., McCall G., and Ricard J. (1988) in Proceedings of theTwelfth International Conference on Plasma Physics and Controlled Nuclear Fusion Research, Nice, France, Paper IAEA-CN50/C-4-4-2.

Haines M.G. (1989) Plasma Phys. Contr. Fusion **31**, 759.

Haines M.G. and Coppins M. (1991) Phys. Rev. Letters **66**, 1462.

Haines M.G. (1994) Dense Z-Pinches, Third International Conference, ed. by M. Haines and A. Knight (AIP Conference Proceedings, vol. 299, p. 472, AIP, New York).

Hammel J. (1976) Report LA-6203-MS, Los Alamos Scientific Laboratory, NM.

Hammel J., Scudder D.W., and Schlachter J.S. (1984) in Proceedings of the First International Conference on Dense Z-Pinches for Fusion, edited by J.D. Sethian and K.A.Gerber (Naval Research Laboratory, Washington, D.C.), p. 13.

Hammel J. and Scudder D.W. (1987) in Proc. of the Fourteenth European Conference on Controlled Fusion and Plasma Physics, Madrid, Spain, 1987, edited by F. Engelmann and J.L. Alvarez Rivaz (European Physical Society, Geneva, Switzerland), p. 450.

Hammel J. (1989) Dense Z-Pinches, ed. by N.R. Pereira, J. Davis and N. Rostoker (AIP Conference Proceedings, vol. 195; AIP, New York), p. 303.

Hammer J.H., Eddleman J.D., Springer P.T., Tabak M., Toor A., Wong K.L., Zimmerman G.B., Denney C., Humphres R., Nash T.J., Sanford T.W.L., Spielman R.B., and De Groot J.S. (1996) Phys. Plasmas 3, 2063.

Han S.J. and Suydam B.R. (1982) Phys. Rev. A **26**, 926.

Hares J.D., Marrs R.E., and Fortner R.J. (1985) J. Phys. D. Appl. Phys. **18**, 627.

Harris E.G. (1961) Phys. Fluids **5**, 1057.

Hartman C.W., Carlson G., Hoffman M., Werner R., and Cheng D.Y. (1977) Nucl. Fusion, **17**, 909.

Hassam A.B. and Lee Y.C. (1984) Phys. Fluids **27**, 438.

Hassam A.B. and Huba J.D. (1987) Geophys. Res. Lett. **14**, 60.

Hassam A.B. and Huba J.D. (1988) Phys. Fluids **31**, 318.

Hattori F., Takabe H., and Mima K. (1986) Phys. Fluids **29**, 1719.

Henshaw M.J. de C., Pert G.J., and Youngs D.L. (1987) Plasma Phys. Contr. Fusion **29**, 405.

Herlach F. and Miura N. (1985) in Strong and Ultrastrong Magnetic Fields, ed. by F.Herlach, Topics in Applied Physics, Vol. 57 (Springer-Verlag, Berlin, 1985), Ch. 6.

Herold H., Kaeppeler H.J., Schmidt H., Shakhatre M., Wong C.S., Deeney C., Choi P. (1988) in Proceedings of the Twelfth International Conference on Plasma Physics and Controlled Nuclear Fusion Research, Nice, France, Paper IAEA-CN-50/C-4-5-3.

Herold H., Jerzykiewicz A., Sadowski M., and Schmidt H. (1989) Nucl. Fusion **29**, 1255.

Hsing W.W. and Porter J.L. (1987) Appl. Phys. Lett., **50**, 1572.

Huba J.D., Lyon J.G., and Hassam A.B. (1987) Phys. Rev. Lett. **59**, 2971.

Huba J.D., Lyon J.G., and Hassam A.B. (1988) Phys. Rev. Lett. **61**, 898.

Huba J.D., Hassam A.B., and Satyanarayana P. (1989) Phys. Fluids B **1**, 931.

Huba J.D., Hassam A.B., and Winske D. (1990) Phys. Fluids B **2**, 1676.

Hussey T.W., Roderick N.F., and Kloc D.A. (1980) J. Appl. Phys. **51**, 1462.

Hussey T.W. (1984) in Ultrahigh Magnetic Fields. Physics. Techniques. Applications, edited by V.M. Titov, G.A. Shvetsov (Moscow: Nauka), p. 208.

Hussey T.W., Matzen M.K., and Roderick N.F. (1986) J. Appl. Phys. **58**, 2677.

Hussey T.W., Roderick N.F., Shumlak U. Spielman R.B., and Deeney C. (1995) Phys. Plasmas **2**, 2055.

Hwang C.S. and Roderick N.F. (1987) J. Appl. Phys. **62**, 95.

Hsing W.W. and Porter J.L (1989) IEEE International Conference on Plasma Science (Cat. No.89CH2760-7), Buffalo, NY, USA, 22-24 May 1989.) New York, NY, USA, p. 113.

Ikuta K. (1988) Jap. J. Appl. Phys. **27**, L266.

Imshennik V.S., Osovets S.M., and Otroschenko I.V. (1973) Zh. Eksp. Teor. Fiz. **64**, 2057 [Sov. Phys. - JETP, in Russian].

Imshennik V.S. et al. (1984) in Proceedings of the Tenth International Conference on Plasma Physics and Controlled Nuclear Fusion Research, London, September, Vol. 2, p. 561.

Imshennik V.S. and Neudachin S.V. (1987) Fizika Plazmy **13**, 1226 [Sov. J. Plasma Phys. (1987) 13, 707].

Imshennik V.S. and Neudachin S.V. (1988) Fizika Plazmy **14**, 668 [Sov. J. Plasma Phys. (1988) 14, 393].

Inogamov N.A. (1985) Zh. Prikl. Mekh. Tekh. Fiz. **5**, 110 [Journal of Applied Mech. and Technical Physics (1985) 26, 702].

Ishii S., Fukuta M., Shimizu K., Hoshina Y., Mineshima T., Kanou M., Liu Y.,and Ogura S. (1989) Dense Z-Pinches, Second International Conference, ed. by N.R. Pereira, J. Davis and N. Rostoker (AIP Conference Proceedings, vol. 195, p. 320, AIP, New York).

Ishii S., Hara T., Sonoda F., Fukuta M., and Hayashi I. (1986) Electr. Eng. Jpn. **106**, 10.

Isichenko M.V., Kulyabin K.L., and Yan'kov V.V. (1989) Fizika Plazmy **15**, 1064 [Sov. J. Plasma Physics (1989) 15, 617].

Jaitly P. and Coppins M. (1994) Dense Z-Pinches, Third International Coonference, ed. by M. Haines and A. Knight (AIP Conference Proceedings, vol. 299, p. 51, AIP, New York).

Jones L.A., Finkin K.H., and Dangor A. (1981) Appl. Phys. Lett. **38**, 522.

Jones L.A. and Kania D.R. (1985) Phys. Rev. Lett. **55**, 1993.

Jones R.D. and Mead W.C. (1986) Nucl. Fusion **26**, 127.

Kadomtsev B.B. (1963) Voprosy teorii plazmy (Reviews of Plasma Physics, in Russian), edited by M.A.Leontovich, Vol. 2, p. 132; English translation (Consultants Bureau, New York, 1966) Vol. 2, p. 153.

Kaeppeler H.J., Hayd A., Maurer M., and Meinke P. (1983) Bericht IPF-83-2, Institut fur Plasmaforschung, Stuttgart.

Kamada Y., Fujita T., Murakami Y., Ohira T., Saitoh K., Fuke Y., Utsumi M., Yoshida Z. and Inoue N. (1989) Nuclear Fusion **29**, 713.

Kanellopoulos M., Coppins M., and Haines M.G. (1988) J. Plasma Phys. **39**, 521.

Katzenstein J. (1981) J. Appl. Phys. **52**, 676.

Kelley H., Garcia G., and Bilbao L. (1989) Plasma Phys. and Contr. Fusion **31**, 1017.

Kidder R.E. (1976) Nucl. Fusion 16, 3.

Kies W. (1988) in Workshop on Z-Pinch and Plasma Focus, Nice, France, 10-11 October.

Kies W., Bachmann H., Baumung K., Bayley J.M., Blaum H., Decker G., Malzig M, Rusch D., Ratajczak W., Stoltz O., van Calker C., Westheide J., and Ziethen G. (1991) J. Appl. Phys. **70**, 7261.

Kim A.A., Kovalchuk B.M., Kokshenev V.A., Kurmaev N.E., Loginov S.V., Fursov F.I., and Jakolev V.P. (1995) in Proceedings of the Tenth IEEE International Pulsed Power Conference, Albuquerque, NM edited by W. Baker and G. Cooperstein (IEEE New York), p. 226.

Kleev A.I. and Velikovich A.L. (1989) XIX International Conference on Phenomena in Ionized Gases, Belgrade, Yugoslavia, 10-14 July, Conributed papers, Vol. 4, p. 840.

Kleev A.I. and Velikovich A.L. (1990) Plasma Phys. Contr. Fusion **32**, 763.

Knoepfel H. (1970) Pulsed High Magnetic Fields (North-Holland Publishing Co., Amsterdam).

Kolb A.C. (1960) Rev. Mod. Phys. **32**, 74.

Kononov E.Ya., Koshelev K.N., Safronova U.I., Sidel'nikov Yu.V., and Churilov S.S. (1980) Pis'ma v Zh. Eksp. Teor. Fiz. **31**, 720 [JETP Letters, in Russian].

Korop E.D., Meierovich B.E., Sidel'nikov Yu.V., and Sukhorukov S.T. (1979) Usp. Fiz. Nauk **129**, 87 [Sov. Phys. - Uspekhi (1979) **22**, 727].

Koshelev K.N., Krauz V.I., Reshetnyak N.G., Salukvadze R.G., Sidel'nikov Yu.V., and Khautaev E.Yu. (1989) Fizika Plazmy **15**, 1068 [Sov. J. Plasma Phys. (1989) **15**, 619].

Krall N.A. and Trivelpiece A.W. (1973) Principles of Plasma Physics (McGraw-Hill, New York).

Kruskal M. and Schwarzschild M. (1954) Proc. Roy. Soc. London A **223**, 348.

Kruskal M. and Oberman C.R. (1958) Phys. Fluids 1, 275.

Kulikovskii A.G. (1957) Dokl. Akad. Nauk USSR 114, 984 [in Russian].

Kulikovskii A.G. (1958) Dokl. Akad. Nauk USSR 120, 984 [in Russian].

Kull H.J. (1991) Physics Reports 206, 197.

Kurchatov I.V. (1957) Nucl. Energy 4, 193.

Lampe M. (1989) Dense Z-Pinches, ed. by N.R. Pereira, J. Davis and N. Rostoker (AIP Conference Proceedings, vol. 195; AIP, New York), p. 252.

Lampe M. (1991) Phys. Fluids B 3, 1521.

Landau L.D. and Lifshitz E.M. (1976) Mechanics, (Pergamon Press, Oxford).

Landau L.D. and Lifshitz E.M. (1977) Quantum Mechanics, Non-relativistic Theory, (Pergamon Press, Oxford).

Landau L.D. and Lifshitz E.M. (1987) Fluid Mechanics, (Pergamon Press, Oxford).

Laval G., Mercier C., and Pellat R. (1965) Nuclear Fusion 5, 156.

Latham R., Nation J.A., Curzon F.L., and Folkierski A. (1960) Nature 186, 624.

Lebert R., Neff W., Holz R., and Richter F. (1989) Dense Z-Pinches, ed. by N.R. Pereira, J. Davis and N. Rostoker (AIP Conference Proceedings, vol. 195; AIP, New York), p. 515.

Lee T.M., McLean E.A., Elton R.C. (1987) Phys. Rev. Lett. 59, 1185.

Lee R.W. and Zigler A., Appl. Phys. Lett.(1988) 53, 2028.

Lee Y. T., and More R. M., Phys. Fluids (1984) 27, 1273.

Lehnert B. (1983) Nucl. Instrum. Methods 207, 233.

Lehnert B. and Scheffel J. (1988) Phys. Rev. Lett. 61, 897.

Lehnert B. and Scheffel J. (1992) Plasma Phys. Control. Fusion 34, 1113.

Leontovich A.M. and Osovets S. M. (1956) Atomnaya Energiya 1, 81 [in Russian].

Lewis D.J. (1950) Proc. Roy. Soc. A 202, 81.

Liberman M.A. and Velikovich A.L. (1984) J. Plasma Phys. 31, 381.

Liberman M.A., Schmaltz, and Velikovich A.L. (1985) Zh. Exper. i Teor. Fiziki, Pis'ma v Redaktsyu 41, 97 [JETP Letters (1985) 41, 116].

Liberman M.A. and Velikovich A.L. (1986a) Physics of Shock Waves in Gases and Plasmas (Springer, Berlin).

Liberman M.A. and Velikovich A.L. (1986b) Nucl. Fusion 26, 709.

Liberman M.A., Velikovich A.L., and Felber F.S. (1987) Megagauss Technology and Pulsed Power Applications, edited by C.M. Fowler, R.S. Caird, and D.J. Erickson (New York: Plenum Press), p. 107.

Liberman M.A., Velikovich A.L., and Felber F.S. (1989) Dense Z-Pinches, ed. by N.R. Pereira, J. Davis and N. Rostoker (AIP Conference Proceedings, vol. 195; AIP, New York), p. 167.

Libin Z. (1927) Zh. Prikl. Fiz. 4, 45 [in Russian].

Lindemuth I.R. and Kirkpatrick R.C. (1984) Atomkernenergie Kerntechik 45, 9.

Lindemuth I.R., McCall G.H., and Nebel R.A. (1989) Phys. Rev. Lett. 62, 264.

Lindemuth I.R. (1989a) Fifth International Conference on Megagauss Magnetic Field Generation and Related Topics, Novosibirsk, July, 3-7, Book of Abstracts, p. 122.

Lindemuth I.R. (1989b) Dense Z-Pinches, ed. by N.R. Pereira, J. Davis and N. Rostoker (AIP Conference Proceedings, vol. 195; AIP, New York), p. 327.

Lindemuth I.R. (1990) Phys. Rev. Letters **65**, 179.

Linhart J.G., Knopfel H., and Gourlan C. (1962) Nuc. Fusion Supplement Part 2 733.

Linhart J.G. (1989) Nucl. Instrum. Methods A **278**, 114.

Lokke, W. A., and Grassberger, W.H., Lawrence Livermore National Laboratory Report No.UCRL-52276, 1977. Copies may be obtained from the National Technical Information Service, Springfield, VA 22161.

Manheimer W.M., Lampe M., and Boris J.P. (1973) Phys. Fluids **16**, 1126.

Manheimer W., Colombant D., and Ott E. (1984) Phys. Fluids **27**, 2164.

Marconi M.C. and Rocca J.J., Appl. Phys. Lett. (1989) **54**, 2180.

Marnachev A.M. (1987) Fizika Plazmy 13, 550 [Sov. J. Plasma Physics (1987) **13**, 312].

Marrs R.E., Dietrich D.D., Fortner R.J., Levine M.A., Price D.F., Stewart R.E., and Young B.K.F. (1983) Appl. Phys. Lett. **42**, 946.

Martin T.H., VanDevender, J.P., Johnson, D.L., McDaniel D.H. and Aker, M. (1975) in Proceedings of the International Conference on Electron Beam Research and Technology, Albuquerque, NM, Vol. 1, p. 450.

Martin T.H., Guenther A.H., and Kristiansen M., J. C. Martin on Pulsed Power, Advances in Pulsed Power Technology Vol. 3, Plenum Press (NY, 1996).

Mather J.W. (1964) Phys. Fluids Suppl. **5**, 28.

Matthews D. et al. (1987) J. Opt. Soc Am. **B4**, 575.

Matzen M.K. and Spielman R.B. (1984) SANDIA Report, Sand 84-1587.

Matzen M.K., Dukart R.J., Hammel B.A., Hanson D.L., Hsing W.W., Hussey T.W., Maguire E.J., Palmer M.A., and Spielman R.B. (1986) J. de Physique **47**, C6-135.

Matzen M.K. (1997) Phys. Plasmas **4**, 1519.

Maxon S., Hagelstein P., Reed K., and Scofield J. (1985) J. Appl. Phys. **57**, 971 (1985).

Maxon S., Hammer J.H., Eddelman J.L., Tabak M., Zimmerman G.B., Alley W.E., Estabrook K.G. Harte J.A., Nash T.J., Sanford T.W.L., and De Groot J.S. (1996) Phys. Plasmas **3**, 1737.

McCall G.H. (1989) Phys. Rev. Lett. **62**, 1986.

Mehlman G., Burkhalter P.G., Stephanakis S.J., Young F.C., and Negel D.J. (1986) J. Appl. Phys. **60**, 3427.

Meierovich B.E. and Sukhorukov S.T. (1975) Zh. Eksp. i Teor. Fiz. 68, 1788 [Sov. Phys. - JETP (1975) **68**, 1783].

Meierovich B.E. (1982) Phys. Reports **92**, 84.

Meierovich B.E. (1983) J. Plasma Phys. **29**, 361.

Meierovich B.E. (1985) Fizika Plazmy 11, 1446 [Sov. J. Plasma Phys. (1985) **11**, 831].

Meierovich B.E. (1986) Usp. Fiz. Nauk **149**, 221[Sov. Phys. - Uspekhi. (1986) **29**, 506].

Meierovich B.E. and Sukhorukov S.T. (1988) Institute of Theoretical and Experimental Physics, Preprint no. 129-89 (in Russian).

Mikaelian K.O. (1982) Phys. Rev. Lett. **48**, 1793.

Mikhailovskii A.B. (1975) Teoriya Plazmennykh Neustoychivostey (Theory of Plasma Instabilities, in Russian) (Atomizdat, Moscow), Vol. 1.

Mikhailovskii A.B. (1977) Teoriya Plazmennykh Neustoychivostey (Theory of Plasma Instabilities, in Russian) (Atomizdat, Moscow), Vol. 2.

Meyer-ter-Vehn J. and Schalk C. (1982) Z. Naturforsch. **A37**, 955.

Miller R. B.(1982) An Introduction to the Physics of Intense Charged Particle Beams (Plenum Press).

Miller G. (1984) Plasma Phys. Contr. Fusion **26**, 1119.

Miyamoto T. (1984) Nucl. Fusion **24**, 337.

Miyamoto T. (1987) Phys. Rev. Lett. A**125**, 5760.

More, R.M., Warren, K.H.,Young, D.A., and Zimmerman, G.B., Phys. Fluids (1988) **31**, 3059.

Mosher D, Stephanakis S.J., Hain K., Dozier C.N., and Young F.C. (1975) Ann. of N.Y. Acad. Sci. **251**, 632.

Murakami M., Shimoide M., and Nishihara K. (1995) Phys. Plasma **2**, 3466.

Nardi V., Bostick W.H., Feugeas J., and Prior W. (1980) Phys. Rev. A **22**, 2211.

Nash T., Deeny C.,Krishnan M., Prasad R.R., Lepell PD., and Warren L. (1990) Quant. Spectrosc. Rad. Transfer **44**, 485.

Nation J. A. (1979) Part. Accel. **10**, 1.

Nash T., Deeney C., Krishnan M., Prasad R.R., Lapell P.D., and Warren L. (1990a) J. Quantum Spectrosc. Radiat. Transfer **44**, 485.

Nash T., Deeney C., Lapell P.D., Prasad R.R.,and Krishnan M. (1990b) Rev.Sci. Instrum. **61**, 2804.

Nash T., Deeney C., Lapell P.D., Prasad R.R.,and Krishnan M. (1990c) Rev.Sci. Instrum. **61**, 2810.

Negus C.R. and Peacock N.J. (1979) J. Phys. D. Appl. Phys. **12**, 91.

Neudachin V.V. and Sasorov P.V. (1991) Nucl. Fusion, **31**, 1053.

Neudachin V.V. and Sasorov P.V. (1988) Fizika Plazmy **14**, 965 [Sov. J. of Plas. Phys. (1988) **14**, 567].

Newcomb W.A. (1960) Ann. Phys. (NY) **10**, 232.

Newcomb W.A. (1961) Phys. Fluids **4**, 391.

Newcomb W.A. (1983) Phys. Fluids **26**, 3246.

Nielson, P. D. and Zimmerman, G. B., Lawrence Livermore National Laboratory Report No. UCRL-53123, 1981. Copies can be obtained from the National Technical Information Services, Springfield VA 22101.

Niffikeer S.L., Beg F.N., Dangor A.E., and Hains M.G. (1994) Dense Z-Pinches, Third International Conference, ed. by M. Haines and A. Knight (AIP Conference Proceedings, vol. 299, p. 495, AIP, New York).

Nuckolls T., Wood L., Thiessen A., and Zimmerman G. (1972) Nature, **239**, 139.

Nycander J. and Wahlberg C. (1984) Nucl. Fusion **24**, 1357.

Oliphant T.A. (1974) Nucl. Fusion **14**, 377.

Orszag S.A. (1984) Physica D **12**, 19.

Ott E. (1972) Phys. Rev. Lett. **29**, 1429.

Parks D. (1983) Phys. Fluids **26**, 448.

Parsons W.M., Ballard E.O., Bartsch R.R., Benage J.F., Bennett G.A., Bowers R.L., Bowman D.W., Brownell J.H., Cochrane J.C., Davis H.A., Ekdahl C.A., Gribble R.F., Griego J.R., Goldstone P.D., Jones M.E., Hinckley W.B., Hosack K.W., Kasik R.J., Lee H., Lopez E.A., Lindemuth I.R., Monroe M.D., Moses Jr. R.W., Ney S.A., Platts D., Reass W.A., Salazar H.R., Sandoval G.M., Scudder D.W., Shlachter J.S., Thompson M.C., Trainor R.J., Valdez G.A., Watt R.G., Wurden G.A., Younger S.M.(1997 IEEE Trans. on Plasma Science **25**, 205.

Pavlovskii A.I., Kolokol'chikov N.P., Dolotenko M.I., Bykov A.I., Karpikov A.A., Mamyshev V.I., Spirov G.M., Tatsenko O.M., Markevtsev I.M., and Sosnin P.V. (1989) Fifth International Conference on Megagauss Magnetic Field Generation and Related Topics, Novosibirsk, July, 3-7, Book of Abstracts, p. 4.

Pavlovskii A.I., Karpikov A.I., Dolotenko M.I., and Mamishev V.I. (1991) Megagauss Technology and Pulsed Power Systems, p.21, ed. by V.M.Titov and G.A.Shvetsov, Nova Sciences Publisher, New York .

Pease R.S. (1957) Proc. Phys. Soc. London Ser. B **70**, 11.

Pereira N.R. and Davis J. (1988) J. Appl. Phys. **64**, R1-27.

Pereira N.R., Rostoker N., and Pearlman J.S. (1984) J. Appl. Phys. **55**, 704.

Pereira N.R., Rostoker N., Riordan J., and Gersten M. (1984) in Proceedings of the First International Conference on Dens Z-Pinches for Fusion, edited by J.D. Sethian and K.A.Gerber (Naval Research Laboratory, Washington, D.C.), p. 71.

Peterson D.L. Bowers R.L., Brownell J.H., Greene A.E., McLenithan K.D., Oliphant T.A., Roderick N.F., and Scannapieco A.J. (1996) Phys. Plasmas **3**, 368.

Porter J.L. (1997) Bull. Am. Phys. Soc. **42**, 1948.

Potter D. (1978) Nuclear Fusion **18**, 813.

Prager S.C. et al. (1990) Phys. Fluids B **2**, 1367.

Prager S.C. (1992) Plasma Phys.Contr. Fusion **34**, 1895.

Rahman H.U., Amendt P., and Rostoker N. (1985) Phys. Fluids **28**, 1528.

Rahman H.U., Felber F.S., Wessel F.J., Liberman M.A., and Velikovich A.L. (1987a) Megagauss Technology and Pulsed Power Applications, ed. by C.M. Fowler, R.S. Caird, and D.J. Erickson (New York: Plenum Press), p. 191.

Rahman H.U., Wessel F.J., Rostoker N., and Fisher A. (1987b) Bull. Am. Phys. Soc. **32**, 1818.

Rahman H.U., Ney P., Wessel F.J., Fisher A., Rostoker N. (1989) Dense Z-Pinches, ed. by N.R.Pereira, J.Davis and N.Rostoker (AIP Conference Proceedings, vol. 195; AIP, New York), p. 351.

Ratakhin N.A., Sorokin S.A., and Chaikovsky S.A. (1988) in Seventh International Conference on High-Power Particle BeamsKarlsruhe, July 4-8, Paper UP6.

Read K.I. (1984) Physica D **12**, 45.

Ripin B.H., McLean E.A., Manka C.K., Pawley C., Stamper J.Peyser T.A., Mostovych A.N., Grun J., Hassam A.B., and Huba J.D. (1987) Phys. Rev. Lett. **59**, 2299.

Robson A.E. (1988) Nucl. Fusion **28**, 2171.

Robson A.E. (1989) Phys. Fluids B **1**, 1834.

Robson A.E. (1991) Phys. Fluids B **3**, 1461.

Roderick, N. F., and Hussey, T.W. J. Appl. Phys. (1984) **56**, 1387.

Roderick N.F. and Hussey T.W. (1989a) Dense Z-Pinches, ed. by N.R. Pereira, J. Davis and N. Rostoker (AIP Conference Proceedings, vol. 195; AIP, New York), p. 157.

Roderick N.F. and Hussey T.W. (1989b) Fifth International Conference on Megagauss Magnetic Field Generation and Related Topics, Novosibirsk, July, 3-7, Book of Abstracts, p. 124.

Rosanov V.B., Rukhadze A.A., and Triger S.A. (1968) Prikl. Mekh. Tekh. Fiz. **5**, 18 [in Russian].

Rosenbluth M.N., Garwin R., and Rosenbluth A. (1954) Los Alamos Scientific Laboratory Report LA-1850, September 1954.

Rosenbluth M.N. (1956) Los Alamos Scientific Laboratory Report LA-2030, April 1956.

Rosenbluth M.N. and Longmire C.L. (1957) Ann. Phys. **1**, 120.

Rosenbluth M.N., Krall N.A., and Rostoker N. (1962) Nucl. Fusion Suppl. Part 1, 143.

Rosenbluth M. N. (1965) "Microinstabilities" in Plasma Physics, p. 458 (International Atomic Energy Agency, Vienna).

Rosenau P., Nebel R.A., and Lewis H.R. (1989) Phys. Fluids B **1**, 1233.

Rostoker N. and Tahsiri H. (1977) Comments Plasma Phys. Contr. Fusion **3**, 39.

Ruden E., Rahman H.U., Fisher A., and Rostoker N. (1987) J. Appl. Phys. **61**, 1311.

Rukhadze A.A. and Triger S.A. (1968) Prikl. Mekh. Tekh. Fiz. **3**, 11 [in Russian].

Sadowski M., Herold H., Schmidt H., Shakhatre H. (1984) Phys. Lett. **105A**, 117.

Saiz E., Gutierrez J., and Cerrato Y. (1992) J.Plasma Phys. **47**, 491.

Sakagami H. and Nishihara K. (1990) Phys. Fluids B **2**, 2715.

Sakharov A.D., Lyudaev R.Z., Smirnov E.N., Plyushchev Yu.I., Pavlovskii A.I., Chernyshev V.K., Feoktistova E.A., Zharinov E.I., and Zysin Yu.A. (1966) Sov. Phys. - Doklady 10, 1045.

Samokhin A.A. (1988) Zh. Prikl. Mat. i Tekh. Fiz., **29**, 89 [Journal of Applied Mech. and Technical Physics (1988) 29, 243].

Sanford T. W. L., Allhouse G. O., Marder B. M., Nash T. J., Mock R. C., Spielman R. B., Seaman J. F., McGurn J. S. , Jobe D., Gilliland T. L., Vargas M., Struve K. W., Stygar W. A., Douglas M. R., Matzen M. K., Hammer J. H., De Groot J. S., Eddleman J. L., Peterson D. L., Mosher D., Whitney K. G., Thornhill J. W., Pulsifer P. E., Apruzese J. P., and Maron Y. (1996) Phys. Rev. Lett. **77**, 5063.

Sanford T. W. L., Nash T. J., Mock R. C., Spielman R. B., Struve K. W., Hammer J. H., De Groot J. S., Whitney K. G., and Apruzese J. P. (1997) Phys. Plasmas **4**, 2188.

Sarfaty M., Maron Y., Krasik Ya. E., Weingarten A., Arad R., Shpitalnik R., Fruchtman A. and Alexiou S. (1995a) Phys. Plasmas **2**, 2122.

Sarfaty M., Shpitalnik R., Arad R., Weingarten A., Krasik Ya. E., Fruchtman A. and Maron Y. (1995b) Phys. Plasmas **2**, 2583.

Sasorov P.V. (1985) Institute of Theoretical and Experimental Physics, Preprint no. 171 (in Russian).

Sasorov P.V. (1991) Fizika Plazmy **17**, 1507 [Sov. J. Plasma Phys. 17 (12), 874].

Scheffel J. and Faghihi M. (1989) J. Plasma Phys. **41**, 427.

Scheffel J., Arber T., and Coppins M. (1994) Dense Z-Pinches, Third International Conference, ed. by M. Haines and A. Knight (AIP Conference Proceedings, vol. 299, p.75, AIP, New York).

Schaper U. (1983) J. Plasma Phys. **29**, 1.

Schmidt H. (1980) Atomkernenergie-Kerntechnik **36**, 161.

Scudder D.W. (1985) Bull. Am. Phys. Soc. **30**, 1408.

Scudder D.W. and Hammel J.E. (1987) IEEE Conference Records Abstracts. 1987 IEEE International Conference on Plasma Science (Arlington, VA, 1-3 June 1987), p. 107.

Scudder D.W., Shlachter J.S., Forman P.R., Riley R.A., and Lovberg, R.H. (1994) Dense Z-Pinches, Third International Conference, ed. by Haines and A. Knoght (AIP Conference Proceedings, vol. 299, p. 503, AIP, New York).

Sedov L.I. (1959) Similarity and Dimensional Methods in Mechanics (Academic, New York).

Sethian J.D., Robson A.E., Gerber K.A., and DeSilva A.W. (1987) Phys. Rev. Lett. **59**, 892.

Sethian J.D., Robson A.E., Gerber K.A., and DeSilva A.W. (1989) Dense Z-Pinches, Second International Conference, ed. by N.R. Pereira, J. Davis and N. Rostoker (AIP Conference Proceedings, vol. 195, p. 308, AIP, New York).

Shafranov V.D. (1956) Atomnaya Energiya **5**, 38.

Shafranov V.D. (1957) Zh. Eksp. Teor. Fiz. **33**, 710 [Sov. Phys. - JETP, in Russian].

Shafranov V.D. (1958a) in Plasma Physics and Problems of Controlled Thermonuclear Reactions (Moscow: Izdatel'stvo AN SSSR; in Russian), Vol. 2, p. 26.

Shafranov V.D. (1958b) in Plasma Physics and Problems of Controlled Thermonuclear Reactions (Moscow: Izdatel'stvo AN SSSR; in Russian), Vol. 2, p. 130.

Shafranov V.D. (1958c) in Plasma Physics and Problems of Controlled Thermonuclear Reactions (Moscow: Izdatel'stvo AN SSSR; in Russian), Vol. 4, p. 61.

Shafranov V.D. (1963) in Reviews of Plasma Physics, edited by M.A.Leontovich, Vol. 2, p. 103, Consultants Bureau, New York.

Shanny R. and Vitkovitsky I. (1969) Naval Research Laboratory Report.

Shaper U. (1983) J. Plasma Physics, 29, 1; **30**, 169.

Sharp D.H. (1984) Physica D **12**, 3.

Shearer J.W. (1976) Phys. Fluids **19**, 1426.

Sheehey P. and Lindemuth I.R. (1994) Dense Z-Pinches, Third International Conference, ed. by M. Haines and A. Knight (AIP Conference Proceedings, vol. 299, p. 157, AIP, New York).

Sheehey P., Hammel J.E., Lindemuth I.R.,Scudder D.W., Shlachter J.S., Lovberg R.H., and Riley R.A. (1992) Phys. Fluids B **4**, 3698.

Shiloh J., Fisher A., and Rostoker N. (1978) Phys. Rev. Letters **40**, 515.

Shiloh J., Fisher A., and Bar-Avraham E. (1979) Appl. Phys. Lett. **35**, 390.

Sincerny P., Ashby S., Childers K., Deeney C., Drury d., Goyer J., Kortbawi D., Roth I., Stallings C., Schlitt L. (1993) in Proceedings of the Ninth IEEE International Pulsed Power Conference, San Diego, CA edited by R. White and W. Rix, p. 880.

Solov'ev L.S. (1984) Fizika Plazmy 10, 1045 [Sov. J. Plasma Phys. (1984) **10**, 602].

Solov'ev L.S. (1984) in Reviews of Plasma Physics, edited by M.A.Leontovich, Vol. 6, p. 239. Consultants Bureau, New York.

Sorokin S.A. and Chaikovsky S.A. (1989) Dense Z-Pinches, ed. by N.R. Pereira, J. Davis and N. Rostoker (AIP Conference Proceedings, vol. 195; AIP, New York), p. 438.

Sorokin S.A. and Chaikovsky S.A. (1994) Dense Z-Pinches, Third International Conference, ed. by M. Haines and A. Knight (AIP Conference Proceedings, vol. 299, p. 83, AIP, New York).

Spielman R.B., Matzen M.K., Palmer M.A., Rand P.B., Hussey T.W., and McDaniel D.H. (1985a) Appl. Phys. Lett. **47**, 229.

Spielman R.B., Hanson D.L., Palmer M.A., Matzen M.K., Hussey T.W. and Peek J.N. (1985b) J. Appl. Phys. **57**, 883.

Spielman R.B., Dukart R.J., Hanson D.L., Hammel B.A., Hsing W.W., Matzen M.K., and Porter J.L. (1989) Dense Z-Pinches, Second International Conference, ed. by N.R. Pereira, J. Davis, and N. Rostoker (AIP Conference Proceedings, vol. 195, p. 3, AIP, New York).

Spielman R.B., Baldwin G.T., Cooper G., Hebron D., Hussey T.W., Landron C., Leeper R.J., Lopez S.F., McGurn J.S., Muron D.J., Porter J.L. , Ruggles L., Ruiz C.L., Schmidlapp A., and Vargas M., "Deuterium Gas Puff and CD_2 Fiber Array Z-Pinch Experiments on Saturn," 1991 IEEE Int. Conf. on Plasma Science, Williamsburg, VA, Session 6A3, (1991).

Spielman R.B. (1992) in Proceedings of the International Conference on Particle Beams, Washington D.C.

Spielman R.B., De Groot J.S., Nash T.J., McGurn J., Ruggles L., Varges M., and Estabrook K.G. (1994) Dense Z-Pinches, Third International Conference, ed. by M. Haines and A. Knight (AIP Conference Proceedings, vol. 299, p. 404, AIP, New York).

Spielman R.B., McGurn J.S., Nash T.J., Ruggles L.E., Seamen J.F., Struve K.W., Gilliland T.L., Jobe D. (1995) Bull. Am. Phys. Soc. **40**, 1845.

Spielman R.B., Stygar W.A., Seamen J.F., Long F., Ives H., Garcia R., Wagoner T., Struve K.W., Mostrom M., Smith I., Spence P., and Corcoran P. (1997) in Proceedings of the Eleventh IEEE Pulsed Power Conference, Baltimore, MD (IEEE New York).

Spielman R.B., Deeney C., Chandler G.A., Douglas M.R., Fehl D.L., Matzen M.K., McDaniel D.H., Nash T.J., Porter J.L., Sanford T.W.L., Seamen J.F., Stygar W.A., Struve K.W., Breeze S.P., McGurn J.S., Torres J.A., Zagar D.M., Gilliland T.L., Jobe D.O., McKenney J.L., Mock R.C., Vargas M., Wagoner T., and Peterson D.L. (1998) Phys. Plasmas 5, 2105

Spies G.O. and Faghihi M. (1987) Phys. Fluids 30, 1724.

Stallings C., Nielson K., and Schneider R. (1976) Appl. Phys. Lett. 29, 404.

Stallings C., Childers K., Ross I., and Schneider R. (1979) Appl. Phys. Lett. 35, 524.

Stein S., Decker G, Kies W., Rowekamp P., Ziethen G., Baumung K., Bluhm H., Ratajczak W., Rusch D., and Bayley J.M. (1994) Dense Z-Pinches, Third International Conference, ed. by M. Haines and A. Knight (AIP Conference Proceedings, vol. 299, p. 509, AIP, New York).

Stephanakis S.J., Apruzese J.P., Burkhalter P.G., Davis J., Meger R.A., McDonald S.W., Mehlman G., Ottinger P.F., and Young F.C. (1986) Appl. Phys. Lett. 48, 829.

Stephanakis S.J. (1989) Second International Conference on High Density Pinches, Laguna Beach, CA, U.S.A., Paper A-3.

Stewart R.E., Dietrich D.D., Egan P.O., Fortner R.J., and Dukart R.J. (1987) J. Appl. Phys. 61, 126.

Suckewer S., Skinner C.H., Keens C., and Voorhees D. (1985) Phys. Rev. Lett. 55, 1753 (1985).

Suydam B.R. (1958) in Proceedings of the Second United National Conference on Peaceful Uses of Atomic Energy (United Nations, Geneva), Vol. 31, p. 157.

Swanekamp S. B., Grossman J.M., Fruchtman A., Oliver B.V. and Ottinger P.F. (1996) Phys. Plasmas 3, 3556.

Tamm I.E. (1956) Osnovy Teorii Elektrichestva (Foundations of the Theory of Electricity, in Russian) (GTTL, Moscow).

Tayler R.J. (1957) Proc. Roy. Soc. B70, 1049.

Tendler M. (1988) Phys. Fluids 31, 3449.

Terletskii Ia.P. (1957) Sov. Phys. - JETP 32, 301.

Terry R. E. and Pereira N. R. (1991) Phys. Fluids B 3, 195.

Thornhill J.W., Whitney K.G., Deeney C., LePell P.D., (1994) Phys. Plasmas 1, 321.

Torricelli-Ciamponi G., Ciampolini V., and Chiuderi C. (1987) J. Plasma Phys. 37, 175.

Town R.P. and Bell A.R. (1991) Phys. Rev. Lett. 67, 1863.

Trubnikov B.A. (1958) in Plasma Physics and Problems of Controlled Thermonuclear Reactions (Moscow: Izdatel'stvo AN SSSR; in Russian), Vol. 1, p. 289.

Trubnikov B.A. (1958) in Plasma Physics and Problems of Controlled Thermonuclear Reactions (Moscow: Izdatel'stvo AN SSSR; in Russian), Vol. 4, p. 87.

Trubnikov B.A. and Zhdanov S.K. (1985) Zh. Exper. i Teor. Fiziki, Pis'ma v Redaktsiyu **41**, 292 [JETP Letters (1985) **41**, 358].

Trubnikov B.A. (1986) Fizika Plazmy **12**, 468 [Sov. J. Plasma Phys. (1986) **12**, 271].

Tuck J.L. (1958) in Proceedings of the Second United National Conference on Peaceful Uses of Atomic Energy (United Nations, Geneva), Vol. 32, p. 3.

Turchi P.J. and Baker W.L. (1973) J. Appl. Phys. **44**, 4936.

Turman B.N., Martin T.H., Neau E.L., Humphreys D.R., Bloomquist D.D., Cook D.L., Goldstein S.A., Schneider L.X., McDaniel D.H., Wilson J.M., Hamil R.A., Barr G.W., and VanDevender J.P. (1985) in Proceedings Of the Fifth IEEE Pulsed Power Conference, Arlington, VA, ed. by P. J. Turchi and M. F. Rose (IEEE, New York,), p. 155.

Turner G. L'E. and Levere T.H. (1973) in Martinus Van Marum Life and Work, Vol. 4, edited by Lefebvre E. and DeBruijn J.G. pp 312-314

Tuszewski M. (1988) Nucl. Fusion **28**, 2033.

Velikovich A.L., Golberg S.M., Liberman M.A., and Felber F.S. (1985) Zh. Eksp. i Teor. Fiz. 88, 445 [Sov. Phys. - JETP (1985) **61**, 261].

Velikovich A.L. (1991) Phys. Fluids B **3**, 492.

Venneri F., Kislev H., and Miley G.H. (1988) Appl. Phys. Lett. **53**, 2269.

Venneri F., Boulais K., and Gerdin G. (1990) Phys. Fluids B **53**, 1613.

Verdon C.P., McCrory R.L., Morse R.L., Baker G.R., Meiron D.I., and Orszag S.A. (1982) Phys. Fluids **25**, 1653.

Vikharev V.D., Zakharov S,V., Smirnov V.P., Starostin A.N., Stepanov A.E., fedulov M.V., and Tsarfin V.Ya. (1991) Zh. Eksp.Teor. Fiz. **99**, 1133 [Sov. Phys. - JETP (1991) **72**, 631].

Vikhrev V.V. and Gureev K.G.(1977) Nucl. Fusion **17**, 291.

Vikhrev V.V. (1978) Pis'ma v Zh. Eksp. Teor. Fiz. 27, 104 [Sov. Phys.- JETP Letters (1978) **27**, 95].

Vikhrev V.V. and Braginskii S.I. (1980), "Dynamics of Z-pinch," in Voprosy teorii plazmy, edited by M.A.Leontovich (Atomizdat, Moscow Vol. 10, p. 243); English translation in Reviews of Plasma Physics, (Consultants Bureau, New York 1986) Vol. 10.

Vikhrev V.V., Ivanov V.V., and Koshelev K.N. (1982) Fizika Plazmy **8**, 1211 [Sov. J. Plasma Phys. (1982) **8**, 688].

Vikhrev V.V. (1986) Fizika Plazmy **12**, 454 [Sov. J. Plasma Phys. (1986) **12**, 262].

Vikhrev V.V., Ivanov V.V., and Rosanova G.A. (1989) Fizika Plazmy **15**, 77 [Sov. J. Plasma Phys. (1989) **15**, 44].

Vikhrev V.V., Ivanov V.V., and Rosanova G.A. (1993) Nucl. Fusion **33**, 311.

Wahlberg C., (1991) Nucl. Fusion **31**, 867.

Wernsman B., Rocca J.J., and Mancini H.L., IEEE Photonics Tech. Lett. (1990) **2**,12.

Wessel F.J., Felber F.S., Wild N.C., Rahman H.U., Fisher A. and Ruden E. (1986) Appl. Phys. Lett. **48**, 1119.

Wessel F.J., Felber F.S., Wild N.C., and Rahman H.U. (1986) Appl. Phys. Lett. **48**, 1119.

Whitham G.B. (1974) Linear and Non-Linear Waves (Wiley - Interscience, New York).

Whitney K.J., Thornhill J.W., Deeney C., LePell P.D., and Coulter M.C. (1992) Proceedings of the 9-th International Conference on High-Power Particle Beams, Washington DC.

Whitney K.J., Thornhill J.W., Spielman R.B., Nash T.J., McGurn J.S., Ruggles L.E., and Coulter M.C. (1994) Dense Z-Pinches, Third International Conference, ed. by M. Haines and A. Knight (AIP Conference Proceedings, vol. 299, p. 429, AIP, New York).

Whitney K.J., Thornhill J.W., Giuliani J.L., Davis J., Miles L.A., Nolting E.E., Kenyon V.L., Spicer W.A., Draper J.A., Parsons C.R., Dang P., Spielman R.B., Nash T.J., McGurn J.S., Ruggles L.E., Deeney C., Prasad R.R., and Warren L. (1994) Phys. Rev. E **50**, 2166.

Wong K.L., Springer P.T., Hammer J.H., Osterheld A., and Denny C. (1995) Bull. Am. Phys. Soc. **40**, 1859.

Yamanaka C. (1983) Nucl. Fusion **23**, 97.

Yan'kov V.V. (1985) I.V.Kurchatov Institute of Atomic Energy, Preprint 4218/7 (in Russian).

Young F.C. et al. (1977) J. Appl. Phys. **48**, 3642.

Youngs D.L. (1984) Physica **12D**, 32.

Zel'dovich Ya.B. and Raizer Yu.P. (1967) Physics of Shock Waves and High-Temperature Hydrodynamic Phenomena (Academic, New York).

Zigler A., Lee R.W., and Mrowka S. (1989) Laser and Particle Beams **7**, 369.

Zimmerman et al.(1973), G. B., Lawrence Livermore National Laboratory Report No. UCRL-74811, 1973. Copies can be obtained from the National Technical Information Services, Springfield VA 22101.

Index

Index